THE GENUS *YERSINIA*

ADVANCES IN EXPERIMENTAL MEDICINE AND BIOLOGY

Recent Volumes in this Series

Volume 521
IMMUNE MECHANISMS IN PAIN AND ANALGESIA
Edited by Halina Machelska and Christoph Stein

Volume 522
NOVEL ANGIOGENIC MECHANISMS: Role of Circulating Progenitor Endothelial Ce
Edited by Nicanor I. Moldovan

Volume 523
ADVANCES IN MODELLING AND CLINICAL APPLICATION OF INTRAVENOUS ANAESTHESIA
Edited by Jaap Vuyk and Stefan Schraag

Volume 524
DIPEPTIDYL AMINOPEPTIDASES IN HEALTH AND DISEASE
Edited by Martin Hildebrandt, Burghard F. Klapp, Torsten Hoffmann, and Hans-Ulrich Demuth

Volume 525
ADVANCES IN PROSTAGLANDIN, LEUKOTRIENE, AND OTHER BIOACTIVE LIPID RESEARCH: Basic Science and Clinical Applications
Edited by Zeliha Yazıcı, Giancarlo Folco, Jeffrey M. Drazen, Santosh Nigam, and Takao Shimizu

Volume 526
TAURINE 5: Beginning the 21st Century
Edited by John B. Lombardini, Stephen W. Schaffer, and Junichi Azuma

Volume 527
DEVELOPMENTS IN TRYPTOPHAN AND SEROTONIN METABOLISM
Edited by Graziella Allegri, Carlo V. L. Costa, Eugenio Ragazzi, Hans Steinhart, and Luigi Varesio

Volume 528
ADAMANTIADES-BEHÇET'S DISEASE
Edited by Christos C. Zouboulis

Volume 529
THE GENUS *YERSINIA*: Entering the Functional Genomic Era
Edited by Mikael Skurnik, José Antonio Bengoechea, and Kaisa Granfors

---

A Continuation Order Plan is available for this series. A continuation order will bring delivery of each new volu immediately upon publication. Volumes are billed only upon actual shipment. For further information please cont the publisher.

THE GENUS *YERSINIA*

Entering the Functional Genomic Era

Edited by

Mikael Skurnik
University of Helsinki
Helsinki, Finland

José Antonio Bengoechea
Hospital Son Dureta
Palma Mallorca, Spain

and

Kaisa Granfors
National Public Health Institute
Turku, Finland

Springer Science+Business Media, LLC

Proceedings of the 8th International Symposium on Yersinia, held September 4–8, 2002, in Turku, Finland

ISSN 0065-2598

DOI 10.1007/978-0-306-48416-2

Originally published by Kluwer Academic / Plenum Publishers, New York in 2003
MyCopy version of the original edition 2003

http://www.wkap.nl/

10 9 8 7 6 5 4 3 2 1

A C.I.P. record for this book is available from the Library of Congress

Contributors

A

Abe, Jun
Abney, Jennifer
Aepfelbacher, Martin
Aftalion, Moshe
Ahrens, Peter
Andor, Andreas
Anisimov, Andrei P.
Appel, Bernd
Ariel, Naomi
Åström, Hanna
Atanassova, Viktoria
Atshabar, Bakyt B.

B

Baba, Kiyosi
Bakanidze, Lela
Baymiev, Andrey H.
Bearden, Scott W.
Bene, Judit
Benedek, Orsolya
Bengoechea, José Antonio
Ber, Raphael
Beridze, Levan
Bino, Tamar
Bogdanovich, Tatiana M.
Boolgakova, Elene
Brade, Lore
Brocchi, Marcelo
Brzezinska, Agnieszka
Bulakh, Elena
Bulgakov, Aleksandr
Bulgakov, Viktor
Butcher, Phillip D.

C

Cabodi, Daniela
Carniel, Elisabeth
Carnoy, Christophe
Chanishvili, Temo
Chanteau, Suzanne
Chesnokova, Margarita V.
Chmyr, Irina A.
Ciebin, Bruce
Cohen, Sara
Collyn, François
Coniglio, Maria A.
Corrêa, Ednéia F. Cristina
Cuccui, Jon

D

Damasko, Christina

Darsavelidze, Marina
Deery, Evelyn
Degtyarev, Sergei
Demakova, Tatiana E.
Dentovskaya, Svetlana V.
Dersch, Petra
Desreumaux, Pierre
Dikhanov, Grigory G.
Dixon, Jack E.
Duchemin, Jean-Bernard
Duplantier, Jean-Marc
Dvalishvili, Megi

E

Eckner, Karl
Edqvist, Petra J.
Eitel, Julia
Eliseikina, Marina
Elizbarashvil, Maia
Elofsson', Mikael
El Tahir, Yasmin
Emödy, Levente

F

Falcão, Deise Pasetto
Falcão, Juliana Pfrimer
Fantino, Czilla
Faveeuw, Christelle
Fetherston, Jacqueline D.
Flashner, Yehuda
Floccari, Mirtha E.
Forsberg', Åke
Forslund, Ola
Foulon, Jeannine
Franzin, Laura
Fredriksson-Ahomaa, Maria
Fukushima, Hiroshi

G

Garbom, Sara
Garcia, Emilio
Gáspár, Erzsébet
Geiger, Anna M.
Goelz, Greta
Goldman, Adrian
Golkocheva, Elitsa
Golubov, Andrey
Gómez, Stella M.
Grangette, Corinne
Gremyakova, Tat'yana A.
Grosfeld, Haim
Guinet, Françoise
Gunasena, Deephti K.
Günþen, Uður (Gunsen, Ugur)
Gur, David
Guzev, Konstantin V.

H

Hägglund, Ulrik
Hallanvuo, Saija
Hammerl, Jens A.
Hayashidani, Hideki
Heesemann, Jürgen
Heroven, Ann Kathrin
Hertwig, Stefan
Hill, James
Hinchliffe, Stewart
Hinds, Jason
Hinnebusch, B. Joseph
Høidal, Hilde Kristin
Holst, Otto
Hoorfar, Jeffrey
Horisaka, Tomoko
Hugenberg, Jutta

I

Imnadze, Paata
Isherwood, Karen E.
Ishikawa, Motoko
Ishiyama, Yuki
Issaeva, Marina P
Ito, Yasuhiko
Iwata, Tutomu
Iwobi, Azuka N.

J

Jaatinen, Silja
Jamieson, Frances
Jensen, Stefan
Joe, Angela
Josefsen, Mathilde H.

K

Kaneko, Ken-Ichi
Kano', Hirotsugu
Kapanadze, Zhana
Karlyshev, Andrey V.
Kaspar, Heike
Kasperkiewicz, Katarzyna
Kato, Yukio
Kauppi, Anna M.
Kekelidze, Merab
Kersley, Joanne E.
Kiljunen, Saija
Kim, Natalya Y.
Kirschning, Carsten J.
Kiseleva, Alla K.
Klein, Iris
Klimov, Victor T.
Knabner, Dorothea
Knight, Stefan D.
Knirel, Yuriy A.
Kocharova, Nina A.
Koh, Susie
Kohsaka, Takao
Komrower, Jenny R.
Konietzny, Antje
Koozmichenko, Inna
Korhonen, Timo K.
Korkeala, Hannu
Korte, Tiina
Kozlovsky, Victor N.
Kukkonen, Maini
Kukleva, Lubov M.
Kuoppa, Kerstin
Kutyrev, Vladimir

L

Lähteenmäki, Kaarina
Lahtinen, Pia
Lång, Hannu
Lassen, Jørgen
Laukkanen, Riikka
Lavander, Moa
Lazar, Shirley
Leary, Sophie E.C.
Lee, Yong
Likhatskaya, Galina N.
Lindner, Buko
Lloyd', Scott A.
Lodri, Czilla
Loiez, Caroline
Lübeck, Peter S.
Lustig, Shlomo

M

MacIntyre, Sheila
Magarlamov, Timyr
Mamroud, Emanuelle
Mangan, Joseph A.
Marceau, Michael
Marranzano, Marina
Martins, Carlos Henrique G.
Matsumoto, Shinichi
Matsuoka, Kentaro
Mauro, Luisa
Mavzutov, Airat R.
Medeiros, Beatriz M.M.
Meka-Mechenko, Tatyana V.
Melegh, Béla
Merckel, Michael C.
Mier, Ildefonso, Jr.
Mishankin, Boris N.
Mishankin, Michail B.
Molineux, Ian J.
Moslehi, Elham
Muszyński, Artur

N

Nagel, Geraldine
Najdenski, Hristo
Nakari, Ulla-Maija
Nedashkovskaya, Elena
Nesbakken, Truls
Neubauer, Heinrich K.J.
Ng, Lee-Ching
Niskanen, Taina
Nogami, Hiroko
Nordfelth, Roland
Novikova, Olga D.
Nummelin, Heli
Nuorti, Pekka

O

Ogawa, Masuo
Ohtomo, Yoshimitsu
Okatani, Tomomitsu A.
Ollikka, Pauli
Oyston, Petra C.F.

P

Pacinelli, Elvia
Pajunen, Maria I.
Parada, Jose L.
Parkhill, Julian
Perry, Robert D.
Petrova, Irina
Pinta, Elise
Plekhova, Natalya G.
Podladchikova, Olga N.
Prentice, Michael B.
Protsenko, Ol'ga

R

Radziejewska-Lebrecht, Joanna
Rahalison, Lila
Rakin, Alexander
Ramos, Orivaldo P.
Ranjalahy, Michel
Rasskazov, Valeri A.
Ratsifasoamanana, Lala
Ravelosaona, Jocelyn
Reeves, Peter R.
Ring, Christian
Robins-Browne, Roy M.
Roggenkamp, Andreas
Rost, Dagmar
Røtterud, Ole-Johan
Roy, Chad
Rüssmann, Holger

S

Saebo, Arve
Saito, Hirohisa
Saito, Masaaki
Savilahti, Harri
Schaudinn, Christoph
Sebbane, Florent
Senchenkova, Sof'ya N.
Shafferman, Avigdor
Shaikhutdinova, Rima Z.
Shao, Feng
Shashkov, Aleksander S.
Shestakova, Irina V.
Shubin, Felix N.
Siitonen, Anja
Silva, Eloisa E.E.
Simbirtsev, Andrey S.
Simonet, Michel
Sing, Andreas
Sjöstedt, Anders
Skinner, Narelle A.
Skurnik, Mikael
Smirnov, George
Smirnov, Igor V.
Solovjeva, Tamara F.
Somova-Isachkova, Larisa M.
Soranummi, Hanna
Sprague, Lisa D.
Stabler, Richard A.
Steinmetz, Ivo
Strauch, Eckhard

Strieder, Thea M.
Suladze, Tamuna
Sundberg, Lena
Suomalainen, Marjo
Svarval, Alena V.
Świerzko, Anna S.

T

Taylor, Victoria L.
Tennant, Sharon M.
Thisted Lambertz, Susanne
Thomson, Nicholas
Tidhar, Avital
Timchenko, Nelly
Titball, Richard W.
Trasak, Claudia
Tsarkov, Vitali V.
Tseneva, Galina Y.
Tsereteli, David
Tvardovskaia, Natalia

U

Underwood, Cindy D.

V

Vasilieva, Galina I.
Velan, Baruch
Velijanashvili, Ioseb
Venho, Reija
Vesselinova, Anna
Vidyaeva, Nadezhda A.
Vilen, Heikki
Vinogradov, Evgeny V.
Virkola, Ritva
Voskressenskaya, Ekaterina A.
Vostrikova, Olga P.

W

Wan, Jeremy
Wang, Lei
Warren, Martin J.
Wenzel, Björn E.
Wiedemann, Agnès
Wiersinga, Wilma M.
Williamson, E. Diane
Witney, Adam A.
Wolf-Watz, Hans
Worsham, Patricia L.
Wren, Brendan W.

Y

Yakushev, Anton V.
Yasnetskaya, Elena
Yoshida, Shin-Ichiro

Z

Zakalashvili, Mariam
Zangaladze, Ekaterine
Zavialov, Anton V.
Zhuravlev, Yuriy
Zudina, Irina V.
Zvi, Anat

Preface

For the eighth time the yersiniologists all over the world gathered together when the 8th International Symposium on *Yersinia* was organized by University of Turku and Turku Microbiology Society in Turku, Finland. Over 250 delegates from 28 countries attended the Symposium.

The Symposium logo (Picture 4, next page) presents a bacteriophage attached to the surface of the bacterium. One can easily imagine that most of the aspects covered in this Symposium are included in the logo: the bacteriophage genome encodes for structural proteins, adhesins and effector proteins that interact with the host cell in most intricate ways to carry out their mission. Life of the bacteriophage depends on the tightly regulated interplay between the phage and the host proteins. This all is also true between *Yersinia* and the different hosts and environments it encounters during its life cycle.

This Symposium Proceedings volume is based on the oral and poster presentations given during the Symposium. The volume has been divided into six parts covering topics such as genomics, surface structures, bacteriophages, molecular and cellular pathogenesis, molecular epidemiology and diagnostics, gene regulation, clinical aspects and vaccines. These topics reflect righteously the present trends in the bacteriology research.

Mikael Skurnik

José Antonio Bengoechea

Kaisa Granfors

8th International Symposium on Yersinia
September 4 - 8, 2002
Turku, Finland

Picture 1. The logo of the Symposium, designed by Jyri Kurkinen.

Acknowledgments

The members of the Scientific Advisory Board of the Symposium: Elisabeth Carniel (Paris), Guy Cornelis (Basel), Åke Forsberg (Umeå), Kaisa Granfors (Turku), Jürgen Heesemann (Munich), Hannu Korkeala (Helsinki), Virginia Miller (St. Louis), Peter Reeves (Sydney), Roy Robins-Browne (Melbourne), Mikael Skurnik (Turku), Richard Titball (Porton Down), Hans Wolf-Watz (Umeå) and Brendan Wren (London), composed a timely and high quality scientific programme to the Symposium.

Important to the success of the Symposium were the sponsors. The scientific sponsors were Academy of Finland, Federation of European Microbiological Societies, Turku University Foundation and The Wellcome Trust. The commercial sponsors were B. Braun Medical Oy, Biofellows Oy, CareerTrax Inc., HK Ruokatalo Oy, Labnet Oy, Leaf, Orion Pharma, Perkin Elmer Life Sciences Wallac Finland Oy, Sarstedt Oy, Scandinavian Airlines System and Thermo Labsystems Oy.

The Symposium delegates were invited to City Reception sponsored by The City of Turku and hosted by Deputy Mayor Kaija Hartiala. Also special thanks are due to Duke John and Princess Catharina Jagiellonica (see Picture 5, page 2) who hosted the Symposium dinner

The practical organising matters were professionally taken care of by the Congress Office of University of Turku and the Local Organising Committee: Mikael Skurnik (President of Symposium), José Antonio Bengoechea, Tatiana Bogdanovich, Kaisa Granfors, Saija Kiljunen, Elise Pinta, Minna Domander and Jaana Lindgren (see Picture 3, next page).

The photographs included in this book were provided by Andrei Anisimov, Lela Bakanidze, Arve Saebo, Tatiana Bogdanovich, Michael Prentice, Tatiana Meka-Mechenko, Alexander Rakin.

Picture 2. The *Yersinia* Research Group in Turku and most of the Local Organising Committee members entering the Symposium Dinner. Front row (from left): Maria Pajunen, Pia Lahtinen, Yasmin El Tahir, Mikael Skurnik, Elise Pinta, Reija Venho and Claire Pacot-Hiriart. Back row (from left): José Antonio Bengoechea, Heli Nummelin, Tatiana Bogdanovich, Saija Kiljunen and Anna-Leena Skurnik.

Contents

PART I

EVOLUTION AND GENOMICS

Picture 5. The hosts of the Syposium dinner Duke John and his lovely spouse Catharina Jagiellonica with Mikael Skurnik (President of the Symposium),

Picture 6. The Mediaeval Symposium Dinner was served from handy trays. Here Matthew Nilles, Scott Lloyd, Ingo Autenrieth (left) and Jürgen Heesemann (right) are being served.

Chapter 1

Evolution of Pathogenic *Yersinia*, Some Lights in the Dark

Elisabeth CARNIEL
Institut Pasteur, Laboratoire des Yersinia, 28 rue du Dr. Roux, 75724 Paris Cedex 15, France

1. EMERGENCE OF PATHOGENIC *Yersinia*

1.1 The different *Yersinia* species

In 1944, Van Loghem proposed that a new genus, designated *Yersinia*, be separated from the genus Pasteurella (Van Loghem, 1944). This proposition became effective in 1974 (Mollaret and Thal, 1974). The first species identified in this genus by Malassez and Vignal in 1883 was *Y. pseudotuberculosis* (Malassez and Vignal, 1883). The second species, *Y. enterocolitica*, was identified in 1939 by Schleifstein and Coleman (Schleifstein and Coleman, 1939). This species was found to be heterogeneous and to contain several related species ("*Y. enterocolitica*-like") (Bercovier *et al.*, 1980a; Brenner *et al.*, 1980b) that were subsequently designated *Y. intermedia* (Brenner *et al.*, 1980a), *Y. kristensenii* (Bercovier *et al.*, 1980b), *Y. frederiksenii* (Ursing *et al.*, 1980), *Y. aldovae* (Bercovier *et al.*, 1984), and *Y. rohdei* (Aleksic *et al.*, 1987). More recently, *Y. mollaretii* and *Y. bercovieri* were also separated from *Y. enterocolitica* (Wauters *et al.*, 1988). Finally, the species *Y. ruckeri* (De Grandis *et al.*, 1988; Ewing *et al.*, 1978) was included in the genus *Yersinia* but its classification in this genus is controversial.

Among the ten true *Yersinia* species, three are pathogenic for mammals: *Y. enterocolitica*, *Y. pseudotuberculosis* and *Y. pestis*, while the others seem to be devoid of virulence-linked properties. How these two subgroups of pathogenic and non-pathogenic species evolved remains a matter of debate.

The Genus Yersinia, Edited by Skurnik *et al.*
Kluwer Academic/Plenum Publishers, New York, 2003

1.2 The *Yersinia* ancestor

The two most classical scenarios for the evolution of pathogenic and non-pathogenic species within a genus are (i) that the ancestor was a pathogenic bacteria which diverged into non-pathogenic species by loss of virulence traits, or (ii) that the ancestor was a non-pathogenic strain, a sub-population of which acquired virulence factors. More and more evidences argue for the second hypothesis, which is currently recognized as the most likely. This hypothesis probably holds true for the *Yersinia* species too. If so, some virulence genes might have been acquired by the pathogenic ancestor, before its divergence into the three species. This may be the case for the pYV plasmid and for some chromosomal genes such as *inv* and *ail* which are common to the three pathogenic species. If acquisition of these genes were ancestral to their branching into the different species, we expect to observe the same accumulation of mutations, especially at non-synonymous sites, in house-keeping genes and in these virulence loci. When a crude BlastN analysis of five house-keeping genes (*thrA, dmsA, tmk, trpE* and *glnA*) from *Y. enterocolitica* and *Y. pseudotuberculosis* was performed, the percentages of nucleotide identity between orthologs was found to vary between 61 and 84%. For the chromosomal *inv* and *ail* loci, these values were in the same range (59 and 61%, respectively), arguing for an acquisition of these genes in a pathogenic *Yersinia* ancestor, before the divergence of the species. In contrast, the pYV plasmid of *Y. pseudotuberculosis* and *Y. enterocolitica* displayed a very high degree of nucleotide conservation (91-99% identity) at the various loci tested (*yopB, lcrV, yadA, yopH* and *yscT*), suggesting a recent and independent acquisition of this plasmid by *Y. enterocolitica* and *Y. pseudotuberculosis*, after their divergence.

The three currently existing pathogenic *Yersinia* species can be separated into two clearly distinct groups based on their clinical and epidemiological features: *Y. enterocolitica* and *Y. pseudotuberculosis* on one hand, and *Y. pestis* on the other hand. These two groups have drastically different modes of transmission (fecal-oral versus flea bite), clinical symptoms (enteritis versus bubonic plague), and outcomes (mild versus highly lethal infections). Such drastic differences raise the question of the evolutionary relationships between these species and of the emergence of a bacterium such as *Y. pestis* which has such a peculiar behavior compared to the enteropathogenic members of the *Enterobacteriaceae* family.

2. ORIGIN OF *Y. pestis*

Strikingly, DNA-DNA hybridization experiments performed in the 1980s showed that *Y. pestis* and *Y. pseudotuberculosis* are genetically almost identical (>90% chromosomal relatedness) (Bercovier *et al.*, 1980a). It was proposed to group them into one single species, *Y. pseudotuberculosis*, composed of two pathovars: *Y. pestis* and *Y. pseudotuberculosis*. This proposal was not adopted because of the potential hazard this would pose for clinical laboratories.

The evolutionary relationship between the three pathogenic species was revisited recently using multilocus sequence typing (MLST). In this technique, sequences of portions of house keeping genes from multiple strains are aligned. A single nucleotide variation at one position in the sequence is considered as a different allele for this locus. Such an MLST analysis was performed on 5 housekeeping genes (*dmsA, glnA, thrA, tmk* and *trpE*), and a gene involved in the synthesis of lipopolysaccharide (*manB*) for 36 *Y. pestis*, 12 *Y. pseudotuberculosis* and 13 *Y. enterocolitica* strains (Achtman *et al.*, 1999). Analysis of the 36 *Y. pestis* isolates, which represented the global diversity of the species, demonstrated a complete absence of variation at 21,881 synonymous sites, indicating that this species is highly clonal. Furthermore, the conserved *Y. pestis* alleles were identical or nearly identical to alleles found in *Y. pseudotuberculosis* but not in *Y. enterocolitica* species, confirming that *Y. pestis* is genetically much closer to *Y. pseudotuberculosis* than to *Y. enterocolitica*. In bacteria where horizontal genetic exchange is rare, sequence polymorphism reflects the accumulation of mutations at a uniform clock rate and correlates with the time elapsed since the existence of a last common ancestor. The use of two clock rates calibrated for *E. coli* allowed to establish that the *Yersinia* common ancestor arose 42 to 187 million years ago and that *Y. pseudotuberculosis* and *Y. enterocolitica* diverged 0.4 to 1.9 million years ago (Achtman *et al.*, 1999). The same analysis established that *Y. pestis* is a clone which emerged very recently, i.e., within the last 1,500 to 20,000 years, from *Y. pseudotuberculosis*.

3. TRANSMISSION OF *Y. pestis* BY FLEAS

Since *Y. pestis* is a very recent descendent of *Y. pseudotuberculosis*, one may wonder how this organism lost the classical fecal-oral route of transmission and acquired the ability to be transmitted by fleas. A putative evolutionary scenario for the stepwise acquisition of this mode of transmission is presented on Figure 1.

3.1 Acquisition of pFra

In this scenario, the first step was the contact, in the rodent intestinal tract, between *Y. pseudotuberculosis* and an enterobacterium that carried a pFra plasmid, and the subsequent transfer of this replicon to *Y. pseudotuberculosis*. A plasmid sharing high sequence identity with a large portion of pFra was found in a *Salmonella enterica* serovar typhi isolate (Prentice *et al.*, 2001), suggesting that such species might have been the donor of this replicon. The 101 kb pFra plasmid, alternatively termed pMT1, carries a phospholipase D encoding *yplD* gene (Hinnebusch *et al.*, 2000; Rudolph *et al.*, 1999). Acquisition of this gene, previously known as the murine toxin gene, was certainly crucial for vector borne transmission since it promotes survival in and colonization of the flea proventriculus (Hinnebusch *et al.*, 2002b).

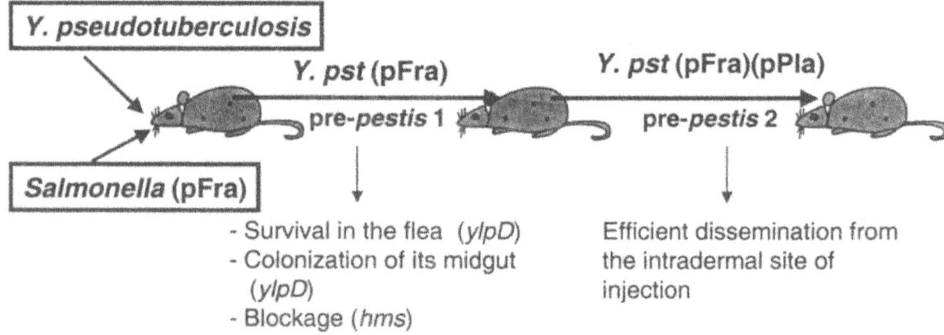

Figure 1. Evolutionary scenario for the transmission of *Y. pestis* by fleas

Furthermore, efficient transmission of *Y. pestis* to a new host by fleas requires that bacteria multiply in the midgut of the insect and form a solid mass that blocks its proventriculus (Bacot and Martin, 1914). During repeated attempts to feed on a new host, the hungry blocked flea is unable to pump blood and subsequently regurgitates the bacteria into the bite wound. The hemin storage locus (*hms*), located on the *Y. pestis* chromosome (Pendrak and Perry, 1991) is responsible for the flea proventriculus blockage (Hinnebusch *et al.*, 1996). This locus is also present on the chromosome of *Y. pseudotuberculosis*. Its presence in the *Y. pestis* progenitor was most likely an essential prerequisite for its transmission by fleas. Any other bacteria carrying the pFra plasmid but lacking the *hms* locus would not have been efficiently transmitted by the insect.

Another prerequisite for the emergence of a flea-borne transmission was the dissemination of the *Y. pseudotuberculosis* (pFra) clone to the bloodstream of the infected rodent, so that the rodent's fleas could ingurgitate the bacteria during their blood meal. This may have indeed been the case since septicemia is a relatively common outcome during *Y. pseudotuberculosis* infections of rodents.

Therefore, the first step in the emergence of a flea-borne transmitted *Y. pestis* progenitor might have required the exceptional conjunction of several factors: (i) coinfection of the intestine of a mammal with a donor bacteria that carried pFra and with the recipient *Y. pseudotuberculosis*, (ii) transfer of this plasmid to the recipient, (iii) presence in the recipient's genome of a locus essential for blockage of the flea proventriculus, and (iv) dissemination of the recipient organism to the bloodstream of the infected animal.

3.2 Acquisition of pPla

The simultaneous presence of the *yplD* and *hms* loci conferred to *Y. pseudotuberculosis* the ability to colonize the flea gut, to block it, and to be efficiently transmitted by the insect. This pre-*pestis* organism was no longer transmitted by the fecal-oral route. Once injected by the flea into the derm of a new host, this newly emerged organism had to deal with an environment drastically different from that of the gut lumen. This is reflected by the much higher median lethal dose (LD_{50}) for mice of *Y. pseudotuberculosis* ($\geq 10^5$ CFU) compared to *Y. pestis* ($\leq$ 10 CFU) upon subcutaneous injection. Subsequent acquisition of a 9.6 kb plasmid unique to *Y. pestis*, alternatively termed pPla, pPCP1, or pPst, has probably been crucial at this stage.

The pPla plasmid carries the plasminogen activator encoding *pla* gene. This factor facilitates *Y. pestis* dissemination from its site of inoculation by cleaving fibrin deposits that trap the organisms and by reducing chemoattraction of inflammatory cells (Sodeinde *et al.,* 1992). Contact between the donor and recipient strains followed by transfer of pPla may have occurred either in the mammalian host or in the flea vector. In the host, the new pre-*pestis* clone now circulated in a normally sterile environment (derm, lymph nodes, spleen, liver, blood), where close contact with a pPla donor organism was unlikely to occur. In contrast, tight contact with potential donor bacteria were more likely to occur in the flea gut. Arguing for this hypothesis is the recent demonstration of genetic transfer of foreign plasmids to *Y. pestis* in the flea midgut environment (Hinnebusch *et al.,* 2002a).

Therefore, the sequential acquisition by *Y. pseudotuberculosis* of the pFra and pPla plasmids along with the pre-existence of the chromosomal *hms* locus have certainly been essential for the emergence of a flea-borne transmitted bacterium such as *Y. pestis*.

4. EXCEPTIONAL PATHOGENICITY OF *Y. pestis*

Some aspects of the particular pathogenicity of *Y. pestis* may also have been the consequence of its new mode of transmission.

4.1 Clinical symptoms

The symptoms caused by enteropathogenic *Yersinia* (fever, abdominal pain and diarrhea) may, at a first glance, seem very different from those of bubonic plague (bubo, fever and septicemia). However, both clinical features result from a tropism of the bacteria for lymph nodes.

In the case of *Y. pseudotuberculosis*, the bacterium is acquired orally and then transits in the intestinal lumen. The draining lymph node by this route of infection is the mesenteric chain whose infection will result in abdominal pains. In the case of *Y. pestis*, the bacteria is injected intradermally and reaches the lymph node draining the cutaneous territory of the flea bite, leading to the well known bubo.

Therefore, although apparently so different, intestinal yersiniosis and plague are both the consequence of the infection of the lymph node draining the site of penetration of the bacterium in the host organism.

4.2 Exceptional level of pathogenicity

In the absence of an adequate treatment, the mortality rate of bubonic plague ranges from 40 to 70% in usually less than a week, while intestinal yersinioses are most often mild and self-limiting. Some explanations for this drastic difference in pathogenicity may once again come from the different modes of transmission.

For an intestinal pathogen such as *Y. pseudotuberculosis*, the most efficient way to be transmitted to new hosts is to cause diarrhea. The bacteria will then be expelled in the environment and will colonize new hosts during the consumption of the contaminated greenery. A severe and lethal infection is not needed for efficient transmission.

The situation is clearly different for *Y. pestis*. This bacterium has to spread to the blood of its host to be transmitted by fleas. The more severe the septicemia, the greater the chances to be taken up by fleas during their blood meal. Therefore, vector-borne transmission of *Y. pestis* most certainly exerted a very strong selective pressure for exacerbating the pathogenicity potential of this bacterium.

4.3 Mechanisms responsible for this exceptional level of pathogenicity

How *Y. pestis* acquired the capacity to cause fulminating septicemia is still an unresolved question. The two *Y. pestis*-specific plasmids, pFra and pPla, clearly play a crucial role in transmission by fleas but they may be, at least in some instances, dispensable for pathogenicity (Davis *et al.*, 1996; Friedlander *et al.*, 1995; Kutyrev *et al.*, 1989; Samoilova *et al.*, 1996; Welkos *et al.*, 1997). Some of the factors essential for *Y. pestis* pathogenicity are most likely chromosomally-encoded.

The sequences of the genomes of two *Y. pestis* strains, CO92 (Parkhill *et al.*, 2001) and KIM (Deng *et al.*, 2002) are now available. These genomes are characterized by a large number of insertion sequences (>130) scattered all over the chromosome and probably responsible for the high genomic plasticity of this species. Another striking feature is the number of non-functional genes (≥149 inactivated coding sequences). IS transposition and mutations that coincidentally increased ectoparasite transmission may have been positively selected. They may also have modified gene functions and/or regulations in such a way that the resulting products would enhance pathogenicity. On the opposite, since proteins necessary for transmission by the fecal-oral route were no longer needed, this may have led to the lack of selective pressure against their mutations. Genes whose inactivation in *Y. pestis* led to increased pathogenicity or transmission by fleas have not yet been identified. Large scale complementation/inactivation experiments will most likely help identifying them.

Another reason for the exceptional pathogenicity of *Y. pestis* may be the acquisition by horizontal transfer of pieces of DNA conferring new pathogenicity potentials. Since *Y. pestis* is a bacterium recently derived from the less virulent progenitor *Y. pseudotuberculosis*, comparative genomics of the two microorganisms may lead to the identification of these newly acquired pieces of foreign DNA. The complete genome sequence of *Y. pseudotuberculosis* has been determined at the Lawrence Livermore National Laboratory in collaboration with our laboratory (http://greengenes.llnl.gov/bbrp/html/microbe.html). *In silico* comparison of *Y. pestis* and *Y. pseudotuberculosis* genomes identified genes and regions present and intact in one species, and absent or mutated in the other species. This comparative genomics along with comparative transcriptome analyses may be valuable tools for identifying horizontally acquired loci or new regulons responsible for the extreme pathogenicity of *Y. pestis*.

5. CONCLUSION

Although there is still a lot of speculation about the emergence and evolution of *Y. pestis*, recent advances made by different laboratories and the promising approach of comparative functional genomics should bring major insights in our knowledge of this field in the near future.

REFERENCES

Achtman, M., Zurth, K., Morelli, C., Torrea, G., Guiyoule, A., and Carniel, E., 1999, *Yersinia pestis*, the cause of plague, is a recently emerged clone of *Yersinia pseudotuberculosis*. Proc. Natl. Acad. Sci. U.S.A. 96: 14043-14048.

Aleksic, S., Steigerwalt, A.G., Bockemuhl, J., Huntley-Carter, G.P., and Brenner, D.J., 1987, *Yersinia rodhei* sp. nov. Isolated from Human and Dog Feces and Surface Water. Int. J. Syst. Bact. 37: 327-332.

Bacot, A.W., and Martin, C.J., 1914, Observations on the mechanisms of the transmission of plague by fleas. J. Hyg. 13 (Plague supplement III): 423-439.

Bercovier, H., Mollaret, H.H., Alonso, J.M., Brault, J., Fanning, G.R., Steigerwalt, A.G., and Brenner, D.J., 1980a, Intra- and interspecies relatedness of *Yersinia pestis* by DNA hybridization and its relationship to *Yersinia pseudotuberculosis*. Curr. Microbiol. 4: 225-229.

Bercovier, H., Steigerwalt, A.G., Guiyoule, A., Huntley-Carter, G., and Brenner, D.J., 1984, *Yersinia aldovae* (Formerly *Yersinia enterocolitica*-Like Group X2): a New Species of *Enterobacteriaceae* Isolated from Aquatic Ecosystems. Int. J. Syst. Bact. 34: 166-175.

Bercovier, H., Ursing, J., Brenner, D.J., Steigerwalt, A.G., Fanning, G.R., Carter, G.P., and Mollaret, H.H., 1980b, *Yersinia kristensenii*: a new species of *Enterobacteriaceae* composed of sucrose-negative strains (formerly called atypical *Yersinia enterocolitica* or *Yersinia enterocolitica*-like). Current Microbiol. 4: 219-224.

Brenner, D.J., Bercovier, H., Ursing, J., Alonso, J.M., Steigerwalt, A.G., Fanning, G.R., Carter, G.P., and Mollaret, H.H., 1980a, Y. *intermedia*: a new species of *Enterobacteriaceae* composed of rhamnose-positive, melibiose-positive, raffinose-positive strains (formerly called atypical *Yersinia enterocolitica* or *Yersinia enterocolitica* -like). Current Microbiol. 4: 207-212.

Brenner, D.J., Ursing, J., Bercovier, H., Steigerwalt, A.G., Fanning, G.R., Alonso, J.M., and Mollaret, H.H., 1980b, Deoxyribonucleic acid relatedness in *Yersinia enterocolitica* and *Yersinia enterocolitica*-like organisms. Curr. Microbiol. 4: 195-200.

Davis, K.J., Fritz, D.L., Pitt, M.L., Welkos, S.L., Worsham, P.L., and Friedlander, A.M., 1996, Pathology of experimental pneumonic plague produced by fraction 1-positive and fraction 1-negative *Yersinia pestis* in African green monkeys (*Cercopithecus aethiops*). Arch. Pathol. Lab. Med. 120: 156-163.

De Grandis, S.A., Krell, P.J., Flett, D.E., and Stevenson, R.M.W., 1988, DNA relatedness of serovars of *Yersinia ruckeri* the enteric Redmouth Bacterium. Int. J. Syst. Bacteriol. 38: 49-55.

Deng, W., Burland, V., Plunkett, G., Boutin, A., Mayhew, G.F., Liss, P., Perna, N.T., Rose, D.J., Mau, B., Zhou, S.G., Schwartz, D.C., Fetherston, J.D., Lindler, L.E., Brubaker, R.R., Plano, G.V., Straley, S.C., McDonough, K.A., Nilles, M.L., Matson, J.S., Blattner, F.R., and Perry, R.D., 2002, Genome sequence of *Yersinia pestis* KIM. J. Bacteriol. 184: 4601-4611.

Ewing, W.H., Ross, A.J., Brenner, D.J., and Fanning, G.R., 1978, *Yersinia ruckeri* sp. nov., the redmouth (RM) Bacterium. Int. J. Syst. Bact. 28: 37-44.

Friedlander, A.M., Welkos, S.L., Worsham, P.L., Andrews, G.P., Heath, D.G., Anderson, G.W., Pitt, M.L.M., Estep, J., and Davis, K., 1995, Relationship between virulence and immunity as revealed in recent studies of the F1 capsule of *Yersinia pestis*. Clin. Infect. Dis. 21: S178-S181.

Hinnebusch, B.J., Perry, R.D., and Schwan, T.G., 1996, Role of the *Yersinia pestis* hemin storage (*hms*) locus in the transmission of plague by fleas. Science 273: 367-370.

Hinnebusch, B.J., Rosso, M.L., Schwan, T.G., and Carniel, E., 2002a, High-frequency conjugative transfer of antibiotic resistance genes to *Yersinia pestis* in the flea midgut. Mol. Microbiol. 46: 349-354.

Hinnebusch, B.J., Rudolph, A.E., Cherepanov, P., Dixon, J.E., Schwan, T.G., and Forsberg, Å., 2002b, Role of *Yersinia* murine toxin in survival of *Yersinia pestis* in the midgut of the flea vector. Science 296: 733-735.

Hinnebusch, J., Cherepanov, P., Du, Y., Rudolph, A., Dixon, J.D., Schwan, T., and Forsberg, A., 2000, Murine toxin of *Yersinia pestis* shows phospholipase D activity but is not required for virulence in mice. Int. J. Med. Microbiol. 290: 483-487.

Kutyrev, V.V., Filippov, A.A., Shavina, N., and Protsenko, O.A., 1989, Genetic analysis and simulation of the virulence of *Yersinia pestis*. Molekuliarnaia Genetika, Mikrobiologia, i Virusologa: 42-7.

Malassez, L., and Vignal, W., 1883, Physiologie pathologique - Tuberculose zoogléique. Académie des Sciences- Comptes Rendus Hebdomadaires des Séances de l'Académie des Sciences 97: 1006-1009.

Mollaret, H.H., and Thal, E., 1974, *Yersinia*. In Bergey's manual of determinative bacteriology Buchanan, R.E.a.G., N. E., (eds), baltimore: Williams and Wilkins, pp. 330-332.

Parkhill, J., Wren, B.W., Thomson, N.R., Titball, R.W., Holden, M.T.G., Prentice, M.B., Sebaihia, M., James, K.D., Churcher, C., Mungall, K.L., Baker, S., Basham, D., Bentley, S.D., Brooks, K., Cerdeño-Tárraga, A.M., Chillingworth, T., Cronin, A., Davies, R.M., Davis, P., Dougan, G., Feltwell, T., Hamlin, N., Holroyd, S., Jagels, K., Karlyshev, A.V., Leather, S., Moule, S., Oyston, P.C.F., Quail, M., Rutherford, K., Simmonds, M., Skelton, J., Stevens, K., Whitehead, S., and Barrell, B.G., 2001, Genome sequence of *Yersinia pestis*, the causative agent of plague. Nature 413: 523-527.

Pendrak, M.L., and Perry, R.D., 1991, Characterization of a hemin-storage locus of *Yersinia pestis*. Biol. Met. 4: 41-47.

Prentice, M.B., James, K.D., Parkhill, J., Baker, S.G., Stevens, K., Simmonds, M.N., Mungall, K.L., Churcher, C., Oyston, P.C.F., Titball, R.W., Wren, B.W., Wain, J., Pickard, D., Hien, T.T., Farrar, J.J., and Dougan, G., 2001, *Yersinia pestis* pFra shows biovar-specific differences and recent common ancestry with a *Salmonella enterica* serovar typhi plasmid. J. Bacteriol. 183: 2586-2594.

Rudolph, A.E., Stuckey, J.A., Zhao, Y., Matthews, H.R., Patton, W.A., Moss, J., and Dixon, J.E., 1999, Expression, characterization, and mutagenesis of the *Yersinia pestis* murine toxin, a phospholipase D superfamily member. J. Biol. Chem. 274: 11824-11831.

Samoilova, S.V., Samoilova, L.V., Yezhov, I.N., Drozdov, I.G., and Anisimov, A.P., 1996, Virulence of pPst(+) and pPst(-) strains of *Yersinia pestis* for guinea-pigs. J. Med. Microbiol. 45: 440-444.

Schleifstein, J.I., and Coleman, M.B., 1939, An unidentified microorganism resembling *B. lignieri* and *Past. pseudotuberculosis*, and pathogenic for man. N.Y. State J. Med. 39: 1749-1753.

Sodeinde, O.A., Subrahmanyam, Y.V.B.K., Stark, K., Quan, T., Bao, Y.D., and Goguen, J.D., 1992, A surface protease and the invasive character of plague. Science 258: 1004-1007.

Ursing, J., Brenner, D.J., Bercovier, H., Fanning, G.R., Steigerwalt, A.G., Brault, J., and Mollaret, H.H., 1980, *Y. frederiksenii*: a new species of *Enterobacteriaceae* composed of rhamnose-positive strains (formerly called atypical *Yersinia enterocolitica* or *Yersinia enterocolitica*-like). Current Microbiol. 4: 213-217.

Van Loghem, J.J., 1944, The classification of the plague bacillus. Antonie van Leeuwenhoek J. Microbiol. Serol. 10: 15-16.

Wauters, G., Janssens, M., Steigerwalt, A.G., and Brenner, D.J., 1988, *Y. mollaretii* sp. nov. and *Y. bercovieri* sp. nov., formerly called *Yersinia enterocolitica* biogroups 3A and 3B. Int. J. Syst. Bacteriol. 38: 424-429.

Welkos, S.L., Friedlander, A.M., and Davis, K.J., 1997, Studies on the role of plasminogen activator in systemic infection by virulent *Yersinia pestis* strain C092. Microb. Pathog. 23: 211-223.

Chapter 2

DNA Adenine Methylation

Victoria L. TAYLOR, Petra C. F. OYSTON and Richard W. TITBALL
Dstl, Chemical and Biological Sciences, Porton Down, Salisbury, Wiltshire, UK

1. INTRODUCTION

Methylation of DNA has many roles in both eukaryotes and prokaryotes. The methylation systems of prokaryotes include restriction/modification systems and solitary methylases. Restriction/modification systems are made up of two enzymes, a methylase and a restriction endonuclease, that recognise the same sequence. Methylation of the target DNA limits accessibility to the restriction endonuclease, thereby preventing "self" restriction. This system allows bacteria to recognise and cleave foreign DNA, due to the different methylation patterns. Solitary methylases do not have a "cognate" restriction endonuclease. They are involved in post-replicative mis-match repair, the synchronisation of DNA replication and the control of gene expression.

There are a number of different solitary methylases and one of the best studied of these is DNA adenine methylase (Dam). The enzyme DNA adenine methyltransferase works by placing a methyl group on to the adenine of the sequence 5'-GATC-3'; the methyl group is donated by S-adenosylmethionine (SAM). Dam can regulate the expression of a number of genes. Some of these genes are regulated by a method termed phase variation. Many molecular switches controlling phase variation involve genomic rearrangements, however phase variation using Dam methylation allows the organism to maintain DNA integrity. The Pap pili of *Escherichia coli* are an example of phase variation controlled by Dam methylation. There are two GATC sites in the *pap* pili promoter region these sites are differentially methylated. When one of the sites is methylated the leucine-responsive regulatory protein (Lrp) binds to the DNA and blocks the

The Genus Yersinia, Edited by Skurnik *et al.*
Kluwer Academic/Plenum Publishers, New York, 2003

promoter from RNA polymerase. If the alternative site is methylated the Lrp binds distally to the promoter allowing RNA polymerase to transcribe the DNA. Post-replicative mis-match repair relies on Dam to distinguish between the parental and daughter strand of DNA. The newly synthesised strand is not methylated for two to four seconds this allows time for the repair mechanism to recognise the errors (Urig *et al*, 2002). The mismatched bases are repaired in the un-methylated strand. Unlike the bulk of DNA the origin of replication, *ori*C, stays hemi-methylated for approximately one third of the cell cycle. The hemi-methylated *ori*C is bound by an outer membrane protein called SeqA (Skarstad *et al*, 2000). Replication cannot begin until SeqA has released the origin of replication and another protein, DnaA, is at high enough levels.

Heithoff *et al* (1999) produced a *dam* mutant in *Salmonella enterica* serovar Typhimurium. The mutants were able to colonise the Peyers patches but were unable to colonise deeper tissue, such as the spleen. The survival in the Peyers patches enabled the host to mount an immune response. The oral LD_{50} for the *dam* knockout mutant was 10,000 times higher than that of the wild-type strain, while the intraperitoneal LD_{50} was 1,000 times higher than that of the wild-type strain. The mutant also provided protection against challenge with wild-type *S. enterica* serovar Typhimurium.

The aims of the project are to construct *dam* mutants of *Yersinia pestis* and *Yersinia pseudotuberculosis*. The effect of this mutation on virulence will be studied and compared to the levels achieved in the *S. enterica* var Typhimurium. Should these mutants be attenuated then the ability of the mutants to protect against a challenge of wild-type *Y. pestis* will then be studied.

2. RESULTS AND DISCUSSION

The *Y. pestis* genome was searched for the presence of a gene homologous to the *dam* gene found in *S. enterica* serovar Typhimurium. A gene was found that had an expect value of $8e^{-98}$ when compared to the *S. enterica dam* gene. The *Y. pseudotuberculosis dam* gene is identical to the *dam* gene of *Y. pestis.* Figure 1 shows the amino acid sequence for the *dam* gene in both *Y. pestis* and *Y. pseudotuberculosis* aligned with the sequence from *S. enterica* serovar Typhimurium.

Restriction digests were used to confirm the presence of DNA adenine methylation. There are three restriction endonucleases that can be used to determine the presence of DNA adenine methylation: *Mbo*I, *Sau*3AI and *Dpn*I. These endonucleases all recognise the nucleotide sequence GATC, which is the same as the sequence recognised by the Dam methylase. The restriction endonuclease *Mbo*I cannot cleave DNA that has the adenine of the recognition sequence methylated (Barbeyron *et al*, 1984). *Sau*3AI cleaves DNA at the recognition sequence regardless of the state of methylation, whereas *Dpn*I will only cleave at the recognition site if the DNA is methylated. In microorganisms that carry out Dam the DNA would not be digestible by *Mbo*I but would be digestible by both *Sau*3AI and *Dpn*I. In microorganisms that do not carry out Dam the DNA would not be digestible by *Dpn*I but would be digestible by both *Sau*3AI and *Mbo*I.

Yersinia	MKKNRAFLKWAGGKYPLVDDIRRHLPAGDCLIEP
S. typhimurium	MKKNRAFLKWAGGKYPLLDDIKRHLPKGECLVEP
Yersinia	FVGAGSVFLNTEFESYILADINNDLINLYNIVKLRTD
S. typhimurium	FVGAGSVFLNTDFSRYILADINSDLISLYNIVKLRTD
Yersinia	DFVRDARVLFTGDFNHSELFYQLRQEFNASTDAYRR
S. typhimurium	EY VQASRELFMPETNQAEVYYQLREEFNTCQDPFRR
Yersinia	ALLFLYLNRHCYNGLCRYNLSGEFNVPFGRYKKRYK
S. typhimurium	AVLFLYLNRYGYNGLCRYNLRGEFNVPFGRYKR-----
Yersinia	KPYFPEAELYWFAEKSQNAVFVCEHYQETLLKAVQGA
S. typhimurium	---PYFPEAELYHFAEKAQNAFFYCESYADSMARADKSS
Yersinia	VVYCDPPYAPLSATANFTAYHTNNFGIADQQNLARLA
S. typhimurium	VVYCDPPYAPLSATANFTAYHTNSF SLT--QQAHLAEI
Yersinia	YQLSTESKVPVLISNHDTELTRNWYHQAASLHVVTAR
S. typhimurium	AENLVSNRI PVLISNHDTALTREWYQLAK- LHVVKVR
Yersinia	RTISRNILGRSKVNELLALYS
S. typhimurium	PSISSNGGTRKKVDELLALYQPGVATPARK

Figure 1. A comparison of the amino acid sequence of the *dam* gene from *Y. pestis* and *Y. pseudotuberculosis* with the *dam* gene from *Salmonella enterica* serovar Typhimurium. Dashes indicate gaps in the aligned sequences.

Genomic DNA from *Y. pestis* strains GB, CO92 and Pexu2 and *Y. pseudotuberculosis* strain 32953 were incubated with either *Mbo*I, *Sau*3AI or no enzyme for four hours at 37 °C in order to determine the methylation state of the DNA. Complete digestion of the DNA by *Sau*3AI and no digestion by *Mbo*I (figure 2) suggested that the DNA of *Y. pestis* and *Y. pseudotuberculosis* is methylated in a dam-dependent manner. This pattern was also achieved with other strains *Y. pestis* that were tested.

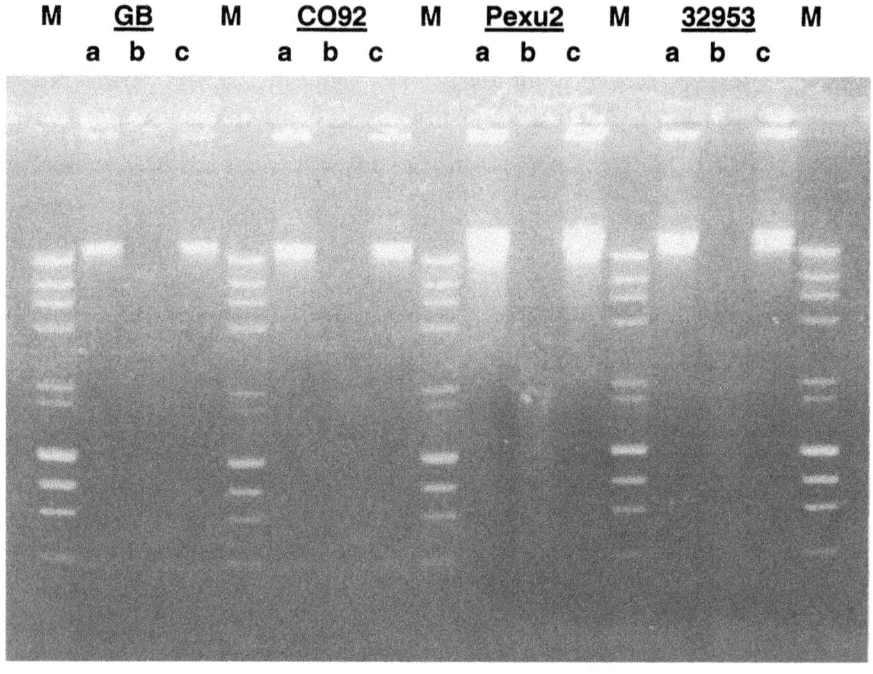

Figure 2. Digestion of *Yersinia* DNA to identify dam methylation. Key: DRIgest III markers (M) (Amersham Biosciences), *Mbo*I (a) (Promega), *Sau*3AI (b) (Promega), no enzyme (c).

The construction of the *dam* knockout mutants was made by amplifying the 5' end of the *dam* gene and the 3' end of the *dam* gene in separate reactions. A kanamycin cassette was inserted between the two incomplete portions of the *dam* gene. The inactivated gene was inserted into a suicide vector and used to construct *dam* mutants of *Y. pestis* and *Y. pseudotuberculosis*.

Preliminary results indicate that both the *Y. pestis* and the *Y. pseudotuberculosis dam* mutants are attenuated in mice. Future work will involve the mutants undergoing further attenuation and protection studies.

REFERENCES

Barbeyron, T., Kean, K. and Forterre, P. 1984. Methylation of GATC sites is required for precise timing between rounds of DNA replication in *Escherichia coli.* Journal of Bacteriology. **160**(2):586-590

Heithoff, D.M., Sinsheimer, R.L., Low, D.A. and Mahan, M.J. 1999. An essential role for DNA adenine methylation in bacterial virulence. Science. **284**:967-970

Skarstad, K., Leuder, G., Lurz, R., Speck, C. and Messer, W. 2000. The *Escherichia coli* SeqA protein binds specifically and co-operatively to two sites in hemimethylated and fully methylated *ori*C. Molecular Microbiology. **36**(6):119-1326

Urig, S., Gowher, H., Hermann, A., Beck, C., Fatemi, M., Humeny, A. and Jeltsch, A. 2002. The *Escherichia coli* dam DNA methyltransferase modifies DNA in a highly processive reaction. Journal of Molecular Biology. **319**:1085-1096

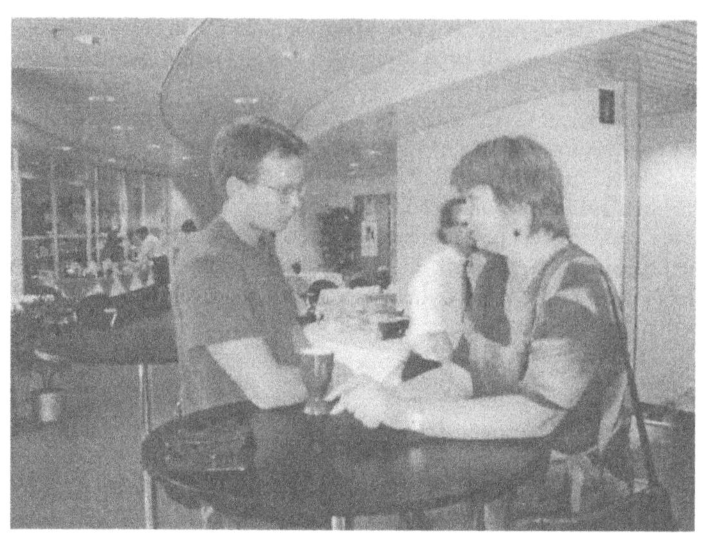

Picture 7. Brooks Wheeler and Tatiana Gremyakova on coffee break.

Picture 8. Emilio Garcia, Mrs. Garcia and Lela Bakanidze.

Chapter 3

Tracing Acquisitions and Losses in *Yersinia* Genomes

Alexander RAKIN*, Andrey GOLUBOV*, Azuka IWOBI, and Jürgen HEESEMANN

Max von Pettenkofer-Institute of Hygiene and Medical Microbiology, University of Munich, Pettenkofer Street 9a, 80336, Munich, Germany.

*A. Rakin and A. Golubov contributed equally to this work.

1. INTRODUCTION

Three species of *Yersinia* are of threat to humans although having a different outcome of infection. *Yersinia pestis* is the causative agent of plague, while *Y. pseudotuberculosis* and *Y. enterocolitica* are enteropathogens. *Y. pestis* and *Y. pseudotuberculosis* belong to the same species, while *Y. enterocolitica* consists of a heterogeneous group of organisms, being less than 50% related to *Y. pseudotuberculosis*. Clinical picture evoked by *Y. enterocolitica* (Bottone, 1997) differs from that of *Y. pseudotuberculosis,* which results mainly in generalized forms of infection. Human pathogenic *Y. enterocolitica* forms two groups, mouse-lethal biogroup (BG) 1B strains including serotypes (ST) O:4, O:8, O:13a, O:13b, O:18, O:20, and O:21 and mouse non-lethal strains (BG 2-4, ST O:3, O:9, O:5,27) (Carter, 1975). The mouse-lethal strains are found to be more destructive in humans, causing extensive ulceration of the gastrointestinal tract and even death. To map out genetic differences that might be responsible for the differences in clinical manifestations in enteropathogenic *Yersinia* we have applied a method of suppression subtractive hybridization.

The Genus Yersinia, Edited by Skurnik *et al.*
'wer Academic/Plenum Publishers, New York, 2003

2. MATERIALS AND METHODS

Y. enterocolitica WA-314 biogroup (BG) 1B serotype (ST) O:8 strain and Y-108 (BG 4, ST O:3), *Y. pseudotuberculosis* Yp346 (ST O:3) and YPIII (ST O:3) were from the strain collection of the Max von Pettenkofer-Institute. Suppression subtractive hybridization (SSH) was performed using the modified protocol from Clontech Laboratories (Diatchenko *et al.*, 1996). Sequencing was performed by AGOWA GmbH as part of BMBF Kompetenzwerk "Genomforschung an pathogenen Bakterien". Analysis of sequences was carried out with the BLAST program from NCBI and the TIGR-CMR program.

3. RESULTS AND DISCUSSION

3.1 Human pathogenic *Y. enterocolitica*

To map out genetic differences between human pathogenic *Y. enterocolitica* we subtracted genomic DNA of *Y. enterocolitica* WA-314, ST O:8, BG 1B representing the mouse lethal group with genomic DNA of Y-108, ST O:3, BG 4 belonging to mouse non-lethal group. Both strains were used each as tester and driver in SSH. By this approach both O:8- and O:3-specific sequences were recognized in hybridization (most significant sequences mapped out by subtractive hybridization are listed in Table 1). The percentage of tester-specific clones was 44% for O:3 strain (in total 83 O:3-specific clones) and 90% - for O:8 isolate (in total 807 O:8-specific clones). It also turned out that 44% *Y. enterocolitica* O:3-specific and 60% *Y. enterocolitica* O:8-specific sequences depicted high similarity (70-95%) to sequences of the completed *Y. pestis* CO92 genome (Parkhill *et al.*, 2001).

Sequences with high similarity to genes encoding Rtx-like cytotoxin of *V. cholerae* (Wei Lin *et al.*, 1999), nematicidal toxin of *Xenorhabdus bovienii*, putative virulence-associated autotransporters YapA and YapG in *Y. pestis* CO92 (Parkhill *et al.*, 2001), as well as biotype-specific *Eco*RII-type restriction-modification system were identified as *Y. enterocolitica* O:3-specific.

Y. enterocolitica O:8-specific sequences show high similarity to genes coding for Pic-like protease with mucinase and hemagglutinin activities encoded on the she-pathogenicity island of *S. flexneri* (Al-Hasani *et al.*, 2001), a putative endochitinase from *V. cholerae* (Folster *et al.*, 2002), a cytotoxic necrotizing factor (Lockman *et al.*, 2002) and *Yen*I restriction-modification system (Kinder *et al.*, 1993).

Sequences with high similarity to known and putative virulence factors uncovered by subtractive hybridization in *Y. enterocolitica* O:8 might be responsible for the high pathogenicity phenotype of *Y. enterocolitica* 1B together with the genes of the high-pathogenicity island that is involved in yersiniabactin mediated iron uptake (Pelludat *et al.*, 2002). On the other hand, Rtx-like cytotoxin represents a good candidate for role of a chromosomally encoded pathogenicity factor of *Y. enterocolitica* O:3 strains.

Table 1. Y. enterocolitica – specific fragments and their characteristics

No	Homologies to predicted gene and encoded protein	Positives (% amino acid)
Y. enterocolitica O:3		
01_05	*wbbX*, lipopolysaccharide O-antigen synthesis protein, *Y. enterocolitica* O:3	162/162 (100%)
01_33	*xnp2*, nematicidal toxin, *Xenorhabdus bovienii*	139/275 (49%)
01_52	*rtxA*, cytotoxin, *V. cholerae*	150/192 (77%)
02_11	*yapG*, putative autotransporter protein, *Y. pestis*	55/67 (81%)
02_72	*yapA*, putative autotransporter protein, *Y. pestis*	79/98 (80%)
04_62	*Eco*RII, DNA-methyltransferase (cytosine-specific), *E. coli*	198/222 (88%)
Y. enterocolitica O:8		
05_19	*yenI*, methyltransferase-endonuclease, *Y. enterocolitica* O:8	39/39 (100%)
05_84	VCA0811, putative chitinase, *V. cholerae*	83/165 (49%)
06_37	*cya*, adenylate cyclase, *Y. enterocolitica* O:8	31/32 (96%)
06_53	*hmsR*, putative hemin binding protein, *Y. pestis*	41/42 (97%)
08_62	*pic*, protease with hemagglutinin activity, *S. flexneri*	67/134 (49%)
12_34	*cnf*, cytotoxic necrotizing factor, *Y. pseudotuberculosis*	64/140 (45%)

3.2 *Y. pseudotuberculosis* serotype O:3

Y. pseudotuberculosis serotype O:3 strains can be divided into two groups (Rakin *et al.*, 1995). Strains of the first group carry a partial high-pathogenicity island (HPI) that lacks the siderophore receptor while strains of the second group have a siderophore system different from the yersiniabactin one. Strains of the latter group are highly virulent to mice in contrast to less virulent ones making up the first group (Fukushima *et al.*, 2001). To map out genomic differences between these two groups, genomic DNAs of siderophore-negative Yp346(pYV346) and siderophore-positive YpIII(pIB1) strains were subtracted. More than 200 YpIII - and 229 Yp346 -

specific sequences were uncovered by SSH. Surprisingly, sequences encoding the cytotoxic effector protein YopT (Zumbihl *et al.*, 1999) and its chaperone SycT were absent from YpIII isolate. On the other hand, several candidates with similarity to iron uptake genes (YPO0776, a putative siderophore biosynthesis protein in *Y. pestis*; *pvsA*, coding for a pyoverdine synthetase A from *Pseudomonas fluorescens* (Mossialos *et al.*, 2002); alr2626, a ferrichrome receptor from *Nostoc sp.*), were uncovered in YpIII strain. Also, sequences with 90% similarity to *rhuM* (encoding a putative cytoplasmic protein) present on the SPI-3 island of *S. typhimurium* and 65% similarity to OrfX of *N. meningitidis* island were present in Yp346 strain. Sequences similar to those coding for putative autotransporters of *Y. pestis* CO92 were differently represented in both O:3 strains: YapA and B were found in Yp346, while YapD and H - in YpIII. Similarly, YapA and G were uncovered only in *Y. enterocolitica* O:8, BG 1B.

Thus method of subtractive hybridization using different tester-driver combinations proved to be effective in elucidating genomic differences that may underlie the observed differences in pathogenesis of medically important *Yersiniae*.

REFERENCES

Al-Hasani, K., Rajakumar, K., Bulach , D., Robins-Browne, R., Adler, B. and Sakellaris, H., 2001, Genetic organization of the she pathogenicity island in *Shigella flexneri* 2a. *Microb. Pathog.* **30**: 1-8.

Bottone, E. J., 1997, *Yersinia enterocolitica*: the charisma continues. *Clin. Microb. Rev.*, **10**:257-276.

Carter, P.B., 1975, Pathogenicity of *Yersinia enterocolitica* for mice. *Infect. Immun.* **11**:164-170.

Diatchenko, L., Lau, Y.-F.C., Campbell, A. P., *et al.*, 1996, Suppression Subtractive Hybridization: A method for generating differentially regulated or tissue-specific cDNA probes and libraries. *Proc. Natl. Acad. Sci. USA* **93**: 6025-6030.

Folster, J.P. and Connell, T.D., 2002, The extracellular transport signal of the *Vibrio cholerae* endochitinase (ChiA) is a structural motif located between amino acids 75 and 555. *J. Bacteriol.* **184**: 2225-2234.

Fukushima H, Matsuda Y, Seki R, *et al.*, 2001, Geographical heterogeneity between Far Eastern and Western countries in prevalence of the virulence plasmid, the superantigen *Yersinia pseudotuberculosis*-derived mitogen, and the high-pathogenicity island among *Yersinia pseudotuberculosis* strains. *J. Clin. Microbiol.* **10:**3541-7.

Kinder, S. A., Badger, J. L., Bryant, G. O., *et al.*, 1993, Cloning of the *Yen*I restriction endonuclease and methyltransferase from *Yersinia enterocolitica* serotype O8 and construction of a transformable R^-M^+ mutant. *Gene*, **136**:271-275.

Lockman, H.A., Gillespie, R.A., Baker, B.D. and Shakhnovich, E., 2002, *Yersinia pseudotuberculosis* produces a cytotoxic necrotizing factor. *Infect. Immun.* **70**: 2708-2714.

Mossialos, D., Ochsner, U., Baysse, C., *et al.*, 2002, Identification of new, conserved, non-ribosomal peptide synthetases from fluorescent pseudomonads involved in the biosynthesis of the siderophore pyoverdine. *Mol. Microbiol.* **45**: 1673-1685.

Parkhill, J., Wren, B. W., Thomson, N. R., Titball, *et al.*, 2001. Genome Sequence of *Yersinia pestis*, the causative agent of plague. *Nature* **413**: 523-527.

Pelludat, C., Hogart, M., and Heesemann, J., 2002, Transfer of the core region genes of the *Yersinia enterocolitica* WA-C serotype O:8 high-pathogenicity island to *Y. enterocolitica* MRS40. *Infec. Immun.* **70**:1832-1841.

Rakin, A., Urbitsch, P., and Heesemann, J., 1995, Evidence for two evolutionary lineages of highly pathogenic *Yersinia* species. *J. Bacteriol.* **177**:2292-2298.

Wei Lin, Fullner, K. J., Clayton, R., Sexton, J. A. *et al.*, 1999. Identification of a *Vibrio cholerae* RTX toxin gene cluster that is tightly linked to the cholera toxin prophage. *Proc. Natl. Acad. Sci. USA*, **96**: 1071-1076.

Zumbihl R, Aepfelbacher M, Andor A, *et al.*, 1999, The cytotoxin YopT of *Yersinia enterocolitica* induces modification and cellular redistribution of the small GTP-binding protein RhoA. *J. Biol. Chem.* **274**: 29289-93.

Picture 9. Alexander Rakin, Mikael Skurnik and Lela Bakanidze on a break during the Blues Session at School

Chapter 4

Subtractive Hybridization Uncovers Novel Pathogenicity-Associated Loci in *Yersinia enterocolitica*

Azuka N. IWOBI[1], Alexander RAKIN[1], Emilio GARCIA[2], and Jürgen HEESEMANN[1]
[1]*Max von Pettenkofer Institute for Hygiene and Medical Microbiology, University of Munich, Pettenkofer street 9a, 80336 Munich, Germany;* [2]*Lawrence Livermore National Laboratory, Human Genome Center L-452, 7000 East Avenue, Livermore, CA 94550, USA*

1. INTRODUCTION

Yersinia enterocolitica species embrace an heterogeneous group of bacteria with different biochemical and antigenic properties. Typically two groups can be found within this species: the non-pathogenic isolates and the pathogenic group. The latter is further divisible into a low-pathogenic, non-mouse lethal group and a highly-pathogenic, mouse-lethal group. The non-pathogenic isolates lack all the virulence markers unique to pathogenic yersiniae, such as the pYV virulence plasmid, the chromosomally encoded Ail (attachment invasion locus) and Inv (invasin) proteins. The low-pathogenic isolates however carry almost all the virulence markers found in the highly-pathogenic isolates, with the exception of the chromosomal locus designated the HPI (high-pathogenicity island) which is involved in the synthesis and transport of the siderophore yersiniabactin. Only the highly-pathogenic, mouse lethal serotypes harbor the HPI (Carniel *et al.,* 1992; Carter, 1975; Pelludat *et al.,* 1998).

In this work, the method of subtractive hybridization was employed to map out novel genetic loci that are unique to the highly-pathogenic, mouse-lethal group of *Y. enterocolitica* species. Here we describe a novel type II secretion system exclusive to the highly-pathogenic *Y. enterocolitica* biotype 1B strains. In addition, a novel IS*10*-like mobile element, IS*1330,* was found to be present in various pathogenic *Y. enterocolitica* species (Iwobi *et al.*, 2002).

2. MATERIALS AND METHODS

2.1 Molecular genetic techniques

Plasmid DNA was prepared by the modified nucleic acid extraction kits from Nucleobond. Chromosomal DNA was prepared as previously described (Ausubel *et al.*, 2000). The cosmid gene bank was constructed using the Supercos 1 cosmid vector (Stratagene). High stringency Southern blot hybridizations were performed with digoxigenin (DIG)- labeled probes at 68°C, following standard procedures.

2.2 Subtractive hybridization

Suppressive subtractive hybridization was based on the PCR-Select Genomic Subtraction system from Clontech. *Y. enterocolitica* NF-O (non-pathogen) was the driver strain and *Y. enterocolitica* WA-314 (highly-pathogenic, mouse-lethal serotype) was the tester strain. DNA sequencing was performed by the dideoxy-chain terminating method on an automated ABI prism 377 DNA Sequencer. The subtracted fragments were analysed by the BLASTN and BLASTX programs provided by NCBI (National Center for Biotechnology Information) and TIGR (The Institute for Genomic Research).

2.3 Construction of a *yts1 E* mutant

The *yts1E* gene was inactivated by insertional mutagenesis by a kanamycin cassette (with transcriptional terminator), which was confirmed by PCR, Southern blot and sequencing.

2.4 Animal experiments

Female BALB/c mice (6 weeks old) were infected (groups of five) with yersiniae orogastrically at a dosage of 5 X 10^9 colony forming units (CFUs). Mice were monitored twice daily and fourty-eight hours after infection they were sacrificed and the average bacterial load from infected organs (liver, spleen and Peyer's patches) was determined. For intravenous inoculation,

10^4 CFUs were inoculated into mice, which were subsequently monitored for 2 days before sacrifice.

3. RESULTS AND DISCUSSION

A total of 200 subtracted clones were analyzed through sequencing and Table 1 gives a summary of some significant homologies.

Table 1. Profile of some subtracted fragments of interest and their corresponding identities to previously described proteins

Clone	Insert size (bp)	Homology	Identity/ Similarity (amino-acids)
o14	1150	Putative transposase in *S. typhimurium*	63/99 (69%)
n35	1100	Putative transporter, permease protein in *Y. pestis*	156/173 (90%)
n5	650	Probable transcriptional regulator in *Pseudomonas aeruginosa*	26/73 (35%)
n16	700	Putative chitinase in *V. cholerae*	75/145 (51%)
n28	850	General secretion pathway protein E in *V. cholerae*	137/226 (60%)
m90	650	LysR family of transcriptional regulators (*Xylella fastidosa*)	80/148 (54%)
a82	500	Putative exported protein in *Y. pestis*	77/145 (53%)
		Haemagglutinin related protein in *N. meningitidis*	64/137 (46%)

3.1 Subtractive hybridization uncovers a novel IS*10*-like element (IS*1330*) unique to pathogenic *Y. enterocolitica* strains.

Through subtractive hybridization, a novel IS*1330* insertion sequence was uncovered. IS*1330* comprises two imperfect 19-bp inverted repeats (IR) and is flanked by a 10-bp duplication of the target sequence. The putative polypeptide, which is predicted to be a 46 kDa protein in size, exhibited a highly basic p*I* of 10.6, common to transposases (Iwobi *et al.*, 2002). Reverse transcription analysis performed to determine whether IS*1330* transposase gene is transcribed was positive. Southern blot analysis revealed that IS*1330* was present only in pathogenic *Y. enterocolitica* serotypes and could therefore be useful in the epidemiological typing of the species.

3.2 A novel type II secretion cluster (Yts1) unique to highly-pathogenic *Y. enterocolitica* serotypes.

The subtracted fragment n28 was highly homologous to the *epsE* gene of *V. cholerae*, part of the type II secretion cluster of this organism. Cosmid n28 carrying this subtracted fragment of interest was sequenced and a unique genetic locus was uncovered, spanning 12kb and containing 13 open reading frames, which potentially encodes a Type II secretion apparatus. The genes were designated *yts1C - S* (*Yersinia* Type II Secretion 1), these letters corresponding to the generally accepted nomenclature for the type II pathway identified in different *Enterobacteriaceae*. Figure 1 presents the genes on this putative type II secretion cluster. A gene, designated *chiY*, flanks downstream the *yts1* cluster and it encodes a putative chitin-binding protein.

Southern blot analysis carried out to determine the distribution of the *yts1* cluster among different *Yersiniae* showed that the cluster is exclusive to the highly-pathogenic *Y. enterocolitica* strains.

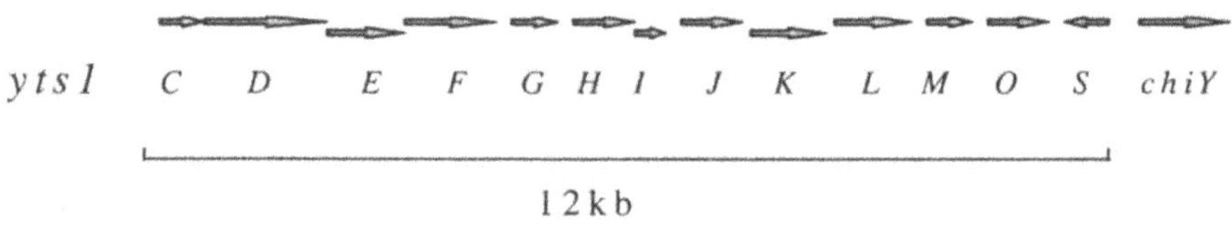

Figure 1. Organization of the *yts1* gene cluster

3.3 Yts1 secreton contributes to the virulence of the highly-pathogenic *Y. enterocolitica* species.

In order to assess the impact of the Yts1 secreton on virulence mouse experiments were carried out with wild type *Y. enterocolitica* WA-314, and the secretion deficient mutant, WA-314 *yts1E*. When the oral route of inoculation was used the secretion-deficient mutant was impaired in liver and spleen colonization compared to the wild type bacteria. In contrast, no significant differences were observed in bacterial counts when the intravenous route of inoculation was used. The Yts1 secreton might thus play a role in the dissemination of the bacteria into deeper tissues (liver and spleen) following the initial invasion of the host.

REFERENCES

Ausubel, F. M., Brent, R., Kingston, R. E., Moore, D. D., Seidman, J. G., Smith, J. A., and Struhl, K. (ed.) 2000. Current Protocols in Molecular Biology. John Wiley & Sons, Inc., USA.

Carniel, E., Guiyoule, A., Guilvout, I. , and Mercereau-Puijalon, O. 1992. Molecular cloning, iron-regulation and mutagenesis of the *irp2* gene encoding HMWP2, a protein specific for the highly-pathogenic *Yersinia*. Mol. Microbiol. **6**: 379-388

Carter, P. B. 1975. Pathogenicity of *Yersinia enterocolitica* for mice. Infect. Immun. **11**: 164-170.

Iwobi, A., Rakin, A., Garcia, E., and Heesemann, J. 2002. Representational Difference Analysis uncovers a novel IS*10*-like Insertion Element unique to pathogenic strains of *Yersinia enterocolitica*. FEMS Microbiol. Lett. **210**: 251-255.

Pelludat, C., Rakin, A., Jacobi, C. A., Schubert, S., and Heesemann J. 1998. The Yersiniabactin Biosynthetic Gene Cluster of *Yersinia enterocolitica*: Organization and Siderophore - Dependent Regulation. J. Bacteriol. **180**: 538-546.

Picture 10. Robert Brubaker, Jürgen Heesemann and Olga Podladchikova on the City Reception.

Picture 11. Igor Domaradsky, Robert Brubaker, Wendy Orent and Vladimir Motin

Chapter 5

Identification of Genes Involved in *Yersinia pestis* Virulence by Signature-Tagged Mutagenesis

Yehuda FLASHNER, Emanuelle MAMROUD, Avital TIDHAR, Raphael BER, Moshe AFTALION, David GUR, Anat ZVI, Naomi ARIEL, Baruch VELAN, Avigdor SHAFFERMAN and Sara COHEN.
Department of Biochemistry and Molecular Genetics, Israel Institute for Biological Research, P.O.Box 19, Ness-Ziona, Israel 74100

1. INTRODUCTION

The three pathogenic *Yersinia* species: *Y. pestis, Y. pseudotuberculosis,* and *Y. enterocolitica* are closely related, but differ in their mode of infection. Both *Y. enterocolitica* and *Y. pseudotuberculosis* are fecal-oral pathogens that cause invasive gastrointestinal diseases. On the other hand, *Y. pestis,* which is transmitted by bites of infected fleas or by uptake of an infective aerosol, is the etiological agent of plague, a fatal disease. The well characterized virulence factors of *Y. pestis* are mainly those encoded by the 70kb plasmid, common to all three *Yersinia* pathogenic species. Only few virulence factors unique to *Y. pestis* are known to reside on the two *Y. pestis* specific plasmids pMT1 and pPCP1. The sequence of the pMT1 plasmid revealed several hypothetical proteins, whose relevance to virulence is yet to be determined. Moreover, the sequencing of the entire *Y. pestis* genome was recently completed, and initial genome analysis indicates the existence of genetic elements, whose hypothetical products are of possible relevance to *Y. pestis* pathogenesis, including several pathogenicity islands. Several methodologies have been developed during the last decade to identify which bacterial genes are required for penetration, colonization and multiplication inside the host. These include IVET (*In-Vivo* Expression Technology), DFI (Differential Fluorescence Induction) and STM (Signature-Tagged

The Genus Yersinia, Edited by Skurnik *et al.*
›wer Academic/Plenum Publishers, New York, 2003

Mutagenesis). These strategies led to the identification of genes expressed preferentially in specific organs, or directly involved in virulence of various pathogenic bacteria.

STM was undertaken to isolate attenuated *Y. enterocolitica* mutants following intraperitoneal inoculation (Darwin and Miller, 1999), or to select attenuated *Y. pseudotuberculosis* strains through orogastric or intravenous infection (Mecsas *et al.*, 2001; Karlyshev *et al.*, 2001).

Here we present the application of STM to select attenuated mutants of *Y. pestis*. For this purpose we used mice infected subcutaneously with the virulent Kimberley53 strain (Ben-Gurion and Hertman, 1958) (LD_{50} = 1CFU). This infection model mimics infection of humans by flea bites, which leads to the development of bubonic plague.

2. RESULTS AND CONCLUSIONS

Application of the STM strategy in mice infected subcutaneously with *Y. pestis* required the establishment of measures for a consistent and reproducible infection conditions. For this purpose development of disease induced by various infection doses was monitored at several time points post infection. Based on these studies the optimal conditions for tagged mutants screening were determined: infection dose of 10^4CFU carrying 20 differently tagged mutants (input pool), and extraction of mutant population from spleen at 48 hours post infection (output pool).

Screening of the virulent *Y. pestis* Kimberley53 mutant library was carried out in two consecutive cycles. Out of 300 mutants screened, 16 were consistently undetectable in spleens following these two cycles. Carrying out infection of mice tests (at infection dose of 100CFU), as well as *in-vivo* competition assays, allowed the grouping of the selected attenuated mutants into four classes: a. defective in ability to induce mortality, b. leading to significant delay in mean time to death, c. not leading to mortality, but having significantly low competition index (CI) values, and d. indistinguishable from wild type.

DNA sequencing of the interrupted sequences of all selected mutants led to identification of genes coding for factors involved in global bacterial physiology (e.g., purine synthesis, DNA metabolism, transcriptional elongation, and response to environmental stress), genes coding for hypothetical polypeptides, whose functions have to be determined, and the already well established virulence gene (*lcrF*). A selected group of mutants are being evaluated as attenuated *Y. pestis* vaccine.

ACKNOWLEDGEMENTS

We would like to thank Dr. D. Holden (UK), and Dr. J.J. Letesson (Belgium) for the gift of tagged pUTminiTn5Km2 plasmids.

REFERENCES

Ben-Gurion, R., and Hertman, I., 1958, Bacteriocin-like material produced by *Pasteurella pestis J. Gen Microbiol.* **19**: 289-297.

Darwin, A.J., and Miller, V.L., 1999, Identification of *Yersinia enterocolitica* genes affecting survival in an animal host using signature-tagged transposon mutagenesis. *Mol. Microbiol.* **32**:51-62.

Karlyshev, A.V., Oyston, C.F., Williams, K., Clark, G.C., Titball, R.W., Winzeler, E.A., and Wren, B.W., 2001, Application of high-density array-based signature-tagged mutagenesis to discover novel *Yersinia* virulence-associated genes. *Infect. Immun.* **69**:7810-7819.

Mecsas, J., Bilis, I., and Falkow, S., 2001, Identification of attenuated *Yersinia pseudotuberculosis* strains and characterization of an orogastric infection in BALB/c mice on day 5 post-infection by signature-tagged mutagenesis. *Infect. Immun.* **67**:2779-2787.

Chapter 6

Characterization of Two Conjugative *Yersinia* Plasmids Mobilizing pYV

Stefan HERTWIG, Iris KLEIN, Jens A. HAMMERL and Bernd APPEL
Robert Koch-Institut, Berlin, Germany

1. INTRODUCTION

The 70 kb virulence plasmid (pYV) residing in pathogenic *Yersinia* strains is known to be non-conjugative. However, it has been demonstrated that plasmid Vwa of *Y. pestis* is mobilizable by F using *E. coli* as donor (Allen *et al.*, 1987). Plasmid pYV of *Y. enterocolitica* could be transmitted by cointegrate formation with a mobilizable vector (Heesemann and Laufs, 1983). These data suggest that the *Yersinia* virulence plasmid might contain an origin of transfer. By molecular cloning and subsequent mating, a 5.5 kb *Bam*HI restriction fragment of pYV has been identified mediating conjugal transfer of pBR328 (Allen *et al.*, 1995). Thus pYV might have evolved from a conjugative plasmid. There are only few reports describing conjugative plasmids in *Yersinia*. Kimura *et al.*, (1975) detected conjugative R plasmids in *Y. enterocolitica* and *Y. pseudotuberculosis* conferring resistance against streptomycin. A Lac+ phenotype was conferred by a self-transmissible plasmid of *Y. enterocolitica* (Cornelis *et al.*, 1976). Here we report on the characterization of two conjugative plasmids (pYcon54 and pYcon966) from *Y. enterocolitica*.

2. RESULTS AND DISCUSSION

The plasmids pYcon54 and pYcon966 were isolated from a biogroup1A strain and a pathogenic O:5,27 strain, respectively. Plasmid pYcon54 has a

The Genus Yersinia, Edited by Skurnik *et al.*
`\uwer Academic/Plenum Publishers, New York, 2003

size of 95.5 kb while the size of pYcon966 is approximately 70 kb. The plasmids were compared by restriction analysis, DNA hybridization, and sequencing indicating that they are closely related. Several putative *tra* genes were compared and showed identical DNA sequences. However, pYcon54 contains two DNA regions which are absent on pYcon966 (Figure 1).

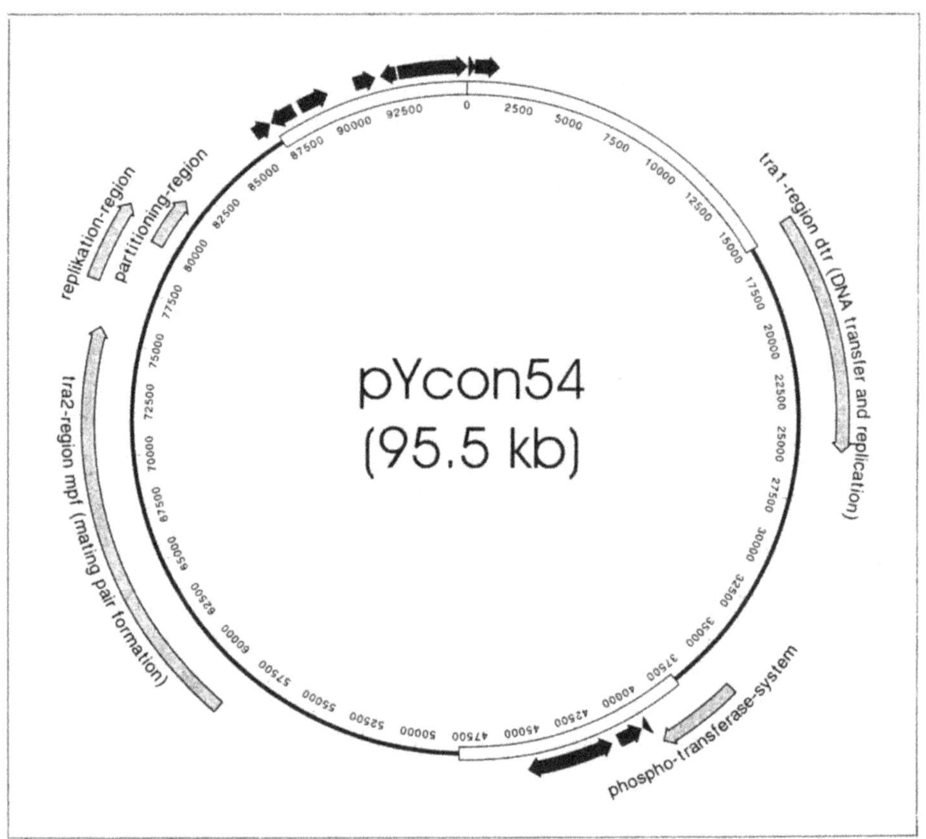

Figure 1. Genetic map of pYcon54. The arrows in dark grey represent genes probably encoding transposases. The white bars indicate the approximate positions of deletions present on pYcon966.

Hybridization studies with *inc* specific DNA probes showed that as the conjugative broad host range plasmid RP4, pYcon54 belongs to the incompatibility group IncP1. In order to study conjugation, the plasmids were marked with a neomycin resistance gene. Mating experiments demonstrated that the plasmids were transferred into a wide range of *Yersinia* strains and into other *Enterobacteriaceae* (*C. freundii, E. coli, S. typhimurium*). The highest transfer frequency obtained was 10^{-2} transconjugants per donor cell. We also investigated the mobilization of pYV. Transmittance of pYV was observed into pathogenic *Yersinia* strains and into a nonpathogenic *Y. enterocolitica* strain with frequencies up to 10^{-3} and 10^{-5} transconjugants per donor, respectively.

The complete DNA sequence of pYcon54 has been determined (95.499 bp). Bioinformatic analysis revealed 155 open reading frames (ORFs) with good coding potential. For most ORFs, probable promoters and Shine-Dalgarno sequences have been identified. Seventy-six predicted products show similarities to known proteins. The strongest homologies were detected to Tra proteins, partitioning proteins and transposases encoded by other plasmids. We have not found any similarities to proteins involved in virulence or antibiotic resistance. Typically for IncP plasmids, pYcon54 contains two *tra* regions, the mating pair formation region (*mpf*) and the DNA transfer and replication region (*dtr*) (Figure 1). The *mpf* region mainly contains genes similarly found on the conjugative plasmids R100, pNL1 and R27 of *E. coli*, *Sphingomonas aromaticvorans*, and *S. typhimurium*, respectively. They code for pilus proteins and are involved in pilus assembly and stabilization. The pYcon54 *dtr* region is closely related to corresponding regions on the *E. coli* plasmids R100 and R388 and on the *Agrobacterium tumefaciens* Ti-plasmid. This region is essential for the formation of the relaxosome. We also identified a locus on pYcon54 probably involved in partitioning. Similar to other low-copy number plasmids, the partitioning locus comprises an operon with two genes and a *cis*-acting centromere-like site.

DNA regions on pYcon54 that are lacking on pYcon966, contain genes not essential for conjugation, e.g. genes for transposases. The comparison of the plasmids suggests that either pYcon54 has acquired additional genes not present on pYcon966 or that pYcon966 is a derivative of pYcon54. Our results indicate that conjugative plasmids might be more widely distributed in *Yersinia* than expected.

REFERENCES

Allen, J.R., Chesbro, W.R., and R.M. Zsigray, 1987, Mobilization of the Vwa plasmid of *Yersinia pestis* by F-containing strains of *Escherichia coli*. *Contrib. Microbiol. Immunol.* Basel, Karger, **vol 9**: 332-341.

Allen, J.R., Kuhnert, W.L., and R.M. Zsigray, 1995, Transfer of the virulence plasmid of *Yersinia pestis* O:19 is promoted by DNA sequences contained within the low calcium response region. *Contrib. Microbiol. Immunol.* Basel, Karger, **vol 13**: 285-289.

Cornelis, G., Bennet, P.M., and J. Grinsted, 1976, Properties of pGC1, a *lac* plasmid originating in *Yersinia enterocolitica* 842. *J. Bacteriol.* **127**: 1058-1062.

Heesemann, J., and R. Laufs, 1983, Construction of a mobilizable *Yersinia enterocolitica* virulence plasmid. *J. Bacteriol.* **155**: 761-767.

Kimura, S., Eda, T., Ikeda, T., and M. Suzuki, 1975, Detection of conjugative R plasmids in genus *Yersinia*. *Japan J. Microbiol.* **19**: 449-451.

Chapter 7

Signature-Tagged Mutagenesis of *Yersinia pestis*

Petra C.F. OYSTON[1], Andrey V. KARLYSHEV[2], Brendan W. WREN[2] and Richard W. TITBALL[1]

Dstl, Chemical and Biological Sciences, Porton Down; London School of Hygiene and Tropical Medicine, University of London

1. INTRODUCTION

STM is a transposon-based approach which allows the simultaneous evaluation of pools of mutants in complex environments (Hensel *et al*, 1995). It has been used to examine a range of pathogens *in vivo*, including *Y. enterocolitica* (Darwin *et al*, 1999). We have modified the standard STM protocol to use high-density oligonucleotide arrays, rather than the usual radioactivity-based method used for the detection of tags. This was first optimised in *Y. pseudotuberculosis* where it was shown to offer significant improvements in terms of performance, efficiency and reliability and allowed quantitative analysis of data (Karlyshev *et al*, 2001).

2. RESULTS

Mutants were created in *Y. pestis* GB using double-tagged mini-Tn5 and assembled into pools of 95 individually tagged mutants. Experiments were then undertaken to identify the optimal level of challenge, the challenge route and a suitable complexity of pool.

Mice were challenged subcutaneously with serial dilutions of a pool of 95 mutants. Overall, the output results were highly variable. The majority of mutants in the pool were missing, and the deletion of mutants appeared to be random. This phenomenon may represent a bottle-neck in the infection process, as was reported previously in co-infection studies (Sebbane *et al*,

The Genus Yersinia, Edited by Skurnik *et al.*
Kluwer Academic/Plenum Publishers, New York, 2003

2001). In general, higher numbers of mutants were recovered from the mice dosed at the higher levels. The best output result was from one mouse receiving $3x10^5$ cfu. Comparison of the output samples from mice challenged by subcutaneous or intra-peritoneal routes of infection showed no improvement in mutant recovery. For other pathogens, simplifying the complexity of the pools resulted in improved results. In STM studies with *Y. pseudotuberculosis* given intra-venously, simplification of the pools reduced variability in outputs (Karlyshev *et al*, 2001). We compared one pool of 96 mutants, two pools of 48 mutants and four pools of 24 mutants. However, no improvement was observed by simplification of the pools. This result would indicate that for future experiments pools of 96 mutants should be used, but with larger numbers of mice, allowing the better output results to be selected.

By comparing the output samples from initial experiments, mutants which were consistently lost on animal passage were identified. DNA was prepared from each of these mutants. The genomic insertion site of the transposon was identified by a single-primer PCR method and sequencing of the PCR product (Karlyshev *et al*, 2000). The results of the sequencing showed two of the mutants had insertions in the *yscB* gene. Insertion of the transposon would have a polar effect in the *yscBCDEFGHIJKL* operon, abolishing Yop secretion and thus attenuating the pathogen. These mutants therefore gave us confidence that the screen had identified attenuated mutants. A third mutant has the transposon inserted into a homolog of the *E. coli hslU* gene encoding a heat shock protein. Inactivation of *hslU* has not been reported previously for any bacterium, but over-expression of HslU has been shown to confer resistance to the SOS inducer nitrofurantoin (Khattar, 1997).

Mice were inoculated with 5 x 10^2 cfu of either the wild type strain GB, or with the *hslU* mutant (2H6) or the unsequencable mutant (2B8). The mutants were not avirulent in pure culture, although mice infected with 2B8 did appear to take slightly longer to die, and thus this mutant was somewhat attenuated *in vivo*.

3. CONCLUSIONS

- A library of signature-tagged transposon *Y. pestis* mutants has been produced
- There are problems with random elimination of mutants from mice
- Some mutants have been identified, although two of these mutants retain virulence for mice.

- Future work will concentrate on refining the *in vivo* screening and identification of other attenuated mutants.

REFERENCES

Hensel M., Shea JE., Gleeson C., Jones MD., Dalton E., Holden DW 1995. Simultaneous identification of bacterial virulence genes by negative selection. Science 269: 400-403

Darwin AJ., Miller VL. 1999. Identification of *Yersinia enterocolitica* genes affecting survival in an animal host using signature-tagged transposon mutagenesis. Mol. Microbiol. 32: 51-62

Karlyshev AV., Oyston PCF., Williams K., Clark G., Titball RW., Winzeler EA., Wren BW. 2001. Application of high-density array-based signature-tagged mutagenesis to discover virulence-associated genes. Infect. Immun. 69: 7810-7819

Sebbane F., Devalckenaere A., Foulon J., Carniel E., Simonet M. 2001. Silencing and reactivation of urease in *Yersinia pestis* id determined by one G residue at a specific position in the *ureD* gene. Infect. Immun. 69: 170-176

Karlyshev AV., Pallen MJ., Wren BW. 2000. Single-primer PCR procedure for rapid identification of transposon insertion sites. BioTechniques. 28: 1078-1082

Khattar MM 1997. Overexpression of the *hslVU* operon supresses SOS-mediated inhibition of cell division in *Escherichia coli*. FEBS Letters. 414: 402-404

Picture 12. Andrey Karlyshev gives a lecture.

Chapter 8

Cobalamin Synthesis in *Yersinia enterocolitica* 8081

Functional Aspects of a Putative Metabolic Island

Michael B. PRENTICE[1], Jon CUCCUI[1], Nick THOMSON[2], Julian PARKHILL[2], Evelyn DEERY[3] and Martin J. WARREN[3]
[1]Bart's and the London Medical School, London EC1A 7BE, UK; [2]Wellcome Trust Sanger Institute, Hinxton, UK; [3]Queen Mary University of London, London, UK

1. INTRODUCTION

The cofactor cobalamin (coenzyme B_{12}) is essential for animals but plays no role in plants or fungi (Roth *et al.*, 1996). Its key structural feature is the cobalt-carbon bond at the heart of a tetrapyrrole ring which generates free radical intermediates for isomerizations of –OH, -NH2 or carbon containing groups (Marsh and Drennan, 2001). Animals need B_{12} to transfer methyl groups. In bacteria it helps rearrange small molecules (ethanolamine, propanediol and glycerol) in anaerobic fermentation (Roth *et al.*, 1996). Bacteria are the ultimate source of all cobalamin. 25 enzymes in 30 different steps are required for synthesis (Roth *et al.*, 1996). *E. coli* has lost the capacity to make B_{12} while *Salmonella enterica* has lost and reacquired it (Roth *et al.*, 1996). An aerobic cobalamin synthesis pathway exists in *Pseudomonas denitrificans* for which all intermediate steps are known (Battersby, 1994) but the *Salmonella* pathway is anaerobic and not fully elucidated (Roth *et al.*, 1996). Other organisms e.g. *Klebsiella pneumoniae, Rhodobacter capsulatum* and *Pseudomonas aeruginosa* can synthesize cobalamin aerobically and anaerobically.

1.1 *Yersinia* and cobalamin

Yersinia species do not ferment glycerol (Bouvet *et al.*, 1995) and cobalamin production or utilisation has not been described in *Yersinia*

The Genus Yersinia, Edited by Skurnik *et al.*
'uwer Academic/Plenum Publishers, New York, 2003

species. However, the *Y. enterocolitica* 8081 genome sequence has a 40 kb region with 75-85% nucleotide identity to the linked *S. enterica cob* (cobalamin synthesis) and *pdu* (propanediol utilisation) operons. This region is not present in the *Y. pestis* genome (Parkhill *et al.*, 2001) although this organism possesses cobalamin uptake genes.

In this preliminary investigation, we present bioassay results suggesting the *Y.enterocolitica* 8081 *cob/pdu* operon is functional.

2. METHODS

A quantitative bioassay was used based on the effect of pelleted lysed bacteria grown in B_{12}-free media on the growth zone of *Salmonella typhimurium* AR2680 *cbiB metE* incorporated into minimal agar medium (Raux *et al.*, 1996). This organism cannot produce B_{12} and requires the B_{12}-dependant methionine synthase encoded by *metH* to grow. A published bioassay (Lawrence and Roth, 1996) based on the limiting cobalt and cobalamin content of standard MacConkey media was used to detect cobalamin-dependent propanediol or ethanolamine fermentation. MacConkey agar with 1% 1,2-propanediol or ethanolamine replacing lactose were used containing either no other additions, $CoCl_2$ at a final concentration of 20 μM or cobinamide at a final concentration of 100nM. Each strain to be tested was subcultured onto a plate of each type incubated aerobically at 37°C. Cobalamin-dependant propanediol fermentation was recorded by a colony colour change colourless-red first on cobinamide then cobalt supplemented and finally the non-supplemented MacConkey media. *Y. enterocolitica* 8081 was a gift from D.A Portnoy, University of California.

3. RESULTS

The *Y. enterocolitica* 8081 *cob/pdu* operon resembles the *Salmonella enterica* serovar Typhimurium operon (Roth *et al.*, 1996). *Y. enterocolitica* 8081 produced cobalamin anaerobically but not aerobically and fermented propanediol in a cobalamin-dependant manner. It did not ferment ethanolamine (no ethanolamine utilisation genes are present in *Y. enterocolitica* 8081). Results for *Salmonella, E.coli* and *Klebsiella* control strains tested were as previously published (Lawrence and Roth, 1996).

4. DISCUSSION

Cobalamin production and propanediol utilisation have not been described in *Yersinia sp*. The 40 kb *cob/pdu* operon confers a major phenotypic difference on *Y. enterocolitica* compared with *Y. pestis*. The distribution of this operon in *Enterobacteriaceae* does not match 16SrRNA phylogeny (data not shown) suggesting multiple horizontal transmissions and/or deletions. This raises several questions for further study:

1. Why is cobalamin produced anaerobically in *Salmonella* and *Y. enterocolitica* 8081 and both anaerobically and aerobically in *Klebsiella pneumoniae*?
2. How is this large operon horizontally transmitted and what selects for its deletion?
3. B_{12} is used for a different function in different species: ethanolamine degradation in *E. coli*, propanediol degradation in *Salmonella* and *Yersinia*, and for glycerol fermentation in *Klebsiella*. How are B_{12} synthetic pathways controlled differently in three closely related taxa to achieve this?

ACKNOWLEDGEMENTS

The Wellcome Trust has funded sequencing of *Y. enterocolitica* 8081 and *Y. pestis*. MBP gratefully acknowledges funding from the Joint Research Board of St Bartholomew's Hospital.

REFERENCES

Battersby A. R., 1994, How nature builds the pigments of life: the conquest of vitamin B12. Science *264:*1551-7.

Bouvet O. M., Lenormand P., Ageron E., and Grimont P. A., 1995, Taxonomic diversity of anaerobic glycerol dissimilation in the Enterobacteriaceae. Res Microbiol *146:*279-90.

Lawrence J. G., and Roth J. R., 1996, Evolution of coenzyme B12 synthesis among enteric bacteria: evidence for loss and reacquisition of a multigene complex. Genetics *142:*11-24.

Marsh E. N., and Drennan C. L., 2001, Adenosylcobalamin-dependent isomerases: new insights into structure and mechanism. Curr Opin Chem Biol *5:*499-505.

Parkhill J., Wren B. W., Thomson N. R., Titball R. W., Holden M. T., Prentice M. B., *et al.*, 2001, Genome sequence of *Yersinia pestis*, the causative agent of plague. Nature *413:*523-7.

Raux E., Lanois A., Levillayer F., Warren M. J., Brody E., Rambach A., *et al.*, 1996, *Salmonella typhimurium* cobalamin (vitamin B12) biosynthetic genes: functional studies in *S. typhimurium* and *Escherichia coli*. J Bacteriol *178:*753-67.

Roth J. R., Lawrence J. G., and Bobik T. A., 1996, Cobalamin (coenzyme B12): synthesis and biological significance. Annu Rev Microbiol *50:*137-81.

Chapter 9

Construction of a *Yersinia pestis* Microarray

Richard A. STABLER[1], Jason HINDS[1], Adam A. WITNEY[1], Karen ISHERWOOD[2], Petra OYSTON[2], Richard TITBALL[2], Brendan WREN[3], Stewart HINCHLIFFE[3], Michael PRENTICE[4], Joseph A. MANGAN[1] and Phillip D. BUTCHER[1]
[1]Bacterial Microarray Group, St. George's Hospital Medical School, Tooting, London; [2]Chemical and biological sciences, Dstl, Porton Down, Wiltshire; [3]London School of Hygiene and Tropical Medicine, Keppel Street, London; [4]Dept of Medical Microbiology, St Bartholomew's and the Royal London Hospital, London.

1. INTRODUCTION

The complete genome DNA sequence of the Gram-negative 'plague' bacterium *Yersinia pestis* CO92 (biovar Orientalis) was published in October 2001 (Parkhill *et al.*, 2001). The genome consisted of 4.6 Mb single circular chromosome containing 4102 coding sequences (including 149 pseudogenes) and three plasmids pMT1 (96.2 kb), pCD1 (70.3 kb) and pPCP1 (9.6 kb). The *Y.* pestis DNA microarray was designed to cover all the genes, including pseudogenes, on the *Y. pestis* CO92 chromosome and the three *Y. pestis* plasmids.

2. ARRAY DESIGN

The approach used was to produce spotted PCR product microarrays. The first step in this process is to design gene-specific primers to amplify a PCR product to represent each of the genes in a the genome. To achieve this Microarray Design (MAD) software performed BLAST analysis of potential PCR products against the genome sequence. The software algorithm selected a PCR product for each gene that was unique and only self-detected in the

The Genus Yersinia, Edited by Skurnik *et al.*
uwer Academic/Plenum Publishers, New York, 2003

BLAST analysis. If a unique PCR product was not possible for a particular gene then the algorithm selected a PCR product from the BLAST analysis that demonstrated minimal cross-hybridisation with other non-target genes.

PCR products were amplified in 96 well plate format using an MWG Biotech RoboAmp 4200, Providing capacity for around 1000 PCR reactions per day. Incorporation into the 96 well plate of non-cross contamination (NCC) design ensures that only a single well is open at any one time. This prevents potential cross-contamination of PCR reactions to maintain PCR product integrity. PCR products were checked on an agarose gel to confirm that single products of the expected size were generated. 5% of PCR products from each plate were sequence verified. The PCR products are gridded at high density on coated glass microscope slides using a BioRobotics MicroGrid II robot. PCR products can be printed from a maximum of 24 384 well plates to 108 slides. Split pin technology allows the robot to dispense the PCR products to multiple slides following a single visit to the sample plate. Combined with SoftTouch technology this pin design aids spotting reproducibility. For the *Y. pestis* whole genome array the spot to spot distance used was 230μm with an average spot diameter of 150μm. This enabled the complete *Y. pestis* microarray to fit into an area of 22×22mm.

3. APPLICATIONS OF THE *Y. pestis* MICROARRAY

The microarray can be used for intra-species genomic analysis between strains and biovars of *Y. pestis* and *Y. pseudotuberculosis*. Data analysis can be used to identify putative deletions, including DFR's (Radnedge *et al.*, 2002), and duplications but does not identify genetic rearrangements. The microarray can also be used for inter-species comparisons between different *Yersinia* species e.g.*Y. pestis* and *Y. enterocolitica*.

The microarray can be used to analyse the transcriptome to asses differences in expression profiles with mutant vs wild-type, environmental responses/shock, growth curve/time course experiments.

4. DATA HANDLING

The development and implementation of a relational database system (BuG@Sbase) is underway to store and track all information relating to the array process and hybridisation data generation. The system will include a web based user interface allowing multi-platform access to registered users. Users will be able to download resources such as gridmap files and protocols and will be able to determine the history of each element on each array. Tools for normalisation and analysis will be implemented to allow manipulation of data within the database. Users will be able to upload their data to the server and make use of these tools. The system will enable users to export MIAME compliant data in the currently evolving MAGE-ML format, such that on publication, it can then be easily uploaded into publicly available microarray data repositories such as ArrayExpress and GEO.

ACKNOWLEDGEMENTS

The Wellcome Trust for funding the BμG@S facility. Dr. R.W. Titball (Defence Science and Technology Laboratory) for DSTL funding of the *Y. pestis* and *Y. pestis/S. typhi* plasmid arrays. Finally, our commercial collaborators, BioRobotics and MWG Biotech. BmG@Sbase is derived from a microarray database schema developed by AAW and colleagues at the Naval Medical Research Center, Maryland, USA, with funding from USAMRMC and the US Department of the Navy.

REFERENCES

Parkhill, J., Wren, B.W., Thomson, N.R., Titball, R.W., Holden, M.T., Prentice, M.B., *et al* (2001) Genome sequence of *Yersinia pestis*, the causative agent of plague. *Nature*. 413: 523-527.

Radnedge, L., Agron, P.G., Worsham, P.L. and Andersen, G.L. (2002) Genome plasticity in *Yersinia pestis*. *Microbiology*. 148: 1687-1698.

Picture 13. Karen Isherwood, Richard Stabler, Stewart Hinchliffe, Victoria Taylor, Sarah Howard and Sharon Tennant are posing on the bank of river Aura.

Chapter 10

A Conjugal Type IV Transfer System in *Yersinia enterocolitica* Strains

Greta GOELZ, Dorothea KNABNER, Bernd APPEL and Eckhard STRAUCH
Robert Koch-Institut, Berlin, Germany

1. INTRODUCTION

Type IV transfer systems have been found in several bacteria and mediate the transfer of large nucleoprotein complexes or proteins. The prototype of type IV transporters is the VirB system of *Agrobacterium tumefaciens* which exports the oncogenic T-DNA across the bacterial membrane into susceptible plants, which leads to the formation of crown galls (Lai and Kado, 2000). The *virB* locus of *A. tumefaciens* consists of 11 genes (*virB1* through *virB11*) coding for a pilus-like mating system and a mating channel that spans the bacterial cell wall. Homologous systems to this transfer system are the conjugal transfer system (Tra) of the conjugative IncN plasmid pKM101 of *Escherichia coli* and the *Bordetella pertussis* toxin exporter Ptl. Increased interest in type IV transfer systems has arisen from the discovery of more type IV systems in a number of pathogenic bacteria likely contributing to the virulence, e.g. *Brucella suis*, *Helicobacter pylori* (Christie, 2001).

By analysing the enterocoliticin producing strain *Y. enterocolitica* 29930 we found a type IV transfer system encoded on a large cryptic plasmid.

The Genus Yersinia, Edited by Skurnik *et al.*
˜uwer Academic/Plenum Publishers, New York, 2003

2. MATERIAL AND METHODS

Y. enterocolitica 29930 (serotype O:7,8 biotype 1A) is a foodborne isolate (Hoffmann *et al.*, 1998). Mating experiments were performed on solid media as described previously (Strauch *et al.*, 2000).

3. RESULTS AND DISCUSSION

A cosmid library from the enterocoliticin producing strain *Y. enterocolitica* 29930 (O:7,8; biotype 1A) was constructed using the cosmid vector SuperCos1 (Stratagene, Amsterdam NL). A cosmid clone (Cos100) was isolated that carries a 40.3 kb fragment of a large cryptic plasmid present in strain 29930.

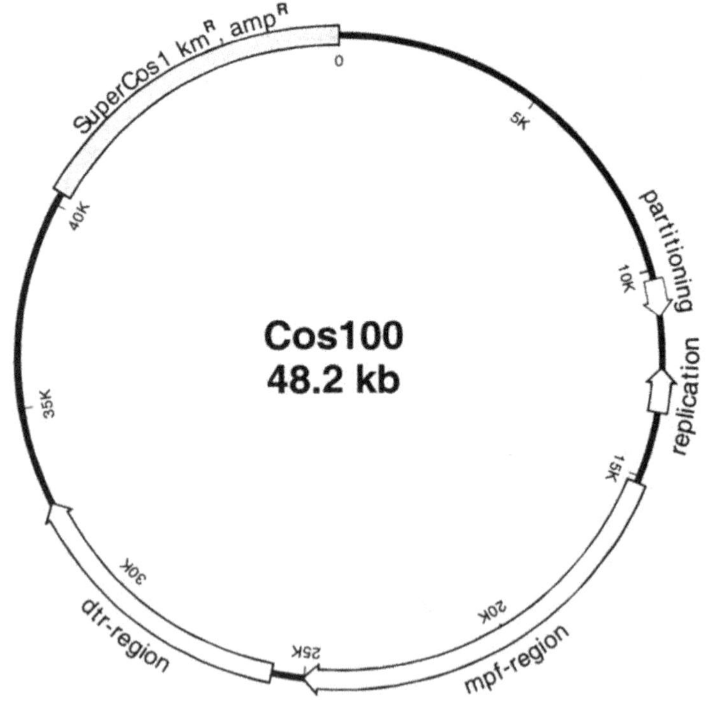

Figure 1. Genetic map of Cos100 carrying a 40.3 kb insert derived from a cryptic plasmid of *Y. enterocolitica* 29930 (see text)

Sequence analysis of the 40.3 kb fragment of the cryptic plasmid (Figure 1) revealed the presence of a complete cluster of *virB* homologous genes (designated *trl1* to *trl11*) encoding a mating pair formation (Mpf) complex. Putative genes (*mobB*, *mobC*, *hns*, *traE*, *ardC*) whose gene products may be involved in DNA processing for the transfer (Dtr complex) are indicated. Using a Tn*5* derived *in vitro* mutagenesis system mutations in some of the *mpf* genes and the *dtr* genes abolished the conjugal DNA transfer.

The sequenced plasmid fragment carries more plasmid specific genes, e.g. a putative replication gene *repA* with the highest homolgy to the *repA* gene of the pYV plasmid of *Y. enterocolitica* 8081 serotype 0:8 indicating that the cryptic plasmid of strain 29930 possesses a replicon of the I-complex plasmids. Furthermore partitioning genes (*parF* /*parG*) are present in the 40.3 kb fragment.

The cosmid was transferable from *E.coli* to different *Yersinia* strains and between *Yersinia* strains at frequencies up to 10^{-3} transconjugants per donor cell. Mobilization of the virulence plasmid pYV was investigated and was detectable between *Y. enterocolitica* strains only at low frequency (10^{-8} transconjugants per donor).

The occurence of genes related to the type IV system of the cryptic plasmid of *Y. enterocolitica* 29930 was investigated in a number of *Yersinia* strains. Hybridization experiments using gene probes derived from the *trl4* and *trl9* genes of plasmid p29330 and PCR reactions were performed. In total, six *Y. enterocolitica* biogroup 1A strains (including strain 29930) were identified which are likely to harbour genes of the type IV transfer system on cryptic plasmids. All six strains were PCR-positive for the genes *trl1*, *trl4, trl8, trl9, trl10* and *trl11* (the sequenced PCR products showed more than 98% sequence identity to the sequence of p29930). The presence of *mobB* and *repA* homologous genes in all strains was verified by hybridization.

REFERENCES

Christie, P.J., 2001, Type IV secretion: intercellular transfer of macromolecules by systems ancestrally related to conjugation machines. *Mol. Microbiol.* **40**:294-305.

Hoffmann, B., E. Strauch, C. Gewinner, H. Nattermann, and B. Appel, 1998, Characterization of plasmid regions of foodborne *Yersinia enterocolitica* biogroup 1A strains hybridizing to the *Yersinia enterocolitica* virulence plasmid. *Syst. Appl. Microbiol.* **21**:201-211.

Lai, E.-M., and C.I. Kado, 2000, The T-pilus of *Agrobacterium tumefaciens. Trends Microbiol.* **8**:361-9.

Strauch, E., I. Voigt, H. Broll, and B. Appel, 2000, Use of a plasmid of a *Yersinia enterocolitica* biogroup 1A strain for the construction of cloning vectors. J. *Biotechnol.* **79**:63-72.

Chapter 11

Transmission Factors: *Yersinia pestis* Genes Required to Infect the Flea Vector of Plague

B. Joseph HINNEBUSCH
**Laboratory of Human Bacterial Pathogenesis, Rocky Mountain Laboratories, National Institute of Allergy and Infectious Diseases, National Institutes of Health, Hamilton, MT 59840 USA*

1. INTRODUCTION

Yersinia pestis is unique among the yersiniae in utilizing an arthropod-borne means of transmission– the flea. This paper will review what is known about the genetic and molecular mechanisms of fleaborne transmission of *Y. pestis*. As is true for other pathogens that cycle between arthropods and mammals, *Y. pestis* relies on distinct sets of genes to infect these two very different hosts. Those *Y. pestis* genes specifically required to colonize and produce a transmission-competent infection in the flea are referred to as transmission factors, to contrast them with virulence factors, the genes required for infection in the mammal.

2. INFECTION OF THE FLEA BY *Y. pestis*

Transmission of the plague bacillus by the rat flea *Xenopsylla cheopis* was first demonstrated by Paul-Louis Simond, working in Vietnam (Simond, 1898). This concept was met with skepticism initially, but Simond's findings were soon amply verified. Many different rodents and other small mammals have been implicated as natural hosts of *Y. pestis*; and as each rodent species tends to have its own flea species, plague circulates within many different flea-rodent complexes throughout the world (Pollitzer, 1954). Thus, although *Y. pestis* can be transmitted among some animals by aerosol and by

'he Genus Yersinia, Edited by Skurnik *et al.*
wer Academic/Plenum Publishers, New York, 2003

ingestion, the existence of plague in nature is considered to depend on rodent-flea-rodent transmission cycles.

Simond and others thought that contamination of the bite site with infected flea feces was the most likely mechanism of plague transmission. The true manner in which fleas transmit plague was discovered by A. W. Bacot, an entomologist working for the Lister Institute's Commission for the Investigation of Plague in India. In their classic paper, Bacot and Martin (1914) described infected *X. cheopis* and *Ceratophyllus fasciatus* fleas that attempted to feed vigorously and persistently, but could not pump blood into their stomachs, although the esophagus became distended with blood. Upon examining these fleas more closely, they found that the lumen of the proventriculus, a muscular valve that connects the esophagus to the midgut, was obstructed with a solid mass of plague bacteria. Investigating further the temporal development of infection in a series of fleas, they observed that during the first week after ingestion in a blood meal, *Y. pestis* grew in the form of small clumps in the lumen of the flea gut that gradually coalesced and grew to form large, brown-colored "jelly-like" cohesive masses. These masses could develop in the proventriculus, as well as the midgut, and lead to proventricular blockage. Based on these observations, Bacot and Martin proposed the following model for plague transmission: When a blocked flea attempts to suck blood, the esophagus fills with blood and becomes distended, but then the blood is driven back into the bite site by elastic recoil of the esophageal wall, carrying with it some of the bacteria from the proventriculus. Bacot later amended the model to propose that incompletely blocked fleas transmit more efficiently than fully blocked fleas, because bacteria from the midgut could flow through the partially obstructed valve into the bite site (Bacot, 1915). This was an important corollary because complete blockage of the proventriculus does not develop readily in some flea species. Nevertheless, according to the Bacot model, *Y. pestis* must infect and interfere with the valvular function of the proventriculus in order for efficient biological transmission to occur, even if the passage of blood into the midgut is not completely blocked.

3. GENETIC AND MOLECULAR MECHANISMS OF *Y. pestis* INFECTION OF THE FLEA

The foregoing description indicates that *Y. pestis* exhibits a distinctive phenotype during the arthropod stage of its life cycle. What is the genetic basis for this phenotype? Many virulence factors of *Y. pestis* that are required in the mammalian host have been identified, including adhesins and the cytotoxic Yop components introduced into immune cells by the Type III secretion system encoded on the *Yersinia* virulence plasmid. In contrast, the analogous genes and mechanisms required to infect the insect have been relatively neglected. Two such transmission factors have been identified, however: the *Yersinia* murine toxin, and the gene products responsible for the pigmentation phenotype of *Y. pestis*.

3.1 Yersinia Murine Toxin: A Flea Gut Colonization Factor

The history of the *Yersinia* murine toxin has been enigmatic. A protein fraction enriched from *Y. pestis* cell lysates that had selective toxicity toward mice was identified in the 1950s and termed *Yersinia* murine toxin (Ymt). Toxicity, manifested by hypotension, tachycardia, and vascular collapse at the terminal stage of septicemic plague was evident in mice and rats, but Ymt was nontoxic to other animals even at high doses. Although a biochemical mechanism of action has not been determined, toxicity has been attributed to a β-adrenergic blocking ability of Ymt (Brown and Montie, 1977). Murine toxicity is distinguishable from endotoxemia, but synergism between Ymt and endotoxin or some other component of *Y. pestis* has not been ruled out (Walker, 1967). Ymt was not a classic secreted exotoxin, but a cytoplasmic or membrane-associated protein whose toxic effects were only seen after being released from lysing bacteria. The nature of murine toxicity remains enigmatic, because recombinant Ymt produced and purified from *Escherichia coli* is nontoxic for mice (unpublished data), but native Ymt purified from *Y. pestis* by affinity chromatography is toxic (Hinnebusch *et al.*, 2000).

The gene for the 61 kDa Ymt protein is encoded on the 100 kb pFra plasmid. It was sequenced and identified as a member of a large family of phospholipase D (PLD) proteins (Cherepanov *et al.*, 1991; Ponting and Kerr, 1996). Members of this PLD family can be found in all kingdoms: bacteria, fungi, plants, animals, and animal viruses. These enzymes are characterized by having two copies of the signature catalytic domain $HXKX_4DX_6GG/S$ (Figure 1). There is limited sequence identity outside of these catalytic domains, but *Y. pestis* Ymt is most similar to the PLDs of *Streptomyces antibioticus* and the plants rice and castor bean. Unlike Ymt, the *Streptomyces* PLD is reportedly a secreted extracellular protein (Iwasaki *et al.*, 1994). Although always presumed to be a virulence factor, the importance of Ymt to virulence was called into question by studies showing that the lethal dose and virulence of *Y. pestis* in mouse infection models are not changed significantly by deletion of *ymt* (Du *et al.*, 1995; Hinnebusch *et al.*, 2000). The first indication that one or more genes carried on the large 100-kb pFra plasmid was important for fleaborne transmission came from an analysis of plasmid-cured derivatives of *Y. pestis* 195/P (Hinnebusch *et al.*, 1998). A strain lacking all or part of pFra was unable to infect or block fleas normally. When this strain was complemented with a recombinant plasmid containing just a single pFra gene (*ymt*), normal infection and blockage of the flea was restored (Figure 2). The same results were observed with a *ymt* deletion mutant of *Y. pestis* KIM, as well as a KIM mutant containing a single amino acid change of H to N in the first Ymt catalytic domain, which eliminated PLD activity (Rudolph *et al.*, 1999; Hinnebusch *et al.*, 2002b).

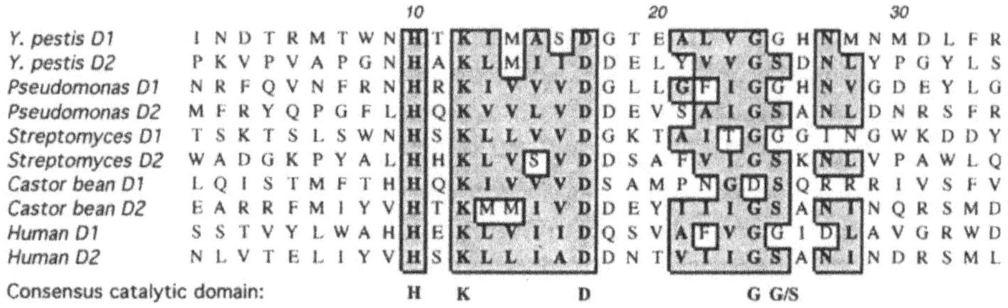

Figure 1. Sequence alignment of catalytic domains 1 and 2 of PLD family members

Interestingly, although these PLD⁻ mutants were eliminated from greater than 95% of fleas within the first few days after infection, a small number of fleas (< 1%) did develop a chronic infection and proventricular blockage. In these rare fleas, however, the infection was confined to the proventriculus; the midgut did not contain bacteria. An explanation for these findings came from observing the phenotype of the PLD⁻ mutant in the flea midgut. During the first 24 hours after being taken up in a blood meal, the morphology of the PLD⁻ mutant changed from the normal bacillary form to a round form, identical in appearance to a spheroplast (Hinnebusch *et al.*, 2002b). Thus, PLD activity of Ymt appears to prevent bacterial membrane damage and lysis in the flea gut.

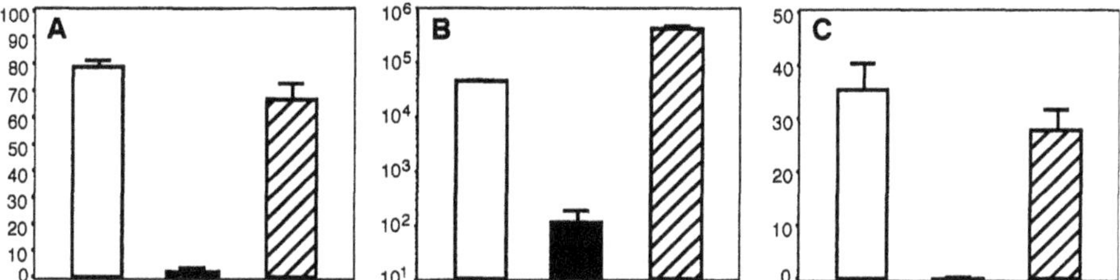

Figure 2. (A) Percentage of fleas infected; (B) average number of *Y. pestis* per infected flea; and (C) percentage of fleas that developed proventricular blockage after infection with *Y. pestis* strains 195/P-(wild type; grey bars), 195/P-3 (cured of pFra; black bars), or 195/P-3 complemented with the *ymt* gene (hatched bars).

3.2 Models for Ymt function in the flea

An intracellular PLD conceivably could protect *Y. pestis* in the flea gut by modifying an endogenous membrane component to make the bacteria impervious to the cytotoxic agent (prophylaxis model). For example, Ymt might modify one or more of the membrane phospholipids. This model would predict that Ymt^- and Ymt^+ *Y. pestis* differ in membrane properties and be differentially effected by conditions in the flea gut. However, quantitative analysis of membrane phospholipids and phosphodiester-linked substitutions of lipid A revealed no differences, and the mutant was also no more susceptible than the parent *Y. pestis* to SDS, EDTA, lysozyme, antibacterial peptides, and other agents that target the Gram-negative outer membrane (unpublished data). In fact, no phenotypic difference between Ymt^- and Ymt^+ *Y. pestis* has been detected outside of the flea gut.

Alternatively, Ymt might act to neutralize a cytotoxic agent in the flea gut after it interacts with the bacteria (antidote model). Neutralization could result from PLD acting directly upon the cytotoxic agent itself, or the action could be indirect. For example, exposure to stressful environments can induce autolysis in some bacteria. If an agent in the flea gut stimulates *Y. pestis* autolysis to cause the observed rapid spheroplast formation, intracellular Ymt activity may provide protection by blocking or redirecting a step in the autolytic pathway. Such a role would be analogous to that of mammalian PLD, which is an intracellular effector in multiple signal transduction cascades.

3.3 The *Y. pestis hms* genes: flea proventriculus colonization factors

After *Y. pestis* has successfully colonized the flea gut, it still has the task of being transmitted to a new host when the flea feeds again. To accomplish this, it must first establish infection in the proventriculus. In 1996, the *Y. pestis hms* genes were discovered to be required to infect and block the proventriculus, but not to infect the midgut, of *X. cheopis* fleas (Hinnebusch *et al.,* 1996). The *hms* (hemin storage) genes are responsible for the well-known pigmentation phenotype of *Y. pestis* colonies grown at < 28°C on media containing hemin (or the structurally analogous Congo red dye) (Jackson and Burrows, 1956). The five *hms* genes required for the pigmentation phenotype occur in two locations on the *Y. pestis* chromosome. An operon consisting of *hmsHFRS* and the *Yersinia* high-pathogenicity island are both contained within the Pgm locus, a 102-kb segment of the *Y. pestis* chromosome that is flanked with IS100 elements and is subject to frequent deletions (Fetherston and Perry, 1994). The unlinked monocistronic *hmsT* gene is outside of the Pgm locus (Hare and McDonough, 1999; Jones *et al.,* 1999). HmsH and HmsF appear to be outer membrane proteins, but the cellular location of the other three Hms proteins has not been determined.

Hms$^-$ mutants lacking either the entire Pgm locus, or a small deletion mutation in *hmsR*, had a striking phenotype in *X. cheopis* fleas. These mutants were able to colonize the flea midgut as well as the wild-type parent strains, but they were completely unable to colonize and block the proventriculus (Hinnebusch *et al.,* 1996). Complementation of the Pgm locus deletion mutant with a plasmid containing *hmsHFRS* was able to restore normal blockage ability, indicating that the high-pathogenicity island, or other genes within the 102-kb Pgm locus, are not required in the flea.

A second Hms-dependent phenotype is autoaggregation, the tendency of *Y. pestis* to form multicellular clumps in certain liquid media. This aspect of the Hms phenotype, rather than pigmentation per se, may be more relevant to the ability of Hms$^+$ *Y. pestis* to block fleas. *Y. pestis* appears to "stick" to the cuticle-covered, hydrophobic proventricular spines by producing a dense cohesive cellular aggregate that is embedded in an extracellular matrix (Hinnebusch *et al.,* 1998; 2002a). This type of infection fits the definition of a bacterial biofilm.

4. CONCLUSION

Y. pestis Ymt and Hms proteins have complementary but distinct roles. Ymt is required for survival in the midgut compartment, but not the proventriculus compartment, of the flea digestive tract. Conversely, the *hms* genes are not required to infect the midgut, but are required to infect the proventriculus. In addition to *ymt* and *hms*, the outer surface plasminogen activator of *Y. pestis* is considered to be important for fleaborne transmission, although this role does not occur in the flea (Hinnebusch *et al.*, 1998). Sodeinde *et al.*, (2002) showed that Pla is important for dissemination from a subcutaneous inoculation site in mice. This finding suggested that *pla* is pertinent to the fleaborne life-cycle of *Y. pestis*, which demands that the bacteria be able to disseminate from the dermis where they are deposited by a flea. Notably, *ymt* and *pla* are located on plasmids that are unique to *Y. pestis*. Thus, acquisition of these two plasmids by horizontal transfer helps to account for the rapid evolutionary transition of *Y. pestis* to fleaborne transmission within the last 20,000 years.

REFERENCES

Bacot, A.W., and Martin, C.J., 1914, Observations on the mechanism of the transmission of plague by fleas. *J. Hygiene* Plague Suppl. **3**: 423-439.

Bacot, A.W., 1915, Further notes on the mechanism of the transmission of plague by fleas. *J. Hygiene* Plague Suppl. **4**: 774-776.

Brown, S.D., and Montie, T.C.,1977, Beta-adrenergic blocking activity of *Yersinia pestis* murine toxin. *Infect. Immun.* **18**: 85-93.

Cherepanov, P.A., Mikhailova, T.G., Karimova, G.A., Zakharova, N.M., Ershov, I.V., and Volkovoi, K.I., 1991, [Cloning and detailed mapping of the fra-ymt region of the *Yersinia pestis* pFra plasmid]. *Mol. Gen. Mikrobiol. Virusol.* **12**: 19-26.

Du, Y., Galyov, E., and Forsberg, A., 1995, Genetic analysis of virulence determinants unique to *Yersinia pestis*. *Contrib Microbiol Immunol* **13**: 321-324.

Fetherston, J.D., and Perry, R.D.,1994, The pigmentation locus of *Yersinia pestis* KIM6+ is flanked by an insertion sequence and includes the structural genes for pesticin sensitivity and HMWP2. *Mol. Microbiol.* **13**: 697-708.

Hare, J.M., and McDonough, K.A., 1999, High-frequency RecA-dependent and -independent mechanisms of Congo red binding mutations in *Yersinia pestis*. *J. Bacteriol.* **181**: 4896-4904.

Hinnebusch, B.J., Perry, R.D., and Schwan, T.G., 1996, Role of the *Yersinia pestis* hemin storage (*hms*) locus in the transmission of plague by fleas. *Science* **273**: 367-370.

Hinnebusch, B.J., Fischer, E.R., and Schwan, T.G., 1998, Evaluation of the role of the *Yersinia pestis* plasminogen activator and other plasmid-encoded factors in temperature-dependent blockage of the flea. *J. Infect. Dis.* **178**: 1406-1415.

Hinnebusch, B.J., Rosso, M.-L., Schwan, T.G., and Carniel, E., 2002a, High-frequency conjugative transfer of antibiotic resistance genes to *Yersinia pestis* in the flea midgut. *Mol. Microbiol.* **46**: 349-354.

Hinnebusch, B.J., Rudolph, A.E., Cherepanov, P., Dixon, J.E., Schwan, T.G., and Forsberg, Å., 2002b, Role of Yersinia murine toxin in survival of *Yersinia pestis* in the midgut of the flea vector. *Science* **296**: 733-735.

Hinnebusch, J., Cherepanov, P., Du, Y., Rudolph, A., Dixon, J.D., Schwan, T., and Forsberg, A., 2000, Murine toxin of *Yersinia pestis* shows phospholipase D activity but is not required for virulence in mice. *Int. J. Med. Microbiol.* **290**: 483-487.

Iwasaki, Y., Nakano, H., and Yamane, T., 1994, Phospholipase D from *Streptomyces antibioticus*: cloning, sequencing, expression, and relationship to other phospholipases. *Appl. Microbiol. Biotechnol.* **42**: 290-299.

Jackson, S., and Burrows, T.W., 1956, The pigmentation of *Pasteurella pestis* on a defined medium containing haemin. *Br. J. Exp. Pathol.* **37**: 570-576.

Jones, H.A., Lillard, J.W., Jr., and Perry, R.D., 1999, HmsT, a protein essential for expression of the haemin storage (Hms+) phenotype of *Yersinia pestis*. *Microbiology* **145**: 2117-2128.

Pollitzer, R., 1954, *Plague.* World Health Organization, Geneva.

Ponting, C.P., and Kerr, I.D., 1996, A novel family of phospholipase D homologues that includes phospholipid synthases and putative endonucleases: identification of duplicated repeats and potential active site residues. *Protein Sci* **5**: 914-922.

Rudolph, A.E., Stuckey, J.A., Zhao, Y., Matthews, H.R., Patton, W.A., Moss, J., and Dixon, J.E., 1999, Expression, characterization, and mutagenesis of the *Yersinia pestis* murine toxin, a phospholipase D superfamily member. *J. Biol. Chem.* **274**: 11824-11831.

Schwan, T.G., and Hinnebusch, B.J., 1998, Bloodstream-versus tick-associated variants of a relapsing fever bacterium. *Science* **280**: 1938-1940.

Simond, P.-L., 1898, La propagation de la peste. *Ann. Inst. Pasteur* **12**: 662-687.

Sodeinde, O.A., Subrahmanyam, Y.V., Stark, K., Quan, T., Bao, Y., and Goguen, J.D., 1992, A surface protease and the invasive character of plague. *Science* **258**: 1004-1007.

Walker, R.V., 1967, Plague toxins- a critical review. *Curr. Top. Microbiol. Immunol.* **41**: 23-42.

PART II

PATHOGENESIS AND HOST INTERACTIONS

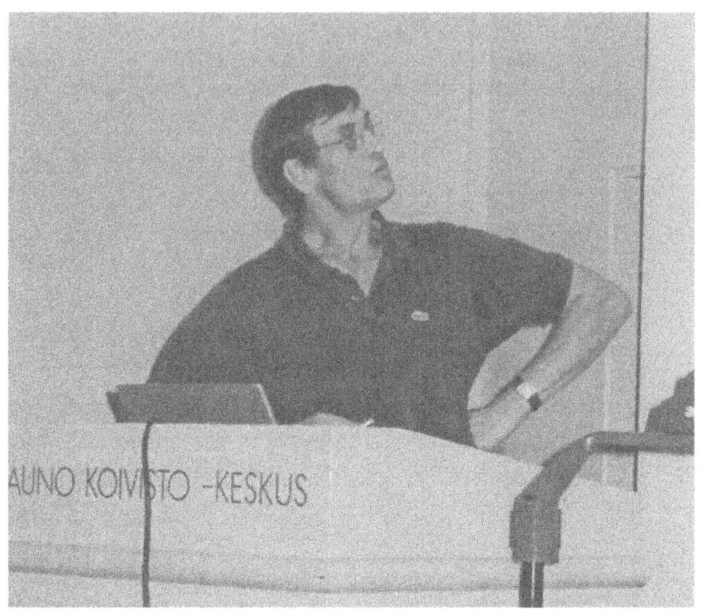

Picture 14. Guy Cornelis gives a lecture.

Picture 15. Jon Goguen, Susan and Joseph Straley and Arthur Friedlander discussing during a coffee break.

Chapter 12

Rho-GTP Binding Proteins in *Yersinia* Target Cell Interaction

Martin AEPFELBACHER, Claudia TRASAK, Agnès WIEDEMANN and Andreas ANDOR
Max von Pettenkofer-Institut für Medizinische Mikrobiologie, LMU München, Pettenkoferstr. 9a, 80336 Munich, Germany

1. INTRODUCTION

Among the more than 10 *Yersinia* species known three are pathogenic for humans: (i) *Y. pestis* is the causative agent of plaque, (ii) *Y. pseudotuberculosis* and (iii) *Y. enterocolitica* are enteropathogenic. The infection process of enteropathogenic yersiniae involves invasive as well as antiphagocytic and other immunomodulatory components. It starts with yersiniae at ambient temperature becoming ingested with contaminated food. A fraction of the infecting bacteria reaches the gut and attaches to the intestinal cells. At this point the chromosomal gene *inv* encoding an outer membrane protein (Invasin) that binds with high affinity to β1 integrin-containing extracellular matrix (ECM) receptors is upregulated (Isberg *et al.*, 2000). Binding of invasin to β1-integrins on the surface of intestinal M-cells triggers bacterial translocation and delivery into subepithelial Peyer's patches. Here *inv* is down-regulated and a variety of genes on the 70 kb virulence plasmid (called pYV in *Y. enterocolitica*) are upregulated (Jacobi *et al.*, 1998). It is well established that the antiphagocytic and immunomodulatory functions of the three pathogenic *Yersinia* species depend on the pYV plasmid which carries genes encoding (i) a protein type III secretion system (TTSS) (ii) a set of at least six effector proteins (*Yersinia* outer proteins, Yop's), (iii) regulators for gene expression and Yop-secretion/translocation, and (iv) a *Yersinia* adhesin (YadA), which among other functions can mediate *Yersinia* attachment to host cells. Contact

The Genus Yersinia, Edited by Skurnik *et al.*
'uwer Academic/Plenum Publishers, New York, 2003

of yersiniae to host cells triggers translocation of the effector Yops (YopH, YopO/YpkA, YopP/YopJ, YopE, YopM and YopT) into the cytoplasm. It has been demonstrated in cell culture studies that the translocated Yops interfere with a variety of cell functions including adhesion, phagocytosis, superoxide anion release, cytokine secretion and the balance of survival vs. apoptosis (Cornelis and Wolf Watz, 1997; Cornelis *et al.,* 1998). Individual deletions of most effector Yops drastically reduce *Yersinia* virulence in the mouse infection model, suggesting that each Yop fulfills a specific function in vivo.

Recent experimental data indicate that GTP-binding proteins of the Rho-family are crucial regulators of *Yersinia* invasion as well as constitute preferred targets for several effector Yops translocated by the *Yersinia* TTSS. The *Yersinia* surface protein invasin-triggered translocation through the M-cells of the intestine may involve Rho GTPases, Rho target proteins of the Wiskott Aldrich syndrome protein (WASp) family and Arp2/3 complex. Three of the *Yersinia* effector Yops, namely YopE, YopT and YopO/YpkA (*Y. pseudotuberculosis*) modulate the function of Rho-GTPases by different mechanisms.

In the following paragraphs we will provide a short review of the molecular mechanisms involved in the invasive and antiphagocytic functions of enteropathogenic *Yersinia* spp. with a focus on Rho GTP-binding proteins.

2. Rho GTPases AND ACTIN IN INVASIN TRIGGERED UPTAKE

Pioneering studies by Isberg el al. demonstrated that invasin-mediated uptake is morphologically similar to zippering phagocytosis and requires high affinity invasin-β1 integrin binding which in the case of *Y. pseudotuberculosis* invasin is further promoted by multimerization (Isberg *et al.,* 2000). The region of invasin necessary and sufficient for integrin binding was mapped to the C-terminal 192 amino acids and was shown to contain a motif resembling the consensus RGD motif of integrin ligands. It has become clear that the integrin triggers signals required for bacterial uptake. Such signals involve tyrosine kinases and lipid kinases and lead to the formation of an actin-rich phagocytic cup essential for the uptake process. Focal adhesion kinase (FAK), a multidomain tyrosine kinase interacting with the cytoplasmic domain of β1 integrins and transferring signals from integrins to growth regulation has also been implicated in invasin-mediated uptake (Isberg *et al.,* 2000). Members of the WASp (Wiskott Aldrich Syndrome protein) family of proteins, which include WASp, N-WASp and three WAVE/Scar isoforms, have received a great deal of attention recently as potential regulators of the actin structures governing bacterial uptake. Upon activation by upstream signalling molecules, which include the Rho GTP-binding proteins Rac and CDC42Hs, phosphatidylinositol-bisphosphate and tyrosine kinases, WASp family proteins expose their C-terminal domains which greatly enhance the actin polymerizing activity of Arp2/3 complex. Arp2/3 complex consists of seven polypeptides and is able to do both, *de novo* nucleate actin filaments and elongate actin filaments at their barbed ends (Higgs and Pollard, 1999). Three recent studies investigated the involvement of WASp family proteins, Arp2/3 complex and RhoGTPases in invasin triggered uptake (Alrutz *et al.,* 2001; McGee *et al.,* 2001; Wiedemann *et al.,* 2001). All three studies concurred that Arp2/3 complex is involved in bacterial uptake stimulated by invasin-integrin binding. The two studies investigating uptake into transformed cell lines (HeLa and COS) found that Rac1 but not CDC42 or Rho is required for uptake, whereas the study investigating invasin-triggered phagocytosis by primary human macrophages concluded that Rac, CDC42 and Rho play a role. Considerable discrepancies were reported concerning the involvement of WASp family proteins in invasin mediated uptake. In WASp negative macrophages formation of actin rich phagocytic cups was drastically reduced but phagocytosis was diminished by only 30 % (Wiedemann *et al.,* 2001). One group found no difference in invasin-promoted uptake in N-WASp negative embryonic fibroblasts compared to wild type fibroblasts (Alrutz *et al.,* 2001). In contrast, another group investigated the effects of inhibitory N-WASp mutants transfected into HeLa cells and concluded that N-WASP is crucially involved in uptake (McGee *et al.,* 2001). The reason for these differences is unclear at the moment but several points need to be considered. First, in macrophages WAVE and N-WASp are expressed and one or both of these could partly take over WASp function (Aepfelbacher *et al.,* unpublished observations). Similarly, in the mutant fibroblasts N-WASp may be substituted by WAVE/Scar which still leaves the possibility that in normal fibroblasts N-WASp does the main job in regulating invasin-mediated uptake. Furthermore, due to the high homology between WASp family proteins, dominant interfering mutants of one protein may well affect the function of its relative. In summary, Arp2/3 complex seems to be involved in invasin-promoted uptake but the upstream regulators may be diverse and redundant. They may include members of the WASp family, RhoGTPases and/or other molecules such as cortactin or actin binding protein.

3. ANTIPHAGOCYTIC FUNCTIONS OF *Yersinia* YOPS

YopH, YopE, YopT and YopO/YpkA, act on the actin cytoskeleton of the host cell. YopE, YopT and YopO/YpkA do so by modulating the function of small GTP-binding proteins of the Rho-family whereas YopH is a tyrosine phosphatase acting on focal adhesion proteins (Aepfelbacher, 2001). The function of YopM is not known at present and YopP/YopJ disrupts MAP-kinase and NF-κB signalling. Below we will shortly describe the most important features of YopE, YopT and YopO/YpkA.

4. YopE IS A RHO-GAP

YopE and YopE-homologous domains of ExoS and ExoT of *Pseudomonas aeruginosa* and of the tyrosine phosphatase SptP of *Salmonella* Typhimurium have been shown to act as GTPase activating proteins (GAP's) for Rho-family proteins (Black and Bliska, 2000; Pawel-Rammingen *et al.,* 2000). Rho GTPases act as molecular switches that are "on" when bound to GTP and "off" when bound to GDP. In their GTP-bound active state Rho proteins can associate with and stimulate various target molecules/effectors including kinases and multidomain scaffolds (Bishop and Hall, 2000). The activation state of Rho GTPases is regulated by i) their intrinsic GTPase activity ii) guanine nucleotide dissociation inhibitors (GDI's) iii) guanine nucleotide exchange factors (GEF's) and iv) GTPase activating proteins (GAP's). GAP's enhance the intrinsic GTPase activity of Ras-like GTPases and thereby convert these from the active into the inactive state. Like all mammalian GAP's, the bacterial Rho-GAP's possess an arginine finger motif, which is required for activity (Scheffzek *et al.,* 1998; Pawel-Rammingen *et al.,* 2000; Black and Bliska, 2000; Andor *et al.,* 2001).

In vitro experiments showed that YopE acts on Rho, Rac and CDC42 (Pawel-Rammingen *et al.,* 2000). The GAP activity of YopE was found to be required for *Yersinia* pathogenicity in a mouse model as well as for disruption of actin stress fibers and antiphagocytosis in HeLa cells. In the same cells constitutively active V12Rac1 but not V14RhoA could counteract the antiphagocytic effect of YopE, suggesting that YopE's effect on Rac is responsible for antiphagocytosis (Black and Bliska, 2000). We have constructed a *Y. enterocolitica* strain that translocates YopE but none of the other effector Yop's to test the intracellular effect of YopE on Rho-, Rac- and CDC42-mediated actin rearrangements in endothelial cells. It was found that YopE does not affect direct Rho-activation by thrombin, direct Rac activation by sphingosine-1-phosphate or direct CDC42 activation by bradykinin in endothelial cells. However, a basal Rac activity required for maintaining cell-cell contacts, as well as Rac activation by CDC42 was blocked by the YopE (Andor *et al.*, 2001). These findings demonstrate that YopE modulates Rho GTPase-dependent signal pathways with a remarkable specificity in primary target cells of *yersiniae.*

5. YopT MODIFIES RHO GTPases THROUGH A NOVEL MECHANISM

The first experiments testing the function of YopT showed that infection of HeLa cells with a *Y. enterocolitica* strain mutated in the effector Yops YopH, O, P, E and M disrupts actin filaments, and this effect was abrogated when YopT was removed also (Iriarte and Cornelis, 1998). Subsequently it was discovered that infection of cells with YopT-producing *Y. enterocolitica* or *Y. pseudotuberculosis* strains causes an increased mobility of RhoA in SDS polyacrylamid gels, an acidic shift of RhoA in isoelectric focussing and removal of RhoA from cellular membranes (Zumbihl *et al.*, 1999). Thus the molecular basis for the YopT effect could be attributed to modification and inactivation of the GTP-binding protein RhoA. Recently it was demonstrated that YopT acts as a cysteine protease leading to C-terminal cleavage of RhoA, Rac1 and CDC42 when these GTPases were overexpressed in cells (Shao *et al.*, 2001). It was convincingly demonstrated that the isoprenoid moiety which becomes posttranslationally attached to the C-termini of most Ras like GTP-binding proteins is removed from RhoA by YopT. A point mutation in the critical cystein residue (C139S) abolished YopT activity and the YopTC139S mutant could be used in coimmunoprecipitation and pull down experiments, likely because of its property to bind to but not dissociate from the RhoGTPases. Data produced in our group using *Yersinia* infected cells indicate that YopT-modified RhoA cannot associate with its cytosolic regulator GDI-1, with cell membranes or with effectors (Aepfelbacher *et al.*, submitted for publication). These YopT effects could be explained by the removal of the isoprenoid geranyl-geranyl group of RhoA, given that these groups are thought to mediate GDI- and membrane interaction (Olofsson, 1999). In contrast to the reported data, however, we could not find an effect of YopT on endogenous Rac or CDC42. Further experiments are warranted to identify the amino acid motif at the C-terminus of Rho GTPases which is preferentially cleaved by the YopT protease. This will enable predictions as to the cellular substrate specificity of YopT.

6. YpkA/YopO A RHO GTPase BINDING KINASE ACTIVATED BY ACTIN

HeLa cells infected with a *Y. pseudotuberculosis* mutant overexpressing YpkA but deficient in expression of YopH, M, E and K were found to display a contractile phenotype associated with pronounced retraction fibers (Hakansson *et al.,* 1996). A similar phenotype could be obtained by overexpressing YpkA using an eukaryotic expression vector in HeLa- and COS cells (Dukuzumuremyi *et al.,* 2000). Biochemically, YpkA/YopO is a serine/threonine protein kinase that has autophosphorylating activity and phosphorylates basic substrates (histones and myelin basic protein) in the presence of eukaryotic activators (Barz *et al.,* 2000; Dukuzumuremyi *et al.,* 2000; Juris *et al.,* 2000). Mutation of a critical lysine residue in YopO (K269A) or aspartic acid residue in YpkA (D270A) largely abolishes kinase activity (Dukuzumuremyi *et al.,* 2000; Juris *et al.,* 2000). One eukaryotic YopO activator was identified as actin and it was demonstrated that removal of the 21 C-terminal amino acids of YopO completely abolishes its actin binding- and autophosphorylating capability. Interestingly, these 21 amino acids show considerable homology to the C-terminus of coronin, an actin bundling protein first discovered in dictyostelium (Juris *et al.,* 2000). Besides binding to actin, YpkA/YopO also associates with the Rho family GTPases Rho and Rac but not CDC42Hs. The YpkA/Rho and YpkA/Rac interactions seem to be largely independent of the nucleotide bound state (GDP or GTP) of the GTPases, although in one report the GDP-bound form of RhoA bound three times more efficiently to YpkA than the GTP-bound form (Dukuzumuremyi *et al.,* 2000). YpkA was able to reduce the level of GTP-bound RhoA in HeLa cells, indicating an inhibitory activity (Barz *et al.,* 2000; Dukuzumuremyi *et al.,* 2000).

In experiments aimed to relate the phosphorylating- and Rho/Rac binding activities of YpkA, it was observed that (i) neither the GDP nor the GTP bound forms of Rho or Rac affect autophosphorylation (ii) autophosphorylated YpkA does not show altered binding to the GDP or GTP bound forms of Rho or Rac, (iii) the kinase dead D270A mutant of YpkA reduces the levels of GTP bound Rho as effectively as wild type YpkA (Barz *et al.,* 2000; Dukuzumuremyi *et al.,* 2000).

Taken together, autophosphorylation and kinase activity, binding to actin and interaction with the Rho GTPases Rho and Rac might point to the mode of action of YopO/YpkA. Whether these activities represent independent or interconnected ways of interference with the host cell cytoskeleton and what their contribution to *Yersinia* virulence is requires further studies.

ACKNOWLEDGEMENTS

We thank Jürgen Heesemann for continuous help and support. M.A.'s work is supported by the Deutsche Forschungsgemeinschaft.

REFERENCES

Alrutz, M.A., Srivastava, A., Wong, K.W., D'Souza-Schorey, C., Tang, M., Ch'Ng L.E., Snapper, S.B., and Isberg, R.R., 2001, Efficient uptake of *Yersinia pseudotuberculosis* via integrin receptors involves a Rac1-Arp 2/3 pathway that bypasses N-WASP function. *Mol. Microbiol.* 42: 689-703.

Aepfelbacher, M., 2001, Modulation of RhoGTPases and the actin cytoskeleton by *Yersinia* outer proteins (Yops). *Int. J. Med. Microbiol.* 291: 269-276.

Andor, A., Trülzsch, K., Essler, M., Wiedemann, A., Roggenkamp, A., Heesemann, J., and Aepfelbacher, M., 2001, YopE of *Yersinia*, a GAP for Rho-GTPases, selectively modulates Rac-dependent actin structures in endothelial cells. *Cell. Microbiol.* 3: 301-310.

Barz, C., Abahji, T.N., Trülzsch, K., and Heesemann, J., 2000, The *Yersinia* Ser/Thr protein kinase YpkA/YopO directly interacts with the small GTPases RhoA and Rac-1. *FEBS Lett.* 482: 139-143.

Bishop, A.L., Hall, A., 2000, Rho GTPases and their effector proteins. *Biochem. J.* 348: 241-255.

Black, D.S., and Bliska, J.B., 2000, The RhoGAP activity of the *Yersinia pseudotuberculosis* cytotoxin YopE is required for antiphagocytic function and virulence. *Mol. Microbiol.* 37: 515-527.

Cornelis, G.R., and Wolf-Watz, H., 1997, The *Yersinia* Yop virulon: a bacterial system for subverting eukaryotic cells. *Mol. Microbiol.* 23: 861-867.

Cornelis, G.R., Boland, A., Boyd, A.P., Geuijen, C., Iriarte, M., Neyt, C., Sory, M.P., and Stainier, I., 1998, The virulence plasmid of *Yersinia*, an antihost genome. *Microbiol. Mol. Biol. Rev.* 62: 1315-1352.

Dukuzumuremyi, J.M., Rosqvist, R., Hallberg, B., Akerstrom, B., Wolf-Watz, H., Schesser, K.: The *Yersinia* protein kinase A is a host-factor inducible RhoA/Rac-binding virulence factor. *J. Biol. Chem.* 275: 35281-35290.

Hakansson, S., Galyov, E.E., Rosqvist, R., and Wolf Watz, H., 1996, The *Yersinia* YpkA Ser/Thr kinase is translocated and subsequently targeted to the inner surface of the HeLa cell plasma membrane. *Mol. Microbiol.* 20: 593-603.

Higgs, H.N., and Pollard, T.D., 1999, Regulation of actin polymerization by Arp2/3 complex and WASp/Scar proteins. *J. Biol. Chem.* 274: 3253-3254.

Iriarte, M., and Cornelis, G.R., 1998, YopT, a new *Yersinia* effector protein, affects the cytoskeleton of host cells. *Mol. Microbiol.* 29: 915-929.

Isberg, R.R., Hamburger, Z., and Dersch, P., 2000, Signaling and invasin-promoted uptake via integrin receptors. *Microbes Infect.* 2: 793-801.

Jacobi, C.A., Roggenkamp, A., Rakin, A., Zumbihl, R., Leitritz, L., and Heesemann, J., 1998, In vitro and in vivo expression studies of *yopE* from *Yersinia enterocolitica* using the *gfp* reporter gene. *Mol. Microbiol.* 30: 865-882.

Juris, S.J., Rudolph, A.E., Huddler, D., Orth, K., and Dixon, J.E., 2000, A distinctive role for the *Yersinia* protein kinase: actin binding, kinase activation, and cytoskeleton disruption. *Proc. Natl. Acad. Sci. U S A* 17: 9431-9436.

McGee K., Zettl, M., Way, M., and Fallman, M., 2001, A role for N-WASP in invasin-promoted internalisation. *FEBS Lett.* 509: 59-65.

Olofsson, B., 1999, Rho Guanine Dissociation Inhibitors: Pivotal Molecules in Cellular Signalling. *Cell. Signal.* 11: 545-554.

Pawel-Rammingen von, U., Telepnev, M.V., Schmidt, G., Aktories, K., Wolf-Watz, H., and Rosqvist, R., 2000, GAP activity of the *Yersinia* YopE cytotoxin specifically targets the Rho pathway: a mechanism for disruption of actin microfilament structure. *Mol. Microbiol.* 36: 737-748.

Scheffzek, K., Ahmadian, M.R., Wittinghofer, A., 1998, GTPase-activating proteins: helping hands to complement an active site. *Trends Biochem. Sci.* 23: 257-262.

Shao, F., Merritt, P., Bao, Z., Innes, R.W., and Dixon, J.E., 2002, A *Yersinia* effector and a *Pseudomonas* avirulence protein define a family of cysteine proteases functioning in bacterial pathogenesis. *Cell* 109: 576-588.

Wiedemann, A., Linder, S., Grassl, G., Albert, M., Autenrieth, I., and Aepfelbacher, M., 2001, *Yersinia enterocolitica* invasin triggers phagocytosis via beta1 integrins, CDC42Hs and WASp in macrophages. *Cell. Microbiol.* 3: 693-702.

Zumbihl, R., Aepfelbacher, M., Andor, A., Jacobi, C.A., Ruckdeschel, K., Rouot, B., and Heesemann, J., 1999, The cytotoxin YopT of *Yersinia enterocolitica* induces modification and cellular redistribution of the small GTP-binding protein RhoA. *J. Biol. Chem.* 274: 29289-29293.

Chapter 13

A Technique of Intradermal Injection of *Yersinia* to Study *Y. pestis* Physiopathology

Françoise GUINET and Elisabeth CARNIEL
Laboratoire des Yersinia, Institut Pasteur, Paris, France

1. INTRODUCTION

Yersinia pestis is one of the most virulent pathogens known. A major development of *Y. pestis*, as compared to *Y. pseudotuberculosis* from which it recently derived, is its ability to be transmitted by a flea vector, which injects the bacteria intradermally. Thus, *Y. pestis* is able to achieve a high degree of virulence from a peripheral entry site. Experimental injections in animals confirmed that this property is specific to *Y. pestis* among the members of the *Yersinia* genus: *Y. pseudotuberculosis* and some *Y. enterocolitica* bioserotypes are highly virulent when injected intravenously, but not by peripheral routes. For these two species, intravenous 50 % lethal dose (LD_{50}) in mice is around 100 CFU while subcutaneous LD_{50} is 10^6 CFU or more. In contrast, LD_{50} of *Y. pestis* injected either intravenously or subcutaneously is < 10 CFU. Mice infected with *Y. pestis* die within 2 to 6 days.

Little is known about the mechanisms by which *Y. pestis* kills its host so effectively and rapidly from the skin compartment. Shortly before death, the host display a high – level septicaemia that is thought to be the ultimate cause of death, through the occurrence of an LPS – induced toxic shock. Septicaemia and death ensure transmission to a new host: fleas feeding on a septicaemic blood become contaminated, and death of the host causes them to migrate to a new host to whom they transmit the disease. How a few units of *Y. pestis*, but not *Y. pseudotuberculosis*, injected at a peripheral site can rapidly generate high – level septicaemias is far from fully understood.

The Genus Yersinia, Edited by Skurnik *et al.*
Kluwer Academic/Plenum Publishers, New York, 2003

To approach this question, we developed a model of intradermal *Yersinia* injection in the mouse in order to monitor events that follow the penetration of bacteria into the skin. Intradermal injection mimics natural infection by a fleabite more closely than does subcutaneous injection. This model, as well as some preliminary results, will be described here.

2. DESCRIPTION OF THE METHOD

The method is adapted from a technique of intradermal injection of *Leishmania* described by Belkaid *et al.,* (Belkaid *et al.,* 1996).

2.1 Injection

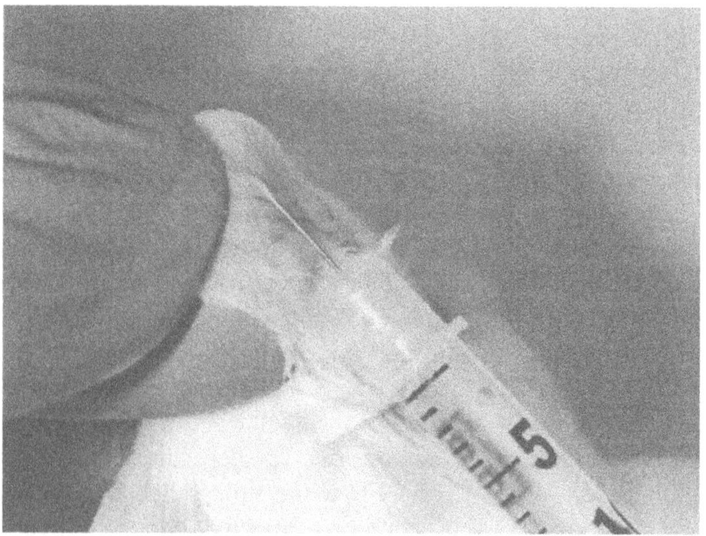

Figure 1. Intradermal injection into the mouse ear

Injection was done between the two leaflets of the mouse external ear. This location, where there is no subcutaneous space, assures that the injection is strictly intradermal. Additionally, the external ear is an anatomically defined compartment easy to collect and drained by one (or a pair of) defined lymph node(s). All experiments were performed on 7 week – old mice.

Prior to injection into the ear, mice were anaesthetised with 0.5 ml intraperitoneal Avertin (1.25 % Tribromoethanol + 2.5 % tert-amyl alcohol).

Three-hundred μl insulin syringes were used to deliver 10 μl into the ear skin. Injection between the two leaflets results in the appearance of a blister, as depicted in Figure 1.

2.2 LD_{50}

LD_{50} determinations were performed with groups of 5 mice that were infected with various bacterial doses and calculated according to the method described by Reed and Muench (Reed and Muench, 1935).

2.3 Harvesting of bacteria from ears and lymph nodes

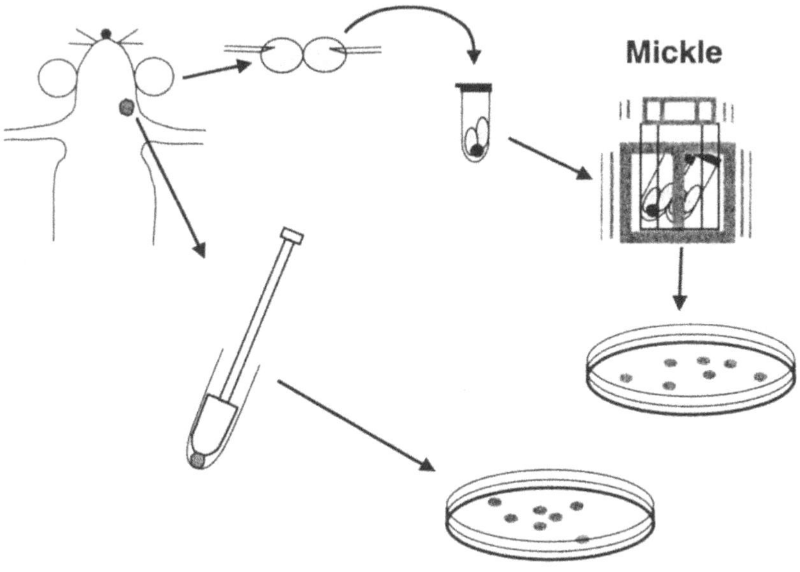

Figure 2. Bacterial recovery from ea.: and lymph node (see text)

Mice were sacrificed, ears were collected and the two leaflets were separated with forceps using sterile conditions. The leaflets were placed in a 2 ml screw-capped tube containing a 5 mm steel bead and were homogenised in a vibratory homogeniser (Mickle®). The resulting suspension was plated for CFU determination. Lymph nodes were homogenised using a hand-driven pestle and the resulting suspension was plated for CFU determination.

3. RESULTS

Using the above-described method, experiments were carried out with *Y. pestis* CO92 and with *Y. pseudotuberculosis* IP32790. The aim was to compare the physiological response of the two species, in order to determine which "behavioural" differences could account for their different virulence potentials. Initial studies focused on bacterial behaviours at the injection site, i.e. the ear skin, and in the draining lymph node.

3.1 LD_{50}

Intradermal injection of serial bacterial loads of either *Y. pestis* or *Y. pseudotuberculosis* confirmed the dramatic difference in virulence potential of the two species by peripheral routes. While the *Y. pestis* LD_{50} was around 20 CFU, that of *Y. pseudotuberculosis* was 5 orders of magnitude higher.

In subsequent experiments, mice were injected with 10^3 bacteria of either species. It was expected that this bacterial load, being 500 X higher than *Y. pestis* LD_{50} and 800 X lower than *Y. pseudotuberculosis* LD_{50}, would result in different outcomes (death or recovery) depending on the species, and thus would yield observable differences in the respective physiopathologies.

3.2 Bacterial counts from ear and lymph nodes

3.2.1 Bacterial recovery

The number of CFU recovered from the ear at time 0 was consistently equivalent to the amount of injected bacteria, thus validating the recovery method.

3.2.2 Follow – up period

Infection with *Y. pseudotuberculosis* was monitored over a period of 72 to 96 hours. With *Y. pestis*, it was not possible to follow bacterial growth over more than 48 hours, because the high mortality rate in the mice past this time precluded reliable conclusions.

3.2.3 Response range

Y. pseudotuberculosis: Two mice were studied for each time point. The number of bacteria recovered from ears and lymph nodes was generally in good agreement within the two mice at each time point. An example is shown in Figure 3. The growth curves were similar from experiment to experiment.

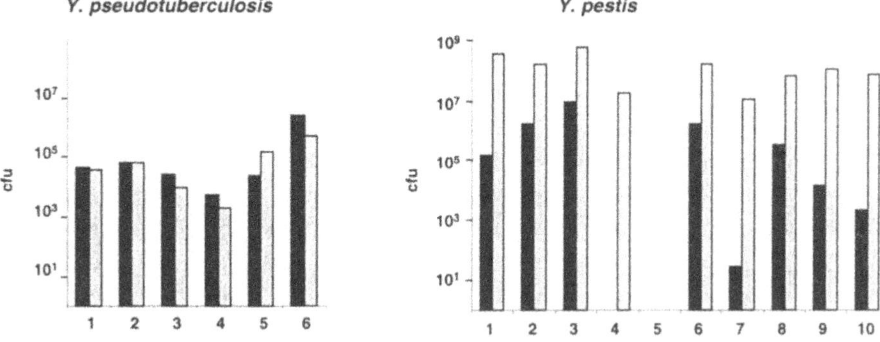

Figure 3. Growth of *Y. pestis* and *Y. pseudotuberculosis* in the ear (black bars) and lymph node samples (grey bars).
Y. pseudotuberculosis: counts in the ear and lymph node samples at 30 h post-injection. Each double bar represents a mouse. Mice 1 and 2, experiment 1; mice 3 and 4, experiment 2; mice 5 and 6, experiment 3.
Y. pestis: counts in the ear and lymph node samples at 48 h post-injection. Each double bar represents a mouse. Mice 1 to 5, experiment 1; mice 6 to 10, experiment 2.

Y. pestis: Experiments were performed where groups of mice to be studied at each time point were defined immediately after injection. At each time point, only mice from the pre-determined corresponding group were studied, even if few were left alive. To ensure a sufficient number of mice for each time point, five mice were initially included in each time-point group. The response range within groups of 5 mice injected with the same bacterial preparation could therefore be observed, and is shown in Figure 3 which combines two experiments. As can be seen, there is a higher degree of bacterial growth variability than with *Y. pseudotuberculosis*. Three mice had little or no bacteria at the injection site at 48 h, while other mice had 10^6 to 10^7 CFU in the ear. The bacterial growth in lymph nodes was more consistent from mouse to mouse, except for one case of no bacterial growth detectable at 48 h in this organ.

3.2.4 Kinetics

Y. pseudotuberculosis: Both in the ear and lymph node, a regular growth to an apparent plateau was found. After an initial delay, growth in the lymph node caught up with that in the ear after about 2 days.

Y. pestis: In the ear, the growth curve was similar to that of *Y. pseudotuberculosis*. By contrast, the amount of bacteria recovered from lymph nodes at 48 h was on average 2 logs higher than that of *Y. pseudotuberculosis*.

4. DISCUSSION

The technique presented here provides a new tool for studying the physiopathology of bubonic plague. With this technique bacteria are delivered into the skin compartment, as they are in natural transmission conditions, and events of bacterial growth or host response can be followed over time at the injection site and in deeper tissues. In this model, we confirmed that *Y. pseudotuberculosis* is much less virulent than the plague agent by the intradermal route. Bacterial counts from the injection site at time 0 proved that the technique of injection and recovery was reliable, the *Y. pseudotuberculosis* growth data that it was reproducible. Therefore, the higher variability observed in the *Y. pestis* behaviour is likely to have real biological significance rather than being a technical artifact. This technique can be used to further study cytological and immunological events taking place in ear and lymph node, which should lead to a better understanding of the physiopathological basis of the plague agent unusual virulence.

REFERENCES

Belkaid Y., Jouin H. and Milon G. A method to recover, enumerate and identify lymphomyeloid cells present in an inflammatory dermal site: a study in laboratory mice. Journal of Immunological Methods. 199:5-25, 1996

Reed LJ and Muench H. A simple method of estimation fifty per cent endpoints. American Journal of Hygiene. 27: 493-497, 1935

Chapter 14

YopT Is a Cysteine Protease Cleaving Rho Family GTPases

Feng SHAO and Jack E. DIXON
Department of Biological Chemistry, University of Michigan Medical School, Ann Arbor, MI 48109-0606, USA

1. INTRODUCTION

YopT is one of the six type III-secreted effector proteins from *Yersinia* (YopE, YopH, YopJ, YopM, YpkA, and YopT) that are delivered into eukaryotic cells to inactivate the host immune response (Cornelis *et al.*, 1998). YopT has been shown to induce a cytotoxic effect in mammalian cells (Iriarte and Cornelis, 1998). This cytotoxicity is characterized by the disruption of the actin cytoskeleton and rounding up of the cells (Iriarte and Cornelis, 1998). YopT is conserved in all three pathogenic *Yersinia* species, suggesting it most likely plays an important role in pathogenesis. Disruption of the host cell cytoskeleton by YopT contributes to the anti-phagocytosis effect of *Yersinia* on macrophages (Grosdent *et al.*, 2002). The biochemical mechanism underlying the cytotoxicity of YopT has not been fully understood. Previous studies have shown that infection of host cells with a mutant *Y. enterocolitica* strain secreting only YopT causes an isoelectric point shift of RhoA, a small GTPase known to regulate the actin cytoskeleton (Zumbihl *et al.*, 1999). In addition, Sorg *et al.*, (Sorg *et al.*, 2001) recently demonstrated that YopT can cause the release of RhoA from cell membranes or artificial vesicles.

The Genus Yersinia, Edited by Skurnik *et al.*
...wer Academic/Plenum Publishers, New York, 2003

2. RESULTS

2.1 The invariant C139, H258 and D274 are essential for YopT cytotoxicity

An exhaustive PSI-BLAST (Altschul *et al.,* 1997) search using YopT sequence as a bait leads to the identification of 19 sequences from both animal and plant bacteria pathogens showing sequence similarity to YopT (see Figure 1 in Shao *et al.,* 2002), suggesting a novel family of proteins involved in bacterial pathogenesis. The overall amino acid sequence identity among the YopT family members is not extensive, yet every member of the family shows several invariant residues, including C139, H258 and D274 (numbered from YopT sequence). In addition, the predicted secondary structure of each YopT family member is similar, with the highest level of structural identity surrounding the conserved C139, H258 and D274 residues.

Previous studies have shown that infection of HeLa cells with a *Y. enterocolitica* strain secreting only YopT leads to a cytotoxic effect (Iriarte and Cornelis, 1998; Zumbihl *et al.,* 1999). We transiently transfected EGFP-YopT fusion constructs into HeLa cells and observed a similar cytotoxic effect. In contrast, cells expressing any of the EGFP-YopT mutants (C139S, H258A and D174A) showed normal morphology and intact actin stress fibers indistinguishable from the control cells. These results demonstrate that the invariant C/H/D residues are required for YopT to disrupt the filamentous actin structure in mammalian cells. In an attempt to use yeast two-hybrid screen to identify the physiological targets of YopT, we found YopT is also cytotoxic to yeast. Using the galactose inducible promoter, we demonstrated that none of the C/H/D mutants were toxic to yeast in contrast to wild type YopT. This observation was consistent with the results observed in HeLa cells. These findings suggest that the invariant C/H/D residues are essential for YopT function in the host cells.

2.2 Rho family GTPases are direct host targets of YopT

We performed a yeast multicopy suppressor screen aiming to identify the physiological target(s) of YopT. The suppressor screen resulted in the isolation of yeast Cdc42 gene, a member of Rho family GTPases, as a suppressor for YopT cytotoxicity. Overexpression of yeast Cdc42 is capable of rescuing the YopT cytotoxicity in yeast. Together with the previous report that the isoelectric point of RhoA was shifted by YopT (Zumbihl *et al.,* 1999), this result indicates that Rho family of small GTPases including

RhoA, Rac and Cdc42 are potential targets of YopT. We wanted to determine if Rho family GTPases would directly interact with YopT. In the yeast two-hybrid assay, YopT (C139S) clearly interacts with the constitutively active RhoA (RhoAL63). Consistently, constitutively active RhoA, Rac and Cdc42 directly bind to YopT when they are co-expressed in mammalian cells. These results indicate that Rho family GTPases are all potential direct targets of YopT.

We were unable to detect any interaction between YopT and Rho GTPases produced in bacteria. Since bacteria are not capable of post-translationally modifying the GTPases, this suggested to us that the interaction of YopT with Rho GTPases could be mediated *via* post-translational modification. RhoA, Rac and Cdc42 are all known to undergo sequential post-translational modifications at their C-terminal CAAX box (C: cysteine, A: aliphatic residue, X: any residue) (Zhang and Casey, 1996). The CAAX box provides the recognition elements for prenylation (geranylgeranylation) of the cysteine, followed by proteolysis of the AAX tri-peptide and methyl esterification of the cysteine. The lipid modification allows for the membrane anchorage of the GTPases (Zhang and Casey, 1996). Indeed, removal of the CAAX box abrogates the two-hybrid interaction between YopT (C139S) and RhoAL63 (Shao *et al.*, 2002). In contrast to the full-length GTPases, CAAX deletion mutants of RhoA, Rac and Cdc42 also fail to bind to YopT (C139S) in the GST pull-down assay.

2.3 YopT is a cysteine protease cleaving post-translationally modifed Rho GTPases

To explore the nature of the prenylation-dependent interaction between YopT and Rho GTPases, we employed the established Triton X-114 partitioning assay which partitions lipid-modified proteins in the detergent phase and non-modified proteins in the aqueous phase (Hancock, 1995). Overexpressing constitutively active form of GST-Rho in HEK293T cells resulted in almost equal amounts of Rho protein partitioning in the detergent phase and aqueous phase. Co-expression of wild type YopT with Rho GTPases resulted in an almost complete loss of the GTPases from the detergent phase. In contrast, substantial amounts of Rho remain in the detergent phase when the YopT mutants (C139S, H258A and D274A) were expresses. This result hints that YopT leads to the loss of the prenyl group of Rho GTPases. To directly assess loss of the prenyl group of Rho GTPases, we expressed GST-RhoAL63 in HEK293T cells in the presence of YopT or YopT (C139S) and metabolically labeled the cells with ^{3}H-mevalonic acid, which is incorporated into the prenyl group of RhoA. Co-expression of wild type YopT, but not YopT (C139S) resulted in a complete loss of the prenyl group from GST-RhoAL63, although the total GST-RhoAL63 remained at a similar level. Furthermore, the recombinant YopT is also capable of liberating the tritium labeled prenyl group from GST-RhoA when it is immobilized onto the GST beads. Taken together, our data suggest that YopT harbors an enzymatic activity towards Rho GTPases that can lead to the loss of their prenyl groups.

The three invariant residues (C/H/D) required for YopT function are also residues that compose the catalytic triad of many cysteine proteases (Rawlings and Barrett, 1994). The YopT family lacks apparent sequence identity to the known families of cysteine proteases, suggesting that YopT might define a new family of cysteine proteases. Barrett and Rawlings have classified all families of cysteine proteases into seven clans, each with a different evolutionary history (Barrett and Rawlings, 2001). We noticed that the predicted secondary structural profiles of the YopT family members resemble those of the so-called CA clan of cysteine proteases. To testing this hypothesis, we carried out the *in vitro* cleavage assay using recombinant YopT preincubated with class-specific protease inhibitors. The activity of YopT to remove the prenyl group from RhoA was unaffected by the aspartyl protease inhibitor pepstatin, the serine protease inhibitor PMSF and metal chelator EDTA. However, The activity of YopT was terminated by the thiol-blocking reagent N-ethylmaleimide and significantly inhibited by the CA clan selective cysteine protease inhibitor, E-64 (Barrett and Rawlings, 2001). These data support our conclusion that YopT is a cysteine protease belonging to the CA clan. To assess the proteolytic activity of YopT directly, we monitored the fate of the prenylated cysteine. GST-RhoAL63, labeled with either ^{3}H-mevalonate or ^{35}S-cysteine, was incubated with wild type or mutant recombinant YopT. Scintillation counting of the chloroform extractable lipid components of the reaction shows that similar amounts of ^{35}S-Cys and ^{3}H-mevalonate were extracted into the chloroform phase. This result clearly demonstrated that the prenylated cysteine together with prenyl group was removed from Rho GTPases as a result of the proteolytic activity of YopT. Indeed, the YopT family of cysteine protease has recently been listed in the MEROPS protease database (http://www.merops.co.uk) as the new C58 family in the CA clan.

2.4 Cleavage of the prenylated cysteine leads to the membrane detachment of Rho GTPases

As a consequence of the removal of the prenyl group, recombinant YopT should be capable of releasing Rho GTPases from membranes into the soluble fraction *in vitro* as recently demonstrated (Sorg *et al.*, 2001). We employed a similar assay to test whether the membrane release of Rho

GTPases is due to the proteolytic activity of YopT. As expected, a significant amount of RhoAL63 was released from the membrane into the soluble fraction upon incubation with recombinant wild type YopT. Incubation with the same amounts of the mutant forms of YopT (C139S, H258A and D274A) produced no soluble GST-RhoAL63. Furthermore, we have shown that YopT can cause the cytoskeleton disruption phenotype in mammalian cells and this depends on the invariant C/H/D residues. To test whether this phenotype is the consequence of the membrane detachment of Rho GTPases induced by YopT *in vitro*, we monitored the membrane distribution of the endogenous RhoA in the presence of YopT. Upon transfection of wild type, but not the protease-deficient YopT mutant (C139S), the endogenous RhoA dissociated from the membranes although the total RhoA expression remains at a similar level. These data suggest that the cytotoxicity of YopT arises from its proteolytic activity which results in the removal of the lipid modification of the endogenous Rho GTPases and their subsequent membrane detachment.

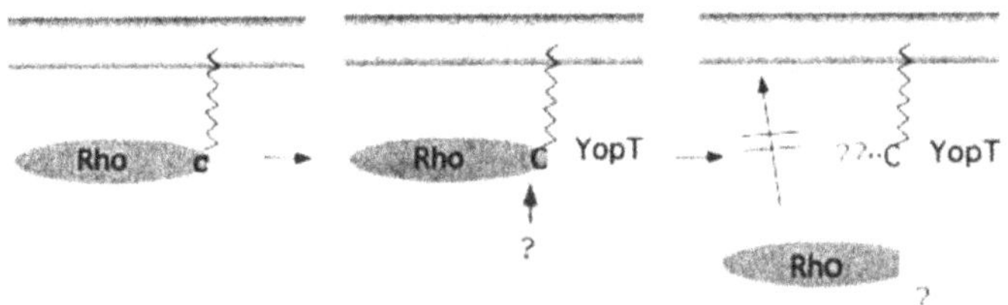

Figure 1. Model of YopT function

3. CONCLUSIONS

We have demonstrated that YopT recognizes the prenylated Rho GTPases (RhoA, Rac and Cdc42) in the host cell and executes a proteolytic cleavage at the C termini of the GTPases. This cleavage leads to the removal of the prenylated cysteine on the GTPases, which results in the membrane detachment of the GTPases. The model of YopT action is depicted in Figure 1. The proteolytic activity towards Rho GTPases described here is responsible for its known cytotoxicity. In addition, the proteolytic triad (C139, H2598 and D274) is conserved in the entire YopT family of bacterial virulence factors, suggesting that the YopT defines a novel family of cysteine protease functioning in bacterial pathogenesis.

ACKNOWLEDGEMENTS

The work in the authors' laboratory is supported by National Institute of Health and the Walther Cancer Institute. F.S. was sponsored by the University of Michigan Rackham Travel Award to participate the symposium.

REFERENCES

Altschul, S.F., Madden, T.L., Schaffer, A.A., Zhang, J., Zhang, Z., Miller, W., and Lipman, D.J., 1997, Gapped BLAST and PSI-BLAST: a new generation of protein database search programs. Nucleic Acids Res 25: 3389-3402.

Barrett, A.J., and Rawlings, N.D., 2001, Evolutionary lines of cysteine peptidases. Biol Chem 382: 727-733.

Cornelis, G.R., Boland, A., Boyd, A.P., Geuijen, C., Iriarte, M., Neyt, C., Sory, M.P., and Stainier, I., 1998, The virulence plasmid of Yersinia, an antihost genome. Microbiol Mol Biol Rev 62: 1315-1352.

Grosdent, N., Maridonneau-Parini, I., Sory, M.P., and Cornelis, G.R., 2002, Role of Yops and adhesins in resistance of Yersinia enterocolitica to phagocytosis. Infect Immun 70: 4165-4176.

Hancock, J.F., 1995, Prenylation and palmitoylation analysis. Methods Enzymol 255: 237-245.

Iriarte, M., and Cornelis, G.R., 1998, YopT, a new Yersinia Yop effector protein, affects the cytoskeleton of host cells. Mol Microbiol 29: 915-929.

Rawlings, N.D., and Barrett, A.J., 1994, Families of cysteine peptidases. Methods Enzymol 244: 461-486.

Shao, F., Merritt, P.M., Bao, Z., Innes, R.W., and Dixon, J.E., 2002, A Yersinia Effector and a Pseudomonas Avirulence Protein Define a Family of Cysteine Proteases Functioning in Bacterial Pathogenesis. Cell 109: 575-588.

Sorg, I., Goehring, U.M., Aktories, K., and Schmidt, G., 2001, Recombinant Yersinia YopT Leads to Uncoupling of RhoA-Effector Interaction. Infect Immun 69: 7535-7543.

Zhang, F.L., and Casey, P.J., 1996, Protein prenylation: molecular mechanisms and functional consequences. Annu Rev Biochem 65: 241-269.

Zumbihl, R., Aepfelbacher, M., Andor, A., Jacobi, C.A., Ruckdeschel, K., Rouot, B., and Heesemann, J., 1999, The cytotoxin YopT of Yersinia enterocolitica induces modification and cellular redistribution of the small GTP-binding protein RhoA. J Biol Chem 27: 29289-29293.

Chapter 15

Structural Studies of *Yersinia* Adhesin YadA

Heli NUMMELIN[1], Michael C. Merckel[1,4], Yasmin EL TAHIR[3], Pauli OLLIKKA[3,5], Mikael SKURNIK[2,3] and Adrian GOLDMAN[1]

[1]*Macromolecular X-ray Crystallography, Structural Biology and Biophysics, Institute of Biotechnology and* [2]*Department of Bacteriology and Immunology, Haartman Institute, University of Helsinki, Helsinki, Finland;* [3]*Department of Medical Biochemistry and Molecular Biology, University of Turku, Turku, Finland;* [4]*Present address: Helsinki Bioenergetics Group, Institute of Biotechnology, University of Helsinki, Helsinki, Finland;* [5]*Present address: The BioTie Therapies Corporation, BioCity, Turku, Finland.*

1. INTRODUCTION

Adhesion of human pathogenic *Yersiniae* to host cells is mediated by several mechanisms, which include the chromosomally encoded proteins invasin and Ail (Isberg & Falkow 1985; Miller & Falkow, 1988) and the *Yersinia* virulence plasmid (pYV) encoded *Yersinia* adhesin, YadA (Heesemann & Grüter, 1987). The expression of the *yadA* gene is temperature regulated in *Y. enterocolitica* and *Y. pseudotuberculosis* (Skurnik & Toivanen, 1992). YadA is not expressed in *Y. pestis*, because of a single-base-pair deletion in the *yadA* gene (Rosqvist *et al.*, 1988). YadA is involved in several virulence-related functions like binding to the extracellular matrix proteins (for a review see El Tahir & Skurnik, 2001). The collagen-binding ability of YadA in *Y. enterocolitica* is directly related to its virulence: loss of the collagen-binding ability of YadA leads to avirulence in mice (Tamm *et al.*, 1993; Roggenkamp *et al.*, 1995).

Wild type YadA of *Y. enterocolitica* serotype O:3 is a 430-amino-acid outer membrane protein, translated with a 25-amino-acid signal peptide that is processed during transportation. On SDS-PAGE YadA forms stable aggregates of molecular weight 160 to 240 kDa suggesting oligomerization of three monomers (Gripenberg-Lerche *et al.*, 1995). Recent electron micrographs show that YadA adopts a lollipop shaped form which consists of a C-terminal membrane anchor domain, a coiled-coil stalk domain and a globular N-terminal head domain (Hoiczyk *et al.*, 2000). The head domain

The Genus Yersinia, Edited by Skurnik *et al.*
Kluwer Academic/Plenum Publishers, New York, 2003

consists of amino acids 26-224 and it includes seven out of eight NSVAIGXXS motifs, which are required for YadA mediated collagen binding (El Tahir *et al.*, 2000).

2. MATERIALS AND METHODS

For the crystallographic studies *yadA*$_{26\text{-}241}$ of *Y. enterocolitica* O:3 was cloned to a vector with an N-terminal His$_6$-tag and transformed to M15(pREP4). For the production of selenomethionyl labelled protein, mutations I130M and I157M were made to NSVAIG--S motifs, which were anticipated to be on the surface of the protein (Nummelin *et al.*, 2002).

YadA was purified from cytoplasmic fraction of broken bacterial cells by a metal chelating affinity column followed by size exclusion chromatography to separate the soluble protein from the aggregated form. For the YadA$_{26\text{-}241}$-SeMet, the bacterial cells were grown in minimal medium with selenomethionine.

Crystal screening was done using sparse matrix screens and a sitting-drop vapor-diffusion method. The crystals were flash frozen and native data were collected at beam line X11 in the DESY, EMBL Hamburg. The MAD data were collected at BM-17, ESRF, Grenoble. Space group assignment and unit cell parameters were defined using DENZO (Otwinowski & Minor 1995).

3. RESULTS

Crystals grew within one week using 12% PEG 8000, 0.1 M Na-cacodylate pH 6.5 and 0.1 M Na-acetate as precipitant and a protein concentration of 8 mg ml^{-1}. The crystals belonged to space group R3 and diffracted to 1.55 Å resolution (Table 1). Expression of a selenomethionyl-incorporated protein led to crystals under similar conditions to native. However, the anomalous signal of arsenic interfered the selenium signal, so the buffer was changed to Tris-HCl. The selenomethionyl-protein was also less soluble and so a lower protein concentration of 2.5 mg ml^{-1} was used (Nummelin *et al.*, 2002). The hexagonal crystals belonged to space group R32 and diffracted to 2.0 Å (Table 1).

Table 1. Data collection statistics for native $YadA_{26-241}$ and selenomethionyl-labelled $YadA_{26-241}$-SeMet data. The values for highest resolution shell 1.61-1.55 Å for native data are on parenthesis.

	$YadA_{26-241}$	$YadA_{26-241}$-SeMet		
Unit cell a, b, c (Å)	67, 67, 222	65, 65, 230		
Completeness (%)	99.7 (98.6)	100.0	99.8	99.8
$1/\sigma(I)$	20.2 (3.1)	21.2	7.7	6.4
R_{merge} (%)	5.2 (39.2)	8.0	8.5	9.6
Phasing power		0.80		
FOM		0.28		

Two selenium atoms per asymmetric unit were found using MAD data, and initial phases and maps were calculated with SOLVE (Terwilliger & Berendzen 1999). The quality of the maps did not correspond to the resolution of the data and the tracing of the polypeptide was difficult. The bad map quality was a result of the location of the selenium atoms very close to one another and in close proximity to crystallographic threefold axis (Figure 1). This resulted in low phasing power and only the central part of the molecule could be traced. The model has been partly refined, and additional phasing power is needed to solve the structure completely.

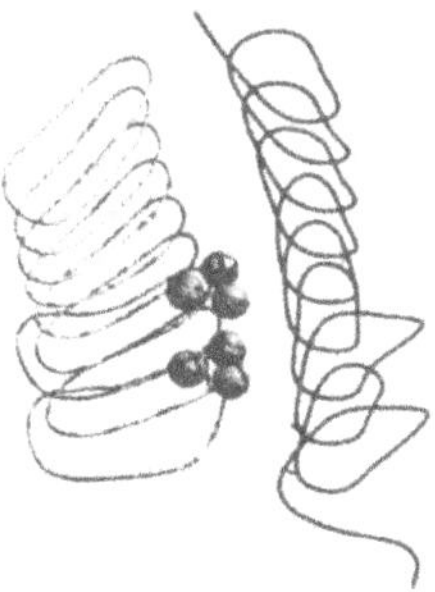

Figure 1. The central part of YadA trimer. In the inside of the trimer Se atoms are seen around the crystallographic threefold axis.

Even with the partial model it can be seen that the central part of the molecule forms a beta solenoid, which has the collagen binding NSVAIG—S motifs inside the trimer. The motifs are thus actually structural motifs that are needed to form a hydrophobic core inside the trimer and most likely they do not interact directly with collagen.

REFERENCES

El Tahir, Y., Kuusela, P., and Skurnik, M., 2000, Functional mapping of the *Yersinia enterocolitica* adhesin YadA. Identification of eight NSVAIG—S motifs in the amino-terminal half of the protein involved in collagen binding., *Mol. Microbiol.* **37**(1), 192-206.

El Tahir, Y., and Skurnik M., 2001, Mini review: YadA, multifaceted *Yersinia* adhesin, *Int. J. Med. Microbiol.* **291**, 209-218.

Gripenberg-Lerche, C., Skurnik, M., and Toivanen P., 1995, Role of YadA-mediated collagen binding in arthritogenicity of *Yersinia enterocolitica* serotype O:8 experimental studies with rats, *Infect. Immun.* **63**, 3222-3226.

Heesemann, J., and Grüter, L., 1987, Genetic evidence that the outer membrane protein YOP1 of *Yersinia enterocolitica* mediates adherence and phagocytosis resistance to human ephithelial cells, *FEMS Microbiol Lett.* **40**, 37-41.

Hoiczyk, E., Roggenkamp, A., Reichenbecher, M., Lupas, A. and Heesemann, J., 2000, Structure and sequence analysis of *Yersinia* YadA and *Moraxella* UspAs reveal a novel class of adhesins, *Embo J.* **19**, 5989-5999.

Isberg, R., and Falkow, S., 1985, A single genetic locus encoded by *Yersinia pseudotuberculosis* permits invasion of cultured animals cells by *Escherichia coli K-12, Nature*, **317**, 262-264.

Miller, V.L., and Falkow, S., 1988, Evidence for two genetic loci in *Yersinia enterocolitica* that can promote invasion of ephithelial cells, *Infect. Immun.* **56**, 1242-1248.

Nummelin, H., El Tahir, Y., Ollikka, P., Skurnik, M., and Goldman A., 2002, Expression, purification and crystallization of a collagen-binding fragment of *Yersinia* adhesin YadA, *Acta Cryst.* **D58**:1042-1044.

Otwinowski, A., and Minor, W., 1995, *DENZO*. Yale University, New Haven, Connecticut, USA.

Roggenkamp, A., Neuberger, H-R., Flügel, A., Schmoll, T., and Heesemann, J., 1995, Substitution of two histidine residues in YadA protein of *Yersinia enterocolitica* abrogates collagen binding, cell adherence and mouse virulence, *Mol. Microbiol.*, **16**, 1207-1219.

Rosqvist, R., Skurnik, M., and Wolf-Watz, H., 1988, Increased virulenve of *Yersinia pseudotuberculosis* by two independent mutations, *Nature* **334**, 522-525.

Skurnik, M., and Toivanen, P., 1992, LcrF is the temperature-regulated activator of the *yadA* gene of *Yersinia entrocolitica* and *Yersinia pseudotuberculosis, J. Bacteriol.* **158**, 1033-1036.

Tamm, A., Tarkkanen, A.M., Korhonen, T., Kuusela, P., Toivanen, P., and Skurnik, M., 1993, Hydrophobic domains affect the collagen-binding specificity and surface polymerization as well as the virulence potential of the YadA protein of *Yerisinia enterocolitica, Mol. Microbiol.* **10**, 995-1011.

Terwilliger, TC., and Berendzen J., 1999, Automated MAD and MIR structure solution, *Acta Cryst.,* **D55,** 849-861.

Chapter 16

Yersinia pseudotuberculosis Harbors a Type IV Pilus Gene Cluster that Contributes to Pathogenicity

François COLLYN, Michael MARCEAU and Michel SIMONET
Equipe Mixte Inserm (E9919)-Université (JE2225)-Institut Pasteur de Lille, Institut de Biologie de Lille, 59021, Lille, France.

1. INTRODUCTION

Type IV pili are appendages emanating from the surface of several Gram-negative bacteria including species pathogenic for animals and plants. These structures, which may be peritrichous or polar on the cell surface sometimes form bundles, and have been implicated in a variety of bacterial functions including cell adhesion, bacteriophage adsorption, plasmid transfer and twitching motility, a form of flagellum-independent locomotion. Pili are composed of pilin subunits, the primary structures of which are conserved. All pilins are produced from prepilin molecules through cleavage of the leader sequence by a prepilin peptidase. There is a long hydrophobic segment (20 to 30 amino acid [aa] residues) at the N-terminal region of the mature pilin protein. Type IV pili are classified into two subclasses, IVA and IVB, on the basis of similarities in the deduced aa sequences of prepilins. Type IVA prepilins have a very short leader sequence (5 to 6 aa) and the N-terminal aa of the mature pilin is a methylated phenylalanine. Type IVB prepilins tend to have longer signal sequences (13 to 30 aa) and the N-terminal aa of the mature protein may be a methionine (methylated in some cases), leucine, tryptophan or serine. The assembly machinery involved in the formation of fimbriae consists of a set of proteins encoded by genes either scattered throughout the bacterial genome (IVA subclass), or organized into operons consisting of 11 to 14 genes (IVB subclass) (Manning and Meyer, 1997).

The Genus Yersinia, Edited by Skurnik *et al.*
Kluwer Academic/Plenum Publishers, New York, 2003

2. *Y. pseudotuberculosis* HARBORS A TYPE IV PILUS GENE CLUSTER FORMING A POLYCISTRONIC UNIT

So far, only one pilus gene cluster has been described in *Y. pseudotuberculosis*: the *psa* gene cluster encoding the pH6 antigen (Yang *et al.,* 1996; Yang and Isberg, 1997). We isolated a recombinant plasmid (pLS91.2) bearing a ~3.4 kb *Eco*RV DNA fragment from *Y. pseudotuberculosis* 32777. Analysis of the sequence of the DNA insert revealed the presence of four open reading frames (*orf1* to *4*), and similarity searches with the deduced amino acid sequences of the proteins encoded by these *orfs* showed that ORF1, ORF2, ORF3 and ORF4 displayed 55, 57, 59 and 55 % similarity to PilL, PilM, PilN and PilO respectively, of *Salmonella enterica*. These Pil proteins are thought to be involved in the biogenesis of type IV pili in this pathogen (Kim and Komano, 1997). To identify the complete type IV pilus gene cluster in *Y. pseudotuberculosis*, a cosmid library was constructed from strain 32777 DNA. Recombinants were further screened by colony blot hybridization under stringent conditions, using PCR-generated DNA probes at the 5' and 3'ends of the pLS91.2 insert. Three positive clones, NM554 (pMM2.1), NM554 (pMM2.A6) and NM554 (pMM3.D6), were isolated.

Cosmid pMM2.1 contained an 11-kb locus composed of eleven *orfs*, the organization of which is shown in Figure 1A. In terms of the arrangement of the first ten genes and their products (see below), the *Y. pseudotuberculosis* gene cluster strongly resembles the *pil* locus recently discovered in *S. enterica* (Kim and Komano, 1997; Zhang *et al.,* 2000). The *pil* cluster was incomplete in cosmid pMM2.A6 and in pMM3.D6, which contained only *pilLMN* and *pilNOPQRSUVW*, respectively.

All *pil orfs* had the same polarity and intergenic spaces ranged from -146 to 179 nucleotides. This suggested that the *pil* gene cluster may be transcribed as a single unit. The polycistronic organization of the locus was confirmed by RT-PCR experiments on total RNA extracted from strain 32777 using 12 sets of primers (1 to 12) (Figure 1A). No amplification product was obtained with primer sets 1 and 12 whereas primer sets 2 to 11 yielded amplicons of expected sizes (Figure 1B).

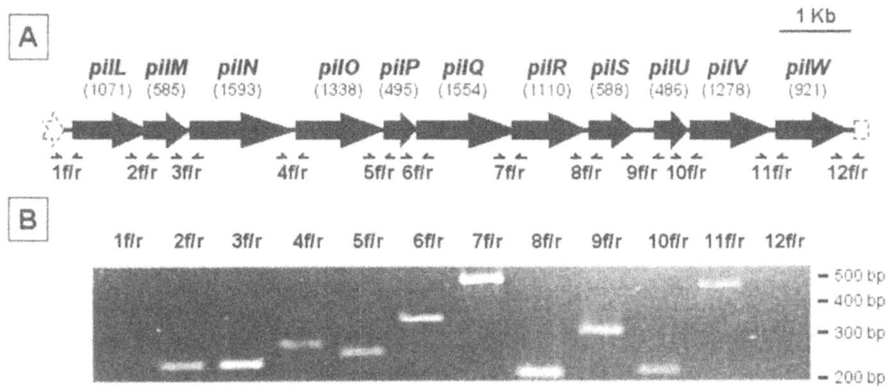

Figure 1. The *Y. pseudotuberculosis pil* operon.
A. Genetic organization of the *pil* gene cluster. The *pil* gene cluster consists of eleven genes with the same polarity and apparently arranged in an operon-like manner (the sizes in base pairs are shown in parentheses).
B. Transcriptional analysis of the *pil* gene cluster. RNA extracted from exponentially growing *Y. pseudotuberculosis* cultures at 37°C were analyzed by RT-PCR using primer sets 1 to 12. Forward (f) and reverse (r) primers were designed within adjacent genes of the *pil* locus and flanking regions (see A). The amplification products were analyzed by agarose gel electrophoresis.

3. THE TYPE IV PILUS GENE CLUSTERS FROM *Y. pseudotuberculosis* AND *S. enterica* ARE HIGHLY HOMOLOGOUS

The deduced aa sequences of the proteins encoded by the *pilLMNOPQRSUV* genes display 30-55 % identity and 46-70 % similarity to PilLMNOPQRSUV proteins involved in the biogenesis of type IV pili in *Salmonella* (Kim and Komano, 1997; Zhang *et al.*, 2000). On the basis of protein similarities, the putative products of *pilS* and *pilV* are thought to be structural prepilins. A putative signal sequence cleavage site has been identified between the 14 and 15th aa of prePilS and the 11 and 12th aa of prePilV. Interestingly, this putative cleavage site for both prePilS and prePilV is between a glycine and a tryptophan based on sequence alignments with all known prepilins from the type IVB family (Table 1). The mature products of *pilS* and *pilV* also have a 20-aa hydrophobic domain in their N-terminal regions whereas in their C-terminal regions, several cysteine residues (two for PilS at positions 140 and 177; eight for PilV at positions 327, 344, 365, 388, 402, 405, 411 and 424) may form intramolecular disulfide bridges, a feature shared by almost all type IV pilins (Manning and Meyer, 1997).

Table 1. Comparison of signal sequences of prepilins of the type IVB family. The pilin monomers are synthesized as precursor proteins with a hydrophilic leader peptide of variable length that is processed at a consensus cleavage site (arrow) by a prepilin peptidase usually recognizing the consensus motif **GXXXXE**. The signal sequences of prepilins and the first 20 aa residues (hydrophobic domain) of the mature proteins are shown.

Prepilins	**Microorganisms**	**Amino acid sequence of the N-terminus of prepilins**[a]	
			▼
prePilS	*Y. pseudotuberculosis*	MLSPVASRKQPHSG	WGILESGGVALVVIVVIAVV
prePilS	*S. enterica* Typhimurium	MLVENINTTLTGNNKKNEPHDKG	WAILEQGTIALVVLFVIVVV
prePilS	*S. enterica* Dublin and Typhi	MKNETEGKMMNEVSTLNPCNRPDRG	MSADAGATALFILVIIGVIA
prePilV	*Y. pseudotuberculosis*	MTFKTRALHRG	WAMMSTGIALLILVIVVIWA
prePilV	*S. enterica* Typhimurium	MKKYDRG	WASLETGAALLIVMLLIAWG
prePilV	*S. enterica* Dublin and Typhi	MKKQKHDGG	FVAMSVGAGLLIVLVMASLA
preBfpA	*E. coli* EPEC	MVSKIMNKKYEKG	LSLIESAMVLALAATVTAGV
preLngA	*E. coli* ETEC	MLSVYNRTQKFKAEARKKIAKYHELRKQRG	MSLLEVIIVLGIIGTIAAGV
preTcpA	*V. cholerae*	MQLLKQLFKKKFVKEEHDKKTGQEG	MTLLEVIIVLGIMGVVSAGV

[a]deduced from prepilin genes (accession numbers in GenBank: AF004308; AF304486; X74730; AF000001; D88588; AF247505)

Y. pseudotuberculosis PilU, the putative product of the gene *pilU*, is homologous to prepilin peptidases; sequence analysis of PilU showed the presence of a pair of aspartate residues at positions 33 and 95, at the carboxyl end. These residues are completely conserved among type IV prepilin peptidases and shown to be essential for enzyme activity (LaPointe and Taylor, 2000). Finally, the last gene of the *Y. pseudotuberculosis pil* cluster, *pilW*, encodes a product showing 61 % identity with a *Y. pestis* putative transposase.

4. EXPRESSION OF THE *Y. pseudotuberculosis pil* OPERON IN *E. coli* K-12 RECONSTITUTES BUNDLE-FORMING PILI

To demonstrate that the *pilLMNOPQRSUV* cluster is responsible for the biogenesis of a type IV pilus, a 10.8 kb *Nco*I fragment from cosmid pMM2.1, containing only the *pil* gene cluster and its putative promoter region, was subcloned into plasmid pACYC184 to yield pMM5. This plasmid was introduced into *E. coli* MC1061 and the bacterial cells were examined by transmission electron microscopy (TEM) (Figure 2). Bundles of filaments protruded from a polar end of strain MC1061 (pMM5) whereas the control strain MC1061 carrying only pACYC184 was nonpiliated (not shown).

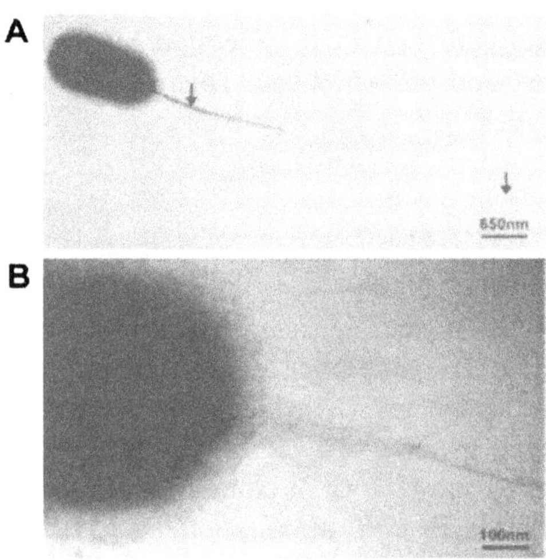

Figure 2. Electron micrograph of *E.coli* containing the recombinant plasmid pMM5. The strain was examined by TEM after uranyl acetate negative staining. A long pilus (6 μm) emanates from one polar position (A, arrows), and is constituted of bundles of fibers with a diameter of 5 nm (B).

5. DELETION OF THE *pil* GENE CLUSTER IN *Y. pseudotuberculosis* REDUCES BACTERIAL PATHOGENICITY IN THE MOUSE MODEL

To determine whether the *pil* gene cluster contributes to *Y. pseudotuberculosis* pathogenicity, an isogenic *pil* mutant was engineered from the wild type strain. A DNA fragment encompassing almost all of the *pil* operon was deleted in the parental strain 32777 and replaced with a kanamycin resistance gene *aphA-Ia*. The virulence of the *pil* mutant was first assessed in an oral model of infection in the BALB/c mouse. The LD_{50} of this strain was slightly higher than that of the parental strain (10^8 versus $10^{7.3}$). After intra-gastric inoculation of a lethal dose of *Yersinia*, death of animals was delayed when the infecting strain was *pil*-deficient. On the other hand, the *pil* mutant was found to be as virulent as the parental strain when administered intra-venously ($LD_{50} < 10^2$).

6. THE *pil* GENE CLUSTER IS PRESENT IN MOST PATHOGENIC *Y. pseudotuberculosis* STRAINS

We investigated the distribution of the *pil* gene cluster in the *Y. pseudotuberculosis* species. We tested 91 *Y. pseudotuberculosis* strains originating from various countries by colony-blot hybridization under stringent conditions with two PCR-generated *pil* probes from strain 32777 DNA: *pilL* (269 bp) and *pilS* (579 bp). Thirty-seven strains hybridized with each of the two probes.

7. CONCLUSIONS

Two features suggest that *Y. pseudotuberculosis* may have acquired the *pil* operon by horizontal gene transfer: (i) the *pil* locus is not uniformly present throughout the species; (ii) whereas the mean G + C % content of the *Y. pseudotuberculosis* genome is 47 %, that of the *pil* operon is 50.8 %. This value is very close to the G + C % content of the genomes of the type IV pilus-producing species *E. coli* and *Salmonella* (51-52 %). This favors the view that *Y. pseudotuberculosis*, which occupies the same ecological niche as these two other enterobacteria, may have acquired the type IV pilus locus from one of these species by genetic exchange in the natural habitat. The two *Salmonella pil* operons have been found located either on a conjugative plasmid (Kim and Komano, 1997) or a large (118 kb) pathogenicity island (Zhang *et al.*, 2000). Other type IVB pilus gene clusters, in enteropathogenic *E. coli* and *V. cholerae*, were also found to be harbored either on a plasmid (Girón *et al.*, 199; Girón *et al.*, 1994) or a pathogenicity island (Karolis *et al.*, 1999). In *Y. pseudotuberculosis* 32777, in which the type IV pilus gene cluster has been discovered, no plasmid other than the virulence plasmid pYV was detected. Thus, given the scenarios in *S. enterica*, *E. coli* and *V. cholerae*, it is tempting to speculate that the *Y. pseudotuberculosis pil* operon could constitute a novel " adaptation-pathogenicity" island, unknown up until now, which in addition to the already described High Pathogenicity Island (HPI) (Buchrieser *et al.*, 1999) contributes to the virulence of *Y. pseudotuberculosis.*

REFERENCES

Buchrieser, C., Rusniok, C., Frangeul, L., Couve, E., Billault, A., Kunst, F., Carniel, E., and Glaser, P., 1999, The 102-kilobase *pgm* locus of *Yersinia pestis*: sequence analysis and

comparison of selected regions among different *Yersinia pestis* and *Yersinia pseudotuberculosis* strains. *Infect. Immun.* **67**: 4851-4861.
Girón, J.A., Ho, A.S., and Schoolnik, G.K., 1991, An inducible bundle-forming pilus of enteropathogenic *Escherichia coli. Science* **254**: 710-713.
Girón, J.A., Levine, M.M., and Kaper, J.B., 1994, Longus: a long pilus ultrastructure produced by human enterotoxigenic *Escherichia coli. Mol. Microbiol.* **12**: 71-82.
Karaolis, D. K., Somara, S.D., Maneval Jr., R., Johnson, J.A., and Kaper, J.B., 1999, A bacteriophage encoding a pathogenicity island, a type-IV pilus and a phage receptor in cholera bacteria. *Nature* **399**: 375-379.
Kim, S.R., and Komano, T., 1997, The plasmid R64 thin pilus identified as a type IV pilus. *J. Bacteriol.* **179**: 3594-3603.
LaPointe, C.F., and Taylor, R.K., 2000, The type 4 prepilin peptidases comprise a novel family of aspartic acid proteases. *J. Biol. Chem.* **275**: 1502-1510.
Manning, P.A., and Meyer, T.F., 1997. Type-4 pili: biogenesis, adhesins, protein export and DNA import. Proceedings of a workshop. *Gene* **192**: 1-198.
Zhang, X.L., Tsui, I.S.C., Yip, M., Fung, A., Wong, W.D., Dai, K.X., Yang, Y., Hackett, J., and Morris C,. 2000, *Salmonella enterica* serovar Typhi uses type IVB pili to enter human intestinal epithelial cells. *Infect. Immun.* **68**: 3067-3073.
Yang, Y., Merriam, J.J., Mueller, J.P., and Isberg, R.R., 1996, The *psa* locus is responsible for thermoinducible binding of *Yersinia pseudotuberculosis* to cultured cells. *Infect. Immun.* **64**: 2483-2489.
Yang, Y., and Isberg, R.R., 1997, Transcriptional regulation of the *Yersinia pseudotuberculosis* pH6 antigen adhesin by two envelope-associated components. *Mol. Microbiol.* **24**: 499-510

Picture 16. Gregory Plano giving a lecture.

Chapter 17

Salicylanilides are Potent Inhibitors of Type III Secretion in *Yersinia*

Anna M. KAUPPI[1], Roland NORDFELTH[2], Ulrik HÄGGLUND[1], Hans WOLF-WATZ[2] and Mikael ELOFSSON[1,2]
[1]Organic Chemistry, Department of Chemistry and [2]Department of Molecular Biology, Umeå University, SE-901 87 Umeå, Sweden

1. INTRODUCTION

The type III secretion systems of gram-negative bacteria constitute attractive targets for development of novel antibacterial agents (Alksne and Projan, 2000; Gauthier and Finlay, 2002). The processes of protein secretion cross the bacterial membranes and translocation into the target cell are crucial for the bacterium to establish an infection. Drugs that inhibit bacterial virulence would be an important addition to the current arsenal of antibacterial drugs that all target growth. Moreover, small molecules that affect the type III secretion machinery can be used as chemical probes to further study the functional details of this important virulence system (Ward *et al.*, 2002). We have used a reporter gene assay in viable *Yersinia pseudotuberculosis* to screen for small molecules that inhibit type III secretion dependent transcription (Kauppi *et al.*, Submitted). Among the compounds identified the *O*-acetyl salicylanilide **1** (Figure 1A) proved to be a potent inhibitor of type III secretion.

2. TYPE III SECRETION INHIBITORS

During infection *Yersinia* spp. will encounter host immune cells and upon direct cell contact the effector proteins, *Yersinia* outer proteins (Yops) are expressed at high levels and secreted cross the bacterial membranes by

The Genus Yersinia, Edited by Skurnik *et al.*
Kluwer Academic/Plenum Publishers, New York, 2003

the *Yersinia secretion* (Ysc) machinery. Subsequently the Yops are translocated into the target cell through a pore in the eukaryotic cell membrane. Inside the target cell the effector proteins effectively inhibit the innate immune response (Cornelis, 2002). To be able to identify small molecules that affect type III secretion we developed a transcriptional reporter assay in viable *Y. pseudotuberculosis*. By fusing the luciferase gene, *luxAB*, from *Vibrio harveyi* (Olsson *et al.*, 1988) to the *yopE* promoter, type III secretion specific transcription can be monitored as a light signal in absence of eukaryotic cells by manipulating the temperature and Ca^{2+} concentration. Screening of a chemical library containing ~9,400 small molecules furnished a number of potent inhibitors of which the *O*-acetyl salicylanilide **1** (Figure 1A) is one of the most potent (Kauppi *et al.*, Submitted). Compound **1** effectively inhibits *yopE* transcription in the reporter strain YPIII(pIB29EL) without detrimental effect on growth at concentration as high as 100 μM (Figure 1B). These results indicate a high selectivity for type III secretion specific processes. Western blot analysis revealed that the inhibitory effect on transcription corresponds to inhibition of Yop secretion (data not shown).

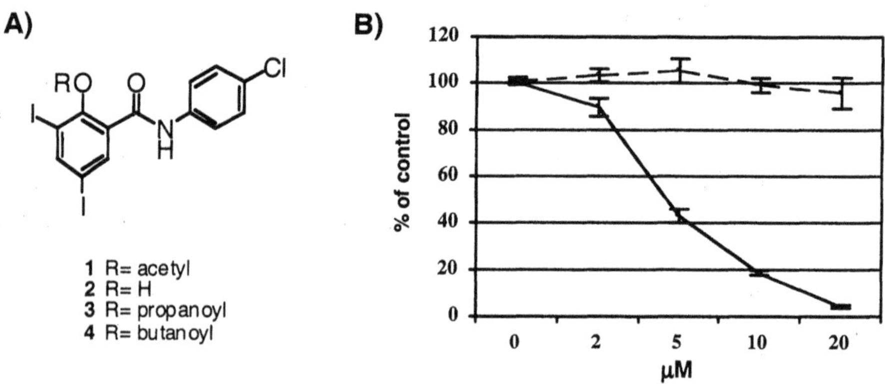

Figure 1. A) Structures for compound 1-4 and B) effect of compound 1 on the reporter gene signal (solid line) and bacterial growth (dashed line) for YPIII(pIB29EL). n=4

Salicylanilide **2** (Figure 1A), the synthetic precursor of **1**, on the other hand proved to be a potent inhibitor of both transcription and bacterial growth (Figure 2A). Lipophilic and weakly acidic salicylanilides such as **2** are known to be potent uncouplers that destroy the proton motive force required for ATP synthesis (Terada *et al.*, 1988). Possibly this effect is the reason for the growth inhibition observed for compound **2** (Figure 2A). Antimicrobial activity has been reported for a number of substituted salicylanilides (Ozawa *et al.*, 1984). Interestingly, substituted salicylanilides have also been shown to inhibit bacterial two-component systems (Macielag *et al.*, 1998). Two-component systems are important regulators of a wide variety of bacterial processes including virulence. It is possible that the acetyl group in **1** abolishes the uncoupling activity while the potential to inhibit two-component systems is maintained. Interestingly no two-component system has been implicated in regulation of the type III secretion machinery in *Yersinia*. Alternatively **1** acts as a pro-drug that is deacetylated to give the active compound **2** at low concentration and perhaps localized to certain compartments within the bacterium. Another possibility is that the acetyl group causes a switch in mode of action and that compound **1** targets some other regulatory component of the type III secretion system.

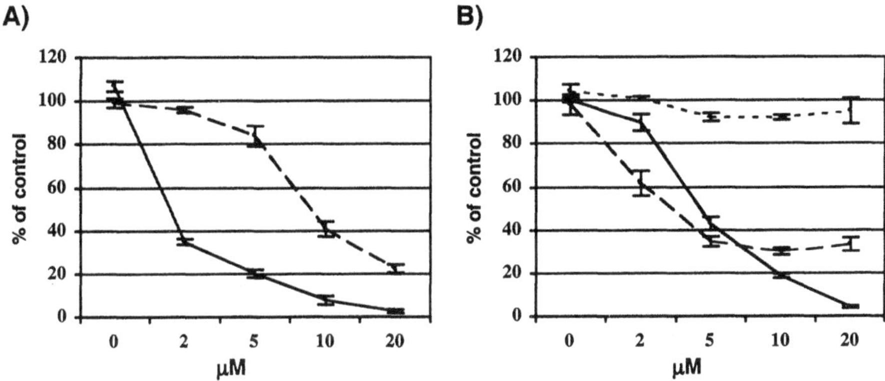

Figure 2. A) Effect of compound 2 on the reporter gene signal (solid line) and bacterial growth (dashed line) for YPIII(pIB29EL). B) Effect of compounds 1 (solid line), 3 (dashed line), and 4 (dotted line) on the reporter gene signal in YPIII(pIB29EL). n=4

These results prompted further studies and compound **3** and **4** (Figure 1A) were prepared in order to study the effect of different acyl groups on the biological activity. The syntheses were carried out essentially according to synthetic protocols reported previously (Ozawa *et al.*, 1984). In terms of inhibition of transcription from the *yopE* promoter compound **3** is a potent inhibitor while **4** proved to be void of any activity (Figure 2B). Derivatives with even larger acyl groups were also found to be inactive (data not shown). Compounds **3** and **4** show no or very low inhibitory effect on bacterial growth in the concentrations range investigated (data not shown). Thus, salicylanilides with small acyl groups e.g. **1** and **3** are potent inhibitors of *yopE* transcription without detrimental effect on growth. However, at this stage we can not conclude whether variation of the acyl group affects direct contact with the target(s) or alters other processes like uptake and metabolism.

3. CONCLUSIONS

We have identified *O*-acyl salicylanilides as potent inhibitors of type III secretion in *Y. pseudotuberculosis*. These compounds can serve as a starting point for development of novel antibacterial agents that target virulence. Furthermore, compounds that affect bacterial virulence can be employed as chemical tools to study and further understand the processes involved in bacterial virulence. We now plan to use these derivatives in *in vitro* and *in vivo* experiments to study type III secretion in *Yersinia* and other Gram-negative bacteria. The *Yersinia* targets of these salicylanilides are unknown and our future work will also focus on target identification and optimisation of promising compounds with the goal to increase potency and selectivity.

ACKNOWLEDGEMENTS

This work was supported by the Swedish National Research Council, The Foundation for Technology Transfer in Umeå, and Innnate Pharmaceuticals AB.

REFERENCES

Alksne, L. E. and Projan, S. J., 2000, Bacterial virulence as a target for antimicrobial chemotherapy. *Curr. Opin. Biotechnol.* **11**, 625-636.

Cornelis, G. R., 2002, *Yersinia* type III secretion: send in the effectors. *J. Cell Biol.* **158**, 401-408.

Gauthier, A. and Finlay, B. B., 2002, Type III secretion inhibitors are potential antimicrobials. *AMS News* **68**, 383-387.

Macielag, M. J. *et al.*, 1998, Substituted salicylanilides as inhibitors of two-component regulatory systems in bacteria. *J. Med. Chem.* **41**, 2939-2945.

Olsson, O., Koncz, C., and Szalay, A. A. 1988, The use of the *luxA* gene of the bacterial luciferase operon as a reporter gene. **215**, 1-9.

Ozawa, I. *et al.*, 1984, Synthesis and antimicrobial activity of salicylanilide derivatives. II. *Chem. Pharm. Bull.* **32**, 305-312.

Terada, H., Goto, S., Yamamoto, K., Takeuchi, I., Hamada, Y., and Miyake, K. 1988, Structural requirements for uncoupling activity in mitochondria: quantitative analysis of structure-uncoupling relationships. *Biochim. Biophys. Acta* **936**, 504-512.

Ward, G. E., Carey, K. L., and Westwood, N. J. 2002, Using small molecules to study big questions in cellular microbiology. *Cell. Microbiol.* **4**, 471-482.

Chapter 18

Mapping of Possible Laminin Binding Sites of *Y. pestis* Plasminogen Activator (Pla) via Phage Display

Orsolya BENEDEK[1], Judit BENE[2], Béla MELEGH[2] and Levente EMŐDY[1]

[1]*Department of Medical Microbiology and Immunology,* [2]*Department of Genetics and Child Development, Medical School, University of Pécs, H-7624 Szigeti út 12., Pécs, Hungary*

1. INTRODUCTION

Yersinia pestis, the causative agent of plague is much more invasive than the two enteropathogenic *Yersiniae*, *Y. pseudotuberculosis* and *Y. enterocolitica*. Despite its high invasiveness *Y. pestis* lacks the well-characterised invasion factors of the two other species namely the *inv* and *ail* gene products and the outer membrane protein YadA. However, it harbours two unique virulence plasmids which might encode candidate proteins responsible for invasion. One of them is a plasminogen activator (Pla) being involved in different steps of invasion. Pla in itself is also able to mediate adhesion to several matrix proteins (Lähteenmäki *et al.,* 1998). In our study we tried to map the laminin binding motifs of Pla by utilising a random phage display library in bacteriophage M13.

2. METHODS AND RESULTS

2.1 Phage display against laminin

A random heptamer phage display library was used for biopanning against laminin based on the manufacturer's instructions. Twenty plaques were purified and sequenced, and eighteen different sequences were obtained.

The Genus Yersinia, Edited by Skurnik *et al.*
…wer Academic/Plenum Publishers, New York, 2003

2.2 Effect of tight-binder phages on Pla mediated laminin binding of *E. coli*

The different phages were tested for their ability to interfere with Pla mediated laminin binding by Pla expressing *E. coli*. We found two independent sequences (WSLLTPA-#5 and YPYIPTL-#14) being able to inhibit laminin binding completely (Figure 1)

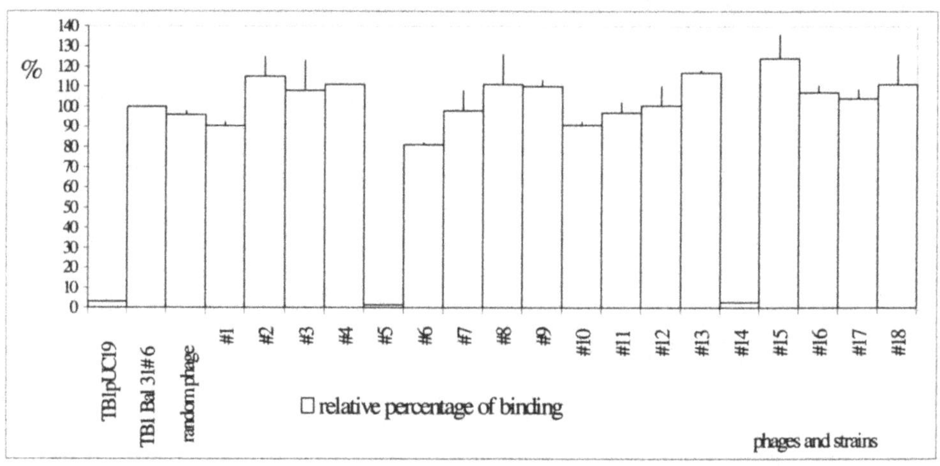

Figure 1. Effect of selected phages on Pla mediated laminin-binding by Pla+ *E.coli*

2.3 Laminin binding by inhibitory phages

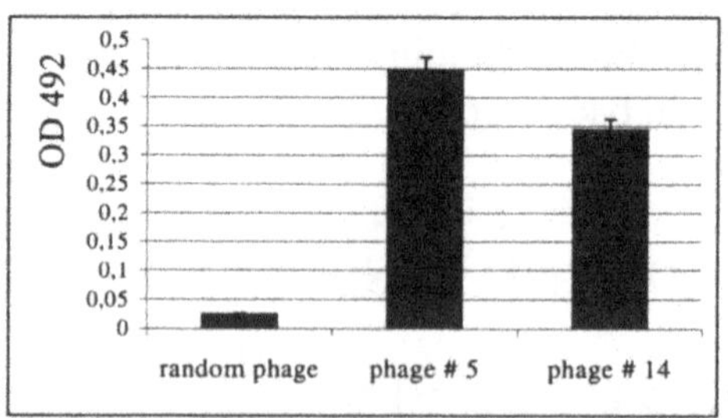

Figure 2. Laminin binding by inhibitory phages in ELISA assay

The two phages displaying strong inhibition on Pla mediated laminin binding bound laminin efficiently (Figure 2) in an ELISA based system using HPO-labelled anti-M13 monoclonal antibodies. WSLLTPA-#5 bound to laminin seventeen times stronger, and YPYIPTL-#14 bound thirteen times stronger than the random phage.

2.4 Fibrinolysis by phages

We checked the fibrinolytic activity of inhibitory phages in Petri dishes. They were spotted on the surface of fibrin films and incubated at 37 °C for 6 hours. Random phages were non-fibrinolytic while phages WSLLTPA-#5 and YPYIPTL-#14 lysed the film in a sharp round spot. (Data not shown.)

2.5 Pattern search

Peptide pattern search was performed with the help of PattinProt Program (Université Claude-Bernard, Lyon, France). Both laminin binding patterns were found outside the putative surface-exposed loops. While YPYIPTL is close to it WSLLTPA is localised at a periplasmic turn.

3. SUMMARY AND CONCLUSIONS

We tried to determine amino acid motifs of *Y. pestis* plasminogen activator (Pla) involved in laminin binding. We selected heptamer peptides using a random phage library which was tested against immobilised laminin. Two sequences seemed to inhibit Pla mediated laminin binding of *E. coli* and exhibited a strong laminin binding capacity revealing a competition with Pla for the same laminin binding site. The motifs are also involved in plasminogen activation because they caused fibrinolysis on fibrin films. The patterns were localised outside the putative surface-exposed loops (Kukkonen *et al.*, 2001).

ACKNOWLEDGEMENTS

This work was supported by OTKA grant T 037833 to L.E.

REFERENCES

Kukkonen, M, Lähteenmäki, K, Suomalainen, M, Kalkkinen, N, Emödy, L, Lång, H, Korhonen, T.K. 2001. Protein regions important for plasminogen activation and inactivation of alpha2-antiplasmin in the surface protease Pla of *Yersinia pestis*. *Mol Microbiol,* **40**:1097-1111

Lähteenmäki, K, Virkola, R, Saren, A, Emödy, L, Korhonen, T. K. 1998. Expression of plasminogen activator Pla of *Yersinia pestis* enhances bacterial attachment to the mammalian extracellular matrix. *Infect Immun*, **66**:5755-5762

Chapter 19

The Fish Pathogen *Yersinia ruckeri* Possesses a TTS System

Deephti K. GUNASENA, Jenny R. KOMROWER and Sheila MACINTYRE
School of AMS, University of Reading, Reading, UK

1. INTRODUCTION

Type III Secretion (TTS) systems are utilized by a number of mammalian and plant pathogens to inject proteins directly into the host cytosol. The set of effector proteins interfere with host cell signalling resulting in a range of phenotypes such as microfilament rearrangement and bacterial adhesion (EPEC *E. coli*), invasion (*Salmonella*, *Shigella*) or host cell death (*Yersinia*). The archetype TTSS has been the plasmid encoded "Yop virulon" shared by the three major species of pathogenic *Yersinia* (Hueck, 1998). However reported studies could find no evidence of this plasmid in the fish pathogen *Y. ruckeri* (Guilvout *et al.*, 1988). Because TTS systems are so critical to the pathogenesis of many Gram-negative bacteria we initiated a study to investigate the significance of TTSS to bacterial pathogens of fish.

2. DEGENERATE OLIGONUCLEOTIDES IDENTIFY TTS ATPase IN *Y. ruckeri*

To screen for TTSS in fish pathogens, a set of degenerate oligonucleotides was designed against the gene for one of the most highly conserved proteins, the type III ATPase. Sequencing of the target PCR fragments (460-470 bp) from positive controls, *S. typhimurium* 12/75 and *Y. pseudotuberculosis* YPIII (pIB102), confirmed amplification of the gene encoding type III ATPase (*ssaN* and *spaL* or *yscN*, respectively) rather than

The Genus Yersinia, Edited by Skurnik *et al.*
'luwer Academic/Plenum Publishers, New York, 2003

the closely associated flagella ATPase genes. The amplified product from *Y. ruckeri* NCIMB 1315 showed significant sequence homology (80% over 149 amino acids) to the chromosomally encoded, low temperature induced type III ATPase gene *ysaN* in *Y. enterocolitica* (Haller *et al.,* 2000).

3. ORGANISATION AND SIGNIFICANCE OF *ysa* OPERON

Genes adjacent to the putative *ysaN* gene of *Y. ruckeri* were probed for and amplified by PCR using degenerate primers designed against the *ysa* operon of *Y. enterocolitica* and the *Y. pestis* chromosomally encoded TTSS (GenBank Acc. No: AF005744 and YPO0255-YPO0273, respectively). The translated sequences of the *Y. ruckeri* genes showed 60%, 51%, 67% and 37 % identity to *ysaV*, *ysaK*, *ysaN* and CDS19 of *Y. enterocolitica*, respectively, confirming presence of a Ysa-like TTSS in *Y. ruckeri* (Figure 1).

Phylogenetic analysis revealed that YsaN of *Y. ruckeri* was clearly distinct from the ATPase in the *Y. pestis* chromosomal TTSS cluster (YPO0267) and also from the plasmid encoded YscN. The complete gene product, YsaN of *Y. ruckeri*, gave 67% identity to YsaN of *Y. enterocolitica*, 55% to InvC of *Sodalis glossinidius*, 51% to InvC of (SPI-I) *S. typhimurium* and 50% to SpaL of *Shigella flexneri*. In contrast, it shared only 45 % identity with *Y. pestis* YPO0267, which was more closely related to *S. typhimurium* SsaN (SPI-2), and 45 % identity with *Y. enterocolitica* YscN. Organization of genes in *Y. ruckeri* also confirmed the close relationship of the *Y. ruckeri* system to that of *Y. enterocolitica ysa* operon (Figure 1).

Interestingly, the Ysa system has recently been shown to be present only in biotype IB strains of *Y. enterocolitica* (Foultier *et al.,* 2002). Since, 16S rDNA sequence of *Y. ruckeri* is closer to that of the *Y. pestis/ pseudotuberculosis* subline than to *Y. enterocolitica* (Ibrahim *et al.,* 1993), this raises an interesting question as to the origin of the Ysa system. The degenerate primers used to screen for type III ATPase did not amplify a *yscN* gene in *Y. ruckeri* consistent with the absence of a pYV type plasmid in this bacterium (Guilvout *et al.,* 1988). Assuming *Y. ruckeri* does possess only this TTSS, this offers an ideal opportunity to study the function and significance of the Ysa system in a bacterium which has evolved without the high virulence Ysc system.

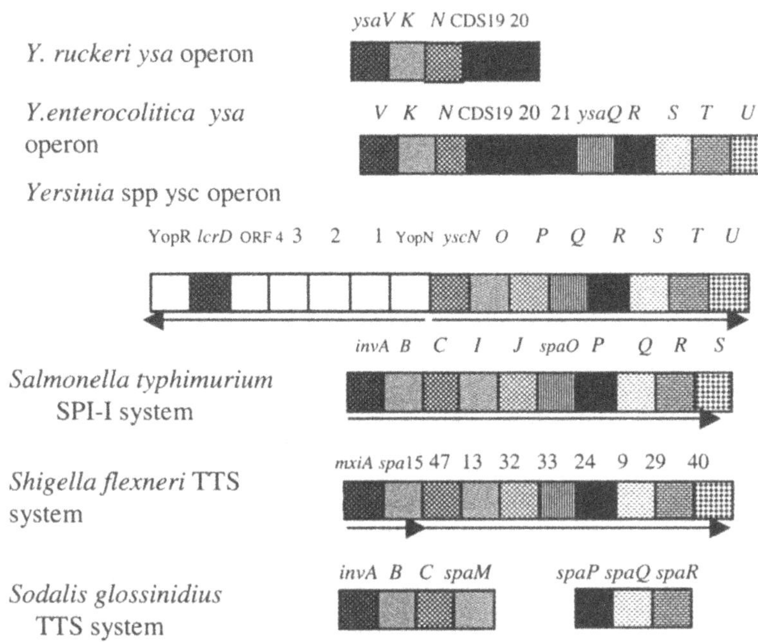

Figure 1. Genetic organization of the type III ATPase and neighboring genes in *Y. ruckeri* and selected bacteria. *Y. ruckeri* genes are named based on their high homology with *Y. enterocolitica ysa* cluster. Homologies of encoded proteins are indicated by by different patterns.

ACKNOWLEDGEMENTS

Work funded by the Sri Lankan Government and the University of Reading, UK.

REFERENCES

Foultier, B., Troisfontaines, P., Muller, S., Opperdoes, F.R. and Cornelis, G.R. (2002) *J Mol Evol* **55**: 37-51.

Guilvout, I., Quilici, M.L., Rabot, S., Lesel, R. and Mazigh, D. (1988) *Applied Env. Microbiol* **54**: 2594-2597.

Haller, J.C., Carlson, S., Pederson, K.J. and Pierson, D.E. (2000) *Mol Microbiol* **36**: 1436-1446

Hueck, C.J.(1998) *Microbiol Mol Biol Rev* **62**: 379-433

Ibrahim, A., Goebel,B.M., Liesack, W. Griffiths, M and Stackebrandt, E. (1993) *FEMS Microbiol Lett* **114**:173-178.

Chapter 20

Characterisation of the Type III Secretion Protein YscU in *Yersinia pseudotuberculosis*

YscU cleavage – dispensable for TTSS but essential for survival

Moa LAVANDER[1], Lena SUNDBERG[1,2], Petra J. EDQVIST[2], Scott A. LLOYD[2,3], Hans WOLF-WATZ[2] and Åke FORSBERG[1,2]

[1]Department of Medical Countermeasures, Division of NBC-Defence, Swedish Defence Research Agency, S-901 82 Umeå, Sweden; [2]Department of Molecular Biology, Umeå University, S-901 87 Umeå, Sweden; [3]Present address: Center for Vaccine Development, Department of Pediatrics, University of Maryland School of Medicine, Baltimore, MD, 21201

1. THE YscU/FlhB PROTEIN FAMILY

The type III secretion system (TTSS) is central for virulence of pathogenic *Yersinia* species and utilised to target the anti-host Yop (*Yersinia* outer proteins) effectors into eukaryotic cells. More than 20 Ysc (Yop secretion) proteins constitute the machinery, which mediates secretion of Yops across the double bacterial membrane (Review: Cornelis *et al.*, 1998). Our focus is *Y. pseudotuberculosis* YscU, one of the core Ysc proteins, which are absolutely required for Yop secretion. In all known TTSSs, members of the YscU protein family are found. These ~40kDa proteins share high amino acid sequence homology and also membrane topology (Figure 1), four domains spanning the bacterial inner membrane and a C-terminal tail protruding into the cytoplasm (Allaoui *et al.*, 1994; Minamino *et al.*, 1994).

The most extensively studied member of the YscU protein family is *Salmonella typhimurium* FlhB of the flagellar TTSS. By expressing the cytoplasmic domain of FlhB in an *E. coli* background, Minamino and Macnab (2000) have shown that the protein is specifically cleaved at residue Proline270. We recently performed a further characterisation of this cleavage (Lavander *et al.*, 2002), investigating whether it occurs in *Yersinia*

The Genus Yersinia, Edited by Skurnik *et al.*
ˈuwer Academic/Plenum Publishers, New York, 2003

YscU aiming to elucidate if it is coupled to type III secretion, or, if not, what an alternate role of cleavage is.

2. ANALYSIS OF YSCU CLEAVAGE

2.1 YscU cleavage is identical to that of FlhB

We verified that the 354 amino acids protein YscU, when expressed in *Y. pseudotuberculosis*, was cleaved in a manner matching that of *Salmonella* FlhB. The amino acid position for YscU cleavage is Proline-264, corresponding to Pro270 of FlhB. Aligning the YscU/FlhB family proteins revealed a strictly conserved site of 4 amino acids, NPTH (Asn263-Pro264-Thr265-His266, numbers referring to amino acid positions in YscU) surrounding the site of cleavage. An in-frame deletion of the NPTH site of YscU was made; resulting in an altered cleavage pattern with a larger C-terminal cleavage product than that of the wild type protein (Figure 1). Moreover, the variant of YscU that lacks the NPTH site was non-functional with respect to Yop secretion. This is possibly due to the altered cleavage disrupting a domain that has to be intact for YscU to function in type III secretion.

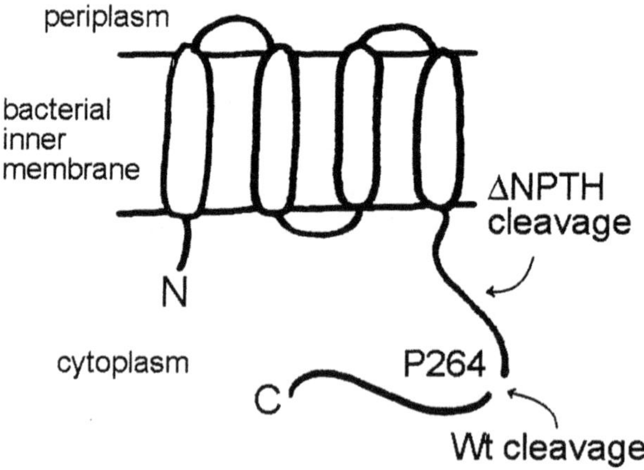

Figure 1. YscU transmembrane structure and cleavage sites for Wt and ΔNPTH YscU

2.2 Cleavage is not required for TTSS but for bacterial survival

No connection between YscU cleavage and type III secretion could be detected by culturing *Yersinia* under conditions known to induce the TTSS. Instead the protein was cleaved as soon as it appeared in the bacterial cells, irrespective of growth conditions. As cleavage occurred also in the TTSS-negative virulence plasmid cured *Yersinia* strain, the protease responsible for cleavage is apparently type III independent. Moreover, single site substitutions at the conserved residues, N263A, P264A or T265A, generated YscU mutant proteins that were not cleaved but still retained partial or full function in Yop secretion. These data determines proteolytic cleavage of YscU to be redundant for functional type III secretion. However, expression of these mutant proteins in *Yersinia* resulted in severe growth inhibition, indicating that the uncleavable variants of YscU had a toxic effect on the bacterial cell.

3. CONCLUSIONS

The YscU/FlhB protein family is very highly conserved, amino acid sequence homologies between the cytoplasmic domains being particularly high. Here we investigated a feature common to YscU and FlhB, i.e. the cleavage of the proteins within the C-terminal cytoplasmic tail. We could conclude that the cleavage is not connected to TTSS regulation. Nor is it required for YscU function in *Yersinia* type III secretion, since the uncleavable YscU variants are able to restore Yop secretion to a *yscU* null mutant strain. However, if YscU is not cleaved within the cytoplasmic tail, it has a toxic effect on the bacteria. The fact that *Yersinia* strains expressing the uncleavable YscU point mutant proteins are seriously inhibited for growth shows that the cleavage is required for survival of the bacterium.

REFERENCES

Allaoui, A., S. Woestyn, C. Sluiters, and G.R. Cornelis. 1994. YscU, a *Yersinia enterocolitica* inner membrane protein involved in Yop secretion. *J Bacteriol.* **176**:4534-42.

Cornelis, G.R., A. Boland, A.P. Boyd, C. Geuijen, M. Iriarte, C. Neyt, M.P. Sory, and I. Stainier. 1998. The virulence plasmid of *Yersinia*, an antihost genome. *Microbiol Mol Biol Rev.* **62**:1315-52.

Lavander, M., L. Sundberg, P.J. Edqvist, S.A. Lloyd, H. Wolf-Watz, and A. Forsberg. 2002. Proteolytic cleavage of the FlhB homologue YscU of *Yersinia pseudotuberculosis* is essential for bacterial survival but not for type III secretion. *J Bacteriol.* **184**:4500-9.

Minamino, T., T. Iino, and K. Kutuskake. 1994. Molecular characterization of the *Salmonella typhimurium flhB* operon and its protein products. *J Bacteriol.* **176**:7630-7.

Minamino, T., and R.M. Macnab. 2000. Domain structure of *Salmonella* FlhB, a flagellar export component responsible for substrate specificity switching. *J Bacteriol.* **182**:4906-14.

Chapter 21

Mutagenesis Elucidates The Assembly Pathway and Structure of *Yersinia pestis* F1 Polymer

Joanne E. KERSLEY[1], Anton V. ZAVIALOV[2], Elham MOSLEHI[1], Stefan D. KNIGHT[2] and Sheila MACINTYRE[1]
[1]*School of Animal and Microbial Sciences, University of Reading, UK,* [2]*Swedish University of Agriculture Sciences, Uppsala, Sweden*

1. INTRODUCTION

The F1 antigen forms an amorphous capsule on the surface of *Yersinia pestis* shortly after infection of the mammalian host. It is antiphagocytic, is highly immunogenic and is a key component of anti-plague vaccines. It is only recently however that any clues to the macromolecular structure of the capsule have been elucidated. Although no defined structure has been resolved by EM, it is known that F1 antigen is assembled via the chaperone-usher pathway (Figure 1) characterised for Pap and Type I pili (Choudhury *et al.*, 1999). We have used mutagenesis to investigate the assembly process and isolate assembly intermediates of Caf1 polymer in an *E. coli* model system (MacIntyre *et al.*, 2001; Zavialov *et al.*, 2002).

2. CHAPERONE MEDIATED SUBUNIT FOLDING

Caf1M belongs to the FGL subfamily of chaperones characterized by the following (see Figure 2) (i) disulphide bond adjacent to the subunit binding site essential for folding of Caf1M but not activity (ii) an extended FGL loop between F1 and G1 β-strands containing hydrophobic residues key to subunit binding (iii) an extended N-terminus upstream of conserved Arg (iv) only one (or in some cases two) subunits in the assembled structure.

The Genus Yersinia, Edited by Skurnik *et al.*
Kluwer Academic/Plenum Publishers, New York, 2003

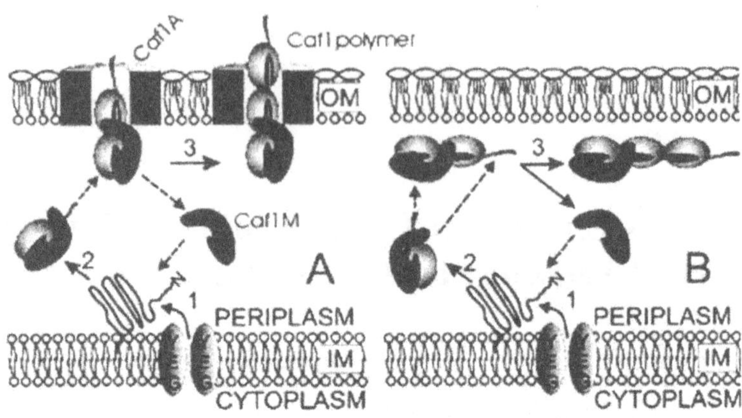

Figure 1. Model of Caf1 folding and polymerisation. (A) Following secretion across the IM the incomplete fold of Caf1 subunit is complemented by donation of an extended G1 β-strand from Caf1M chaperone (donor strand complementation) (2). Free donor strand on chaperone-bound subunit replaces chaperone G1 strand resulting in subunit:subunit complementation (2-3). Growing polymer assembles at the cell surface, via interaction with Caf1A usher, or in periplasm (B) in the absence of Caf1A (adapted from Zavialov *et al.*, 2002).

```
Chaperone           F1 β-strand       FGL 'loop'          G1 β-strand
           20  90 ==========>                     126      ============>
Caf1M (19)R. DKESLKWLCVKGIPPKDEDIWDDATNKQKFNPDKDVGVFVQFAINNCIKLLVRP
PAPd    (7)R .DRESLFYFNLREIPPRSEK-----------------ANVLQIALQTKIKLFYRP
```

Figure 2. G1 binding strand of Caf1M (FGL) and PapD (FGS) chaperones. Conserved Arg (R20 in Caf1M), anchors the COOH residue of subunits in Type I pili in the binding cleft.

Sequence identity between members of this family (mostly >30%) suggests a similar fold to the immunoglobulin fold of FimC (Choudhury *et al.*,1999). With a Caf1MR20S mutant, subunit was rapidly degraded and surface assembled F1 was only 13% of that assembled with wild-type chaperone. An even more dramatic effect was observed with FGL residues V126 and V128 mutated to Ser; Caf1 was again degraded and no surface assembled F1 could be detected. On the basis of these results it appeared that Caf1M chaperone stabilises Caf1 monomer via the basic principle of donor strand complementation, but with a longer extended G1 binding strand implying an extended hydrophobic surface on the Caf1 subunit (MacIntyre *et al.*, 2001).

3. SUBUNIT POLYMERISATION

Analysis of periplasmic assembled F1 polymer (Figure 1B) provided evidence that F1 assembles in a Caf1M:Caf1 (n+1) configuration. To test if subunit polymerization occurred via donor strand complementation (Figure 1) as proposed for Type I pili assembly (Choudhury *et al.,* 1999), three potential donor motifs within the N-terminus of Caf1 subunit were individually mutated (Zavialov *et al.,* 2002). All mutants formed binary complex with chaperone, but mutations within the 3rd motif were incompletely folded as indicated by their sensitivity to proteases. Deletion of the 1st motif had no effect on subunit folding or polymerization. Mutation of the 2nd motif led to stable binary (Caf1ΔA9-V14:Caf1M) or ternary ($A9RCaf1_2$:Caf1M) complex but polymerization was blocked. The essential donor strand was identified as T8ATATL13. Study of the susceptibility of $A9RCaf1_2$:Caf1M to trypsin and wild type ternary complex to proteinase K showed that half of the subunit was cleaved within the donor strand. This shows that the donor strand of one subunit becomes buried during subunit: subunit interaction.

4. CONCLUSIONS AND APPLICATIONS

This study shows that apparently strikingly different architectures can be achieved by the same general principle of donor strand complementation. Perhaps the most significant aspect of this study, however, has been the ability to block polymerisation and obtain stable assembly intermediates. These are currently being used to obtain high resolution structural information on donor strand complementation between chaperone and subunit as well as subunit:subunit interaction and F1 antigen structure. Finally, the fact that each Caf1 monomer is an incomplete β-barrel would explain why Caf1 cannot be isolated as a stable monomer and why Caf1 expressed in the cytoplasm (in the absence of chaperone) is less protective than surface derived F1 (Titball and Williamson, 2001).

ACKNOWLEDGEMENTS

Work supported by BBSRC grant 45/B16926 and Swedish Research Council

REFERENCES

Choudhury, D., Thompson, A., Stojanoff, V., Langermann, S., Pickner, J., Hultgren, S.J., and Knight, S.D. *et al.* (1999) X-ray structure of the FimC-FimH chaperone-adhesin complex from uropathogenic *Escherichia coli. Science* 285: 1061-1066.

MacIntyre, S., Zyrianova, I.M., Chernovskaya, T.V., Leonard, M., Rudenko, E.G., Zav'yalov, V.P. and Chapman D.A.G. (2001) An extended hydrophobic interactive surface of *Yersinia pestis* Caf1M chaperone is essential for subunit binding and F1 capsule assembly. *Mol Microbiol* 39: 12-25

Titball, R.W. and Williamson, E.D. (2001) Vaccination against bubonic and pneumonic plague. *Vaccine* 19: 4175-4184

Zavialov, A.V., Kersley, J., Korpela, T., Zav'yalov, V.P., MacIntyre, S. and Knight, S. (2002) Donor strand complementation mechanism in the biogenesis of non-pilus systems. *Mol Microbiol* 45: 983-995

Chapter 22

Characterization of Infections with Wild and Mutant *Yersinia pseudotuberculosis* Strains in Rabbit Oral Model

Hristo NAJDENSKI[1], Anna VESSELINOVA[1], Elitsa GOLKOCHEVA[1], Sara GARBOM[2] and Hans WOLF-WATZ[2]
[1]*Department of Pathogenic Bacteria, The Stephan Angeloff Institute of Microbiology, Bulgarian Academy of Sciences, 1113 Sofia, Bulgaria;* [2]*Department of Cell and Molecular Biology, University of Umeå, S-901 87 Umea, Sweden.*

1. INTRODUCTION

The expression of virulence in *Y. pseudotuberculosis* requires the presence of a common 70-kb virulence plasmid that encodes a number of secreted proteins called Yops (Heesemann *et al.,* 1984; Bölin *et al.,* 1985). By specific mutagenesis, several of the Yops have been identified as virulence determinants in experimental infection mainly in mouse model. One of these proteins is YpkA (Ser/Thr protein kinase), a virulence factor with the potential to interfere with the signal transduction of eukaryotic cells. Another secreted 82-kDa protein - YopK is involved in translocation of Yop effectors into eukaryotic cells and required for *Y. pseudotuberculosis* to establish a systemic infection in mice after i.p. and oral inoculation (Holmström *et al.,* 1995).

In the present study we compared the course of infection of the virulent *Y. pseudotuberculosis* strain to that of the ΔyopK and ΔypkA mutants in rabbit model and compared the clinical signs, bacterial load, IgG and IgM antibody responses and histopathology.

The Genus Yersinia, Edited by Skurnik *et al.*
'uwer Academic/Plenum Publishers, New York, 2003

2. MATERIALS AND METHODS

Bacterial strains. *Y. pseudotuberculosis* serotype O:3 wild type strain (pIB102) and two mutant strains (pIB155,ΔyopK and pIB44,ΔypkA) were cultivated in BHI broth (Difco) as described by Holmström *et al.,* (1995).

Experimental animals and infection. Fifty-five six-month old New Zealand rabbits,. weighing 2.5 ± 0.2 kg were used. Animals were orally infected as follows: 15 rabbits with the wild strain pIB102 in a dose of 1.6 x 10^{10} CFU; 15 with ΔyopK mutant strain - dose of 2.3 x 10^{10} CFU and 15 – with the mutant strain ΔypkA - dose of 2.0 x 10^{10} CFU. Three infected and two control animals were sedated and killed by intravenous injection of 50 to 80 mg/kg sodium pentobarbital (Pharmachim, Bulgaria) at each time point.

Bacterial clearance, antibody response and histopathology were carried out as previously described (Najdenski *et al.,* 1998).

3. RESULTS AND DISCUSSION

Clinical observations during the course of yersiniosis revealed higher rates of body temperatures up to day 14 p.i. with the wild strain (39.5 – 39.7 °C) in comparison with the control animals (38.6 ± 0.2 °C). Significantly augmented body temperature was found at day 7 p.i. in animals infected with the mutant ΔypkA strain (39.2 – 39.5 °C) and at day 5 in animals infected with the ΔyopK strain (39.2 – 39.6 °C). During this period most animals were weakly dispirited but retained their appetite. No signs of diarrhea or loss of weight were observed.

The cultures of all specimens (excepting feces) from rabbits infected with the wild strain were positive up to day 30 p.i. The infection rate progressed during that period especially in tonsils, brain, lung, spleen, and small intestines persisting in the former till the last period of examination. Brain, tonsil, liver, spleen, MLN and SI of rabbits infected with the mutant ΔyopK strain were not colonized whereas these organs were colonized by the ΔypkA strain. At the next intervals (days 45 and 60 p.i.) bacteria were not found in brain and viscera of rabbits infected with both mutant strains. During the first two weeks at the onset of infection stools of all experimental animals gave positive cultures.

Sera obtained from animals infected with wild and ΔyopK strains showed in the last two intervals four-fold increase in IgG antibody titer (Table 1). Two fold increase in IgM titer was observed in both experimental groups starting from day 30 p.i. Lower titers for IgG and IgM (1:50 to 1:100) were found after ΔypkA infection.

Table 1. IgG and IgM antibody responses screened by ELISA test

Days p.i.	Yersinia pseudotuberculosis strains					
	pIB102		pIB155,ΔyopK		pIB44,ΔypkA	
	IgG	IgM	IgG	IgM	IgG	IgM
7	-	-	-	1:50	-	-
14	1:50	1:50	1:100	1:50	1:50	1:50
30	1:100	1:100	1:100	1:100	1:100	1:100
45	1:200	1:100	1:200	1:100	1:100	1:50
60	1:200	1:50	1:100	1:100	1:100	1:50

Purulent meningoencephalitis was found in the brain of animals infected with the wild and ΔypkA mutant strains. In rabbits infected with the wild strain a catarrhal pneumonia and catarrhal enteritis were observed while after ΔyopK and ΔypkA infections only lymphadenitis and lienitis, respectively, were established. Interesting finding was the observed hyperplasia of lymphoid tissue in the lungs of experimental animals infected with all of three pathogens. Clearly demonstrated hyperplasia of lymphoid tissue was established in the Peyer`s patches especially after ΔyopK infection (Figure 1).

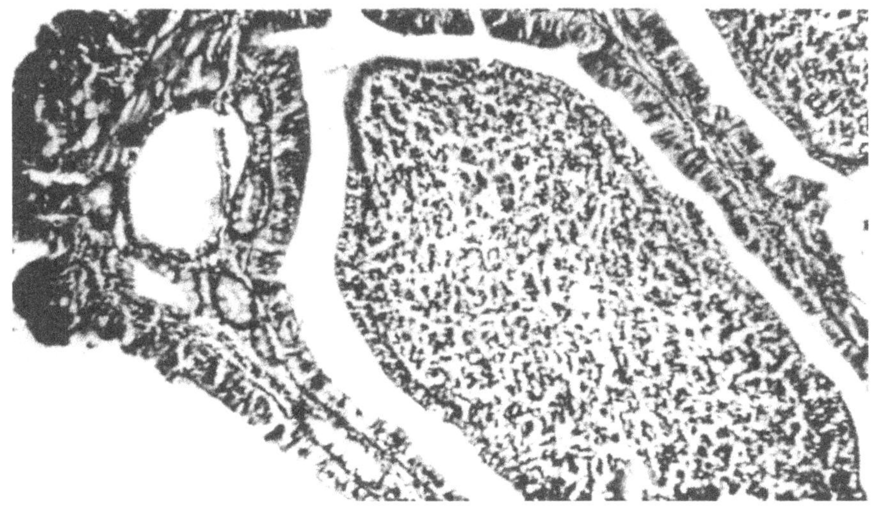

Figure 1. Small intestine of rabbit infected with pIB155,ΔyopK mutant strain of *Y. pseudotuberculosis* O:3, day 7 p.i. Hyperplasia of lymphoid tissue in lamina propria. H.E. staining. Magnification 100x.

The course of examined infections was progressive with a transient hypertermia and dissemination of bacteria in brain and viscera till day 30 p.i. (pIB102 and ΔypkA), respectively day 14 p.i. (ΔyopK). Evidently, ΔyopK mutant cells are markedly impaired in their ability to disseminate into brain and viscera of infected rabbits although they are able to colonize the Peyer's patches and induce significant IgG and IgM antibody titers. The observed significant leucocytosis with granulocytosis, moderate monocytosis, transient lymphopenia and enhanced specific IgG and IgM antibody responses after infection with the mutant strain ΔyopK was in correlation with the low virulence for mice (unpublished data). In contrast, the ΔypkA mutant cells were as rapidly cleared from the brain and viscera as the wild type strain.

The histological analysis of the lymph tissues of rabbits infected with both mutant strains demonstrated immunomorphological changes as a sign of ongoing immune responses.

The comparative study showed that genetically constructed pIB155,ΔyopK mutant strain is significantly attenuated and has a great potential to serve as a safe and effective live carrier vaccine strain.

ACKNOWLEDGEMENTS

This work was supported by the European Union (contract QLK2-CT-1999-00780).

REFERENCES

Bölin, I., Portnoy, D.A., and Wolf-Watz, H., 1985, Expression of the temperature-inducible outer membrane proteins of *yersiniae*. *Infect. Immun.* **48**: 234-240.

Heesemann, J., Algermissen, B., and Laufs, R., 1984, Genetically manipulated virulence of *Yersinia enterocolitica*, *Infect. Immun.* **46**: 105-110.

Holmström, A., Rosqvist, R., Wolf-Watz, H., and Forsberg, A., 1995, Virulence plasmid – encoded YopK is essential for *Yersinia pseudotuberculosis* to cause systemic infection in mice, *Infect. Immun.* **63:** 2269-2276.

Najdenski, H., Nikolova, S., Vesselinova, A., Nejkov, P., 1998, Studies of *Yersinia enterocolitica* 0:3 experimental infection in pigs. *J. Vet. Med. B* **45**: 59-64.

Chapter 23

Identification of *Yersinia pestis* Pigment Receptor

Olga N. PODLADCHIKOVA and Grigory G. DIKHANOV
Research Institute for Plague Control, 117 Gorky str., Rostov-on Don, Russia

1. INTRODUCTION

The ability to adsorb pigments (e.g. hemin or Congo red dye) during growth at 28°C (Hms$^+$ phenotype) is an essential characteristic of *Yersinia pestis* that was shown to be required for plague transmission by fleas (Hinnebusch et al, 1996). Molecular basis of the Hms$^+$ phenotype is currently unknown. Although a number of genes have been found whose deletion or inactivation (hmsHFRS and hmsT) leads to the loss of pigmentation (Jones *et al.*, 1999; Hare and McDonough, 1999), the pigment receptor (PR) has not yet been identified. The goal of this investigation was to purify and characterize PR with the help of different biochemical methods. To achieve this goal, experimentally obtained strain *VEV*, an Hms$^+$ Caf1$^-$ Cad$^-$ derivative of *Y.pestis EV76* (Podladchikova *et al.*, 2002), was used. Hms$^-$ mutants of this strain retained *hmsHFRS* and *hmsT* and with a high frequency (10^{-3}) formed revertants with various degrees of Hms$^+$ expression suggesting that they carry "leaky" mutations in some regulatory gene(s).

2. RESULTS

The comparative analysis of Hms$^+$ and Hms$^-$ cells of *Y.pestis VEV* showed that aqueous extracts of Hms$^+$ cells, in contrast to those of Hms$^-$, contained material able to adsorb hemin and Congo red upon immobilization on nitrocellulose filters. Using this dot-blot dye adsorption assay for

The Genus Yersinia, Edited by Skurnik *et al.*
uwer Academic/Plenum Publishers, New York, 2003

monitoring pigment binding activity, we elaborated the following purification procedure of PR:

1. Preparation of acetone-dried bacterial mass from the cells grown on casamino acid medium at 28°C for 48 h and double-washed with PBS.
2. Extraction of acetone-dried cells with 10 mM NaOH.
3. Precipitation of PR with neutralized ammonium sulfate (20%).
4. Dialysis of the precipitate against 10 mM NaOH.
5. Precipitation of PR at pH 4.6.
6. Precipitation of PR with three volumes of acetone.

When the procedure was applied to the Hms$^-$ mutant, roughly the same amount of the material (about 10%) was isolated but it could not adsorb pigments. The preparations purified from Hms$^+$ and Hms$^-$ cells and designated PR$^+$ and PR$^-$ were compared with the help of different biochemical methods. In SDS-PAGE, PR$^+$ and PR$^-$ presented one dominant protein band of *ca* 17 kDa and multiple minor slowly migrating bands. However, HPLC analysis revealed that both the PR$^+$ and PR$^-$ preparations have only three protein fractions which based on amino acid compositional analysis apparently contained one single protein. Gel filtration of the PR$^+$ and PR$^-$ preparations heated at 100°C for 5 min (Figure 1) revealed that both had a high molecular weight fraction, which appeared to be 17 kDa protein, and a low molecular weight fraction which was different in the two preparations. In PR$^-$ it had negative values of absorbance at 220 nm that might be caused by the fluorescence.

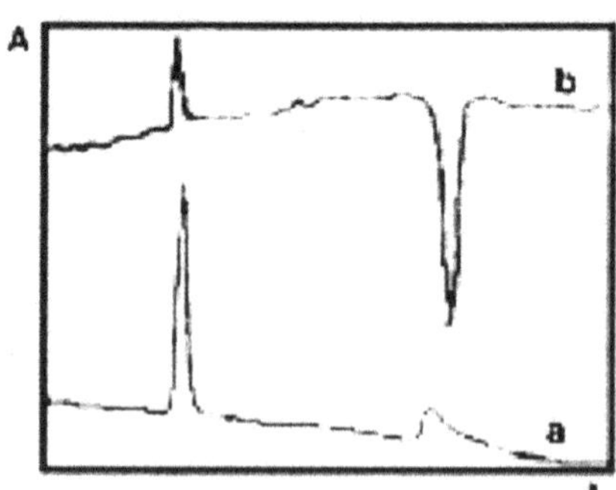

Figure 1. Gel filtration profiles of PR+ (a) and PR- (b) preparations heated at 100°C for 5 min. Bio-gel TSK 20 column. Elution by 50 mM acetic acid. Detection at 220 nm. A – absorbance, t – elution time.

Thin layer chromatography of the low molecular weight fractions (Figure 2) showed that PR^+ contained two components with Rf of 0.7 and 0.8, while PR^- contained an additional one, with Rf of 0.9. The three components had different properties. Rf 0.7 component was stained with iodine but not with ninhydrin and precipitated from the aqueous solutions under acidic conditions suggesting that it was negatively charged (LC^-). Rf 0.8 component was stained with ninhydrin but not with iodine and precipitated under alkaline conditions indicating its positive charge (LC^+). Rf 0.9 component was characterized by bright blue fluorescence in UV, was stained with iodine and ninhydrin, and dissolved in nonpolar organic solvents suggesting a hydrophobic nature. The chemical structure and physiological function of the three low molecular weight components need future investigation.

Thus, PR^+ appeared to contain three components: 17 kDa protein, LC^+, and LC^-. Dot-blot dye adsorption assay revealed that neither of them by itself could adsorb pigments. However, when they were mixed in proportions present in native PR^+, the pigment binding activity was restored. If the proportion was not maintained, the pigment binding was not observed. Moreover, the addition to the native or reconstituted PR^+ of the hydrophobic component purified from PR^- led to the loss of the pigment binding activity suggesting that this component was a pigmentation inhibitor.

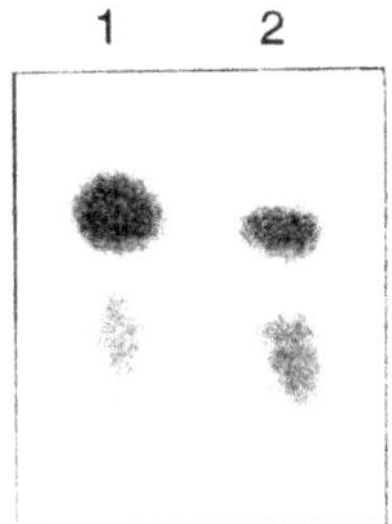

Figure 2. Thin layer chromatography of the low molecular weight fractions of PR+ (1) and PR- (2) preparations. Solid phase – silica gel plates. Mobile phase - 80% ethanol. Detection by ninhydrin followed by iodine.

3. CONCLUSIONS

In the present work the PR of *Y. pestis* was identified as a complex of a 17 kDa protein associated with two oppositely charged low molecular weight components that cooperatively form a pigment binding domain of PR. The pigment binding activity of PR is dependent on an appropriate proportion of the low molecular weight components and on the absence of the pigmentation inhibitor.

ACKNOWLEDGEMENTS

We would like to thank Dr. L. Baratova for amino acid analysis of PR preparations and Dr. J. Fetherston for a critical discussion of the manuscript.

REFERENCES

Hare, J.M., and McDonough, K.A., 1999, High-freguency RecA-dependent and independent mechanisms of Congo red binding mutations in *Yersinia pestis. J. Bacteriol.* **181**: 4896-4904.

Hinnebusch, B.J., Perry, R.D., Schwan, T.G., 1996, Role of the *Yersinia pestis* hemin storage (hms) locus in the transmission of plague by fleas. *Science* **273**: 367-370.

Jones, H.A., Lillard, J.W., and Perry, R.D., 1999, HmsT, a protein essential for expression of the haemin storage (Hms^+) phenotype of *Yersinia pestis. Microbiology* **145**: 2117-2128.

Podladchikova, O.N., Rykova, V.A., Ivanova, V.S., Eremenko, N.S., and Lebedeva, S.A., 2002, Study of Pgm mutation mechanism in *Yersinia pestis* vaccine strain EV76. *.Mol. Genet., Microbiol., Virusol.* (Russian). **2**: 14-19.

Chapter 24

Yersinia enterocolitica Biotype 1A: Not as Harmless as You Think

Sharon M. TENNANT, Narelle A. SKINNER, Angela JOE and Roy M. ROBINS-BROWNE
Department of Microbiology and Immunology, University of Melbourne, and Microbiological Research Unit, Murdoch Childrens Research Institute, Victoria 3010, Australia

1. INTRODUCTION

Yersinia enterocolitica biotype 1A strains are often regarded to be avirulent as they lack the phenotypic and genotypic markers of the pYV-bearing strains. However, there is growing epidemiological and experimental evidence to suggest that some biotype 1A strains can cause disease.

1.1 Epidemiological evidence

Strains of *Y. enterocolitica* biotype 1A have been isolated from patients with gastrointestinal illness from diverse geographic locations, including Australia, New Zealand, South Africa, Chile, Switzerland, Canada, and the United States of America. In a prospective case-control study in Chile, biotype 1A strains were only isolated from children with diarrhoea, whereas, pYV-bearing strains were isolated at similar frequencies from symptomatic and asymptomatic children (Morris *et al.*, 1991). Furthermore, a Swiss study found that the duration and severity of symptoms attributable to biotype 1A infections was indistinguishable from that due to classical, pYV-bearing strains (Burnens *et al.*, 1996).

The Genus Yersinia, Edited by Skurnik *et al.*
'uwer Academic/Plenum Publishers, New York, 2003

1.2 *In vitro* evidence

Investigations of biotype 1A strains of *Y. enterocolitica* for virulence-associated properties have shown that compared to strains isolated from the environment, biotype 1A strains of clinical origin are:

1. significantly more invasive for epithelial cells in vitro and for rabbit enterocytes in vivo,
2. significantly more resistant to killing by macrophages,
3. are able to exocytose from epithelial cells and macrophages without killing these cells, and
4. are able to colonise mice for significantly longer periods (Grant *et al.*, 1998, 1999).

These results indicate that there may be two subgroups of biotype 1A strains: pathogenic strains of clinical origin and non-pathogenic strains that occur in the environment.

2. GENOMIC SUBTRACTIVE HYBRIDISATION

We have used genomic subtractive hybridisation to identify potential virulence genes of pathogenic strains of *Y. enterocolitica* biotype 1A.

2.1 Strains

A biotype 1A strain, T83 (serotype O:5), was obtained from a patient with diarrhoea and possesses all the virulence-associated properties listed in section 1.2, whereas strain IP2222 (serotype O:36) was isolated from food and lacks these properties.

2.2 Subtractive hybridisation

We used the Clontech PCR-Select Bacterial Genome Subtraction Kit to identify sequences in strain T83 that were absent from strain IP2222. The subtracted DNA fragments were cloned and 173 clones were tested for T83-specificity by dot-blot hybridisation. 54 T83-specific fragments were identified and sequenced. The sequences were then analysed using the NCBI BLASTX program (Table 1).

Table 1. Breakdown of 54 T83-specific sequences.

Category	Number of clones
Significant protein matches in database	
Known proteins	17
Phage-encoded proteins	8
Hypothetical proteins	8
No protein match in database	21

2.3 Sequences with homology to known genes

Four sequences showed homology to three insecticidal toxin complex (*tc*) genes first identified in *Photorhabdus luminescens* (Bowen *et al.,* 1998). These genes are a *tcdA*-like gene, a *tcaC*-like gene and a *tccC*-like gene. Therefore, strain T83 possesses at least 1 copy of each of three genetic elements believed to be necessary for oral toxicity of *P. luminescens* for insects.

3. CONCLUSIONS

We have used subtractive hybridisation to identify genomic sequences in a clinical biotype 1A isolate that are absent from a non-clinical biotype 1A strain. We are currently investigating whether these differences contribute to the virulence of the biotype 1A strain.

ACKNOWLEDGEMENTS

This work was supported by a grant from the Australian National Health and Medical Research Council.

REFERENCES

Bowen, D., Rocheleau, T.A., Blackburn, M., Andreev, O., Golubeva, E., Bhartia, R., and ffrench-Constant, R. H., 1998, Insecticidal toxins from the bacterium *Photorhabdus luminescens*. *Science* 280: 2129-2132.

Burnens, A.P., Frey, A., and Nicolet, J, 1996, Association between clinical presentation, biogroups and virulence attributes of *Yersinia enterocolitica* strains in human diarrhoeal disease. *Epidemiol. Infect.* 116: 27-34.

Grant, T., Bennett-Wood, V., and Robins-Browne, R.M., 1998, Identification of virulence-associated characteristics in clinical isolates of *Yersinia enterocolitica* lacking classical virulence markers. *Infect. Immun.* 66: 1113-1120.

Grant, T., Bennett-Wood, V., and Robins-Browne, R.M., 1999, Characterization of the interaction between *Yersinia entoercolitica* biotype 1A and phagocytes and epithelial cells in vitro. *Infect. Immun.* 67: 4367-4375.

Morris, J.G., Jr., Prado, V., Ferreccio, C., Robins-Browne, R.M., Bordun, A.-M., Cayazzo, M., Kay, B.A., and Levine, M.M., 1991, *Yersinia enterocolitica* isolated from two cohorts of young children in Santiago, Chile: incidence of and lack of correlation between illness and proposed virulence factors. *J. Clin. Microbiol.* 29: 2784-2788.

Chapter 25

Pestoides F, a *Yersinia pestis* Strain Lacking Plasminogen Activator, is Virulent by the Aerosol Route

Patricia L. WORSHAM and Chad ROY
United States Army Medical Research Institute of Infectious Diseases, Fort Detrick, MD 21702-5011 USA

Unlike the majority of *Yersinia pestis* strains, the Pestoides subgroup of *Y. pestis* is characterized by the ability to ferment rhamnose and melibiose. In addition to this unusual fermentation pattern, they are atypical in that they carry larger variants of the plasmid pFra than other strains in our collection (Worsham and Hunter, 1998). Molecular fingerprinting studies have revealed that *Y. pestis* is a recently evolved clone of *Y. pseudotuberculosis* and there is growing genetic evidence to suggest that the Pestoides group is the most ancient phylogenetically of *Y. pestis* strains (Motin *et al.*, 2002). Thus, this is a particularly important group to study in a species that lacks the genetic diversity found in many other pathogens. Some isolates, such as Pestoides F, lack the plasmid pPst that encodes both the bacteriocin pesticin and the invasion-associated virulence factor plasminogen activator. Pestoides F is fully virulent by the parenteral route despite the lack of plasminogen activator. In contrast, loss of this gene product in our virulent type strain of *Y. pestis*, CO92, leads to attenuation, as assessed by both parenteral and aerosol challenges (Welkos *et al.*, 1997).

Y. pestis generally carries three plasmids: pFra (pMT), which encodes capsule and murine toxin; pLcr (pYV), which is responsible for the low calcium response; and pPst (pPCP), which encodes pesticin, immunity to pesticin, and plasminogen activator. Pestoides F lacks the small pPst plasmid present in typical *Y. pestis* strains. We confirmed the absence of pPst by PCR using primers specific for *pim* (pesticin immunity), *pst* (pesticin), and *pla* (plasminogen activator). Due to the lack of the *pim* and *pst* gene products, this strain is non-pesticinogenic and is senstitive to this bacteriocin. In other respects, such as production of the F1 capsular antigen and the low-calcium response, Pestoides F resembles other virulent *Y. pestis* strains.

Previously, Pestoides F was shown to be fully virulent in mice challenged s.c. (Welkos *et al.,* 1997). We extended these observations by determining the virulence of this strain by the aerosol route of infection. Challenge organisms were grown at 30 C in heart infusion broth (HIB) containing 0.2% xylose. Cells were harvested by centrifugation, resuspended, and serially diluted in HIB for the challenge inoculum. BALB/c mice (6-7 weeks) were challenged using a 1 μm-generating Collison nebulizer contained within a Class III biocabinet. Mice were exposed in a temperature and humidity-controlled whole-body exposure chamber. The aerosol was continuously sampled by an all glass impinger (AGI) containing HIB. Dose was determined from serial dilution and culture of AGI samples according to the formula determined by Guyton. We found that Pestoides F was comparable in LD_{50} (7.3 X 10^4) to our type strain CO92 (6.0 X 10^4). Thus, it appears that Pla is not a necessary virulence factor for Pestoides F in the aerosol mouse model.

In additional studies, we focused on physiological differences between Pestoides F and "typical" *Y. pestis* strains. When a number of complex growth media were compared, we found that Pestoides F was remarkably more uniform in its plating efficiency than CO92. While the latter demonstrated slow and inconsistent growth on some complex media, such as tryptose blood agar base, Pestoides F was uniformly robust in its growth on all media tested. In some cases, the plating efficiency was approximately 100-fold greater with Pestoides F than with CO92, although this was somewhat batch-dependent.

Recent molecular genetic analyses such as multiple-locus, variable-number tandem repeat analysis (MLVA), IS100-based fingerprinting, and sequencing of housekeeping genes suggest that the Pestoides are the most ancient group of *Y. pestis* strains and may actually represent transitional strains between *Y. pseudotuberculosis* and *Y. pestis.* Since *Y. pseudotuberculosis* is a more robustly growing species, the superior plating efficiency of the Pestoides group strain is consistent with this hypothesis. However, additional strains of typical and Pestoides-type strains need to be examined to determine if these phenotypic differences between the groups hold true.

Several diagnostic and detection systems have been developed using *pla* to identify *Y. pestis.* Although the vast majority of strains of *Y. pestis* worldwide are Pla^+, there are regions where Pla^- isolates are not uncommon. Considering the aerosol virulence of some Pla^- isolates and the threat of bioterrorism, it appears dangerous to rely solely on Pla-based *Y. pestis* detection systems.

Plasminogen activator is a major virulence factor of *Y. pestis*; it is thought to facilitate the spread of organisms from peripheral sites of infection. However, our results suggest that plasminogen activator is not required for aerosol virulence in certain strains and that other factors present in Pestoides F may compensate for the lack of Pla activity. We are currently examining the disease pathology of animals infected with Pestoides strains to compare it to that produced by more typical strains, as well as examining the relative invasiveness of Pestoides F and more typical Pla$^+$ strains in in vitro models.

NOTES

The views of the authors do not purport to reflect the positions of the Department of the Army or the Department of Defense.

REFERENCES

Motin VL, Georgescu AM, Elliott JM, Hu P, Worsham PL, Ott LL, Slezak TR, Sokhansanj BA, Regala WM, Brubaker RR, Garcia E. 2002. Genetic variabiliy of *Yersinia pestis* isolates as predicted by PCR-based genotyping and analysis of structural genes encoding glycerol-3-phospate dehydrogenase (glpD). J. Bacteriol. 184: 1019-1027.

Welkos SL, Friedlander AM, Davis KJ. 1997. Studies on the role of plaminogen activator in systemic infection by virulent *Yersinia pestis* strain CO92. Microb. Pathogen. 23:211-223.

Worsham PL, Hunter M. 1998. Characterization of Pestoides F, an atypical strain of *Yersinia pestis*. Med Microbiol. 6(Suppl. II):34-35.

Chapter 26

Impact of the *Yersinia pseudotuberculosis* -Derived Mitogen (YPM) on the Murine Immune System

Christophe CARNOY[1,2], Caroline LOIEZ[3], Christelle FAVEEUW[4], Corinne GRANGETTE[5], Pierre DESREUMAUX[6] and Michel SIMONET[1]
[1]*Inserm E9919;* [2]*Faculté de Pharmacie de Lille;* [3]*Lab. Bactériologie, CHRU de Lille;* [4]*Inserm U547;* [5]*Lab. des Ecosystèmes, Institut Pasteur de Lille,* [6]*Inserm E0114, Lille France.*

Superantigens are viral and bacterial toxins exhibiting a strong polyclonal T-cell activation in human and animal species. *Y. pseudotuberculosis* is so far the only Gram-negative microorganism to produce a superantigenic toxin. This molecule, designated YPM for *Y. pseudotuberculosis*-derived mitogen, induces a major histocompatibility complex (MHC)-dependent proliferation of human T cells bearing Vβ3, Vβ9, Vβ13.1 and Vβ13.2 T-cell receptor (TcR) domains (Abe *et al.*, 1993). In mice, YPM stimulates T cells carrying Vβ7 and Vβ8 variable regions on their TcR (Miyoshi-Akiyama *et al.*, 1997). The YPM toxin might be involved in inflammatory post-infection complications (reactive arthritis, erythema nodosum) and in Kawasaki syndrome (Baba *et al.*, 1991; Konishi *et al.*, 1997). We recently demonstrated that YPM was responsible for the exacerbation of the virulence of *Y. pseudotuberculosis* in mice (Carnoy *et al.*, 2000).

1. YPM INDUCES IL-4 PRODUCTION

To assess the impact of YPM on the host immune system, the production of inflammatory (TNFα, IFNγ, IL-2, IL-6), immunoregulatory (IL-4), and anti-inflammatory (IL-10) cytokines was followed during the course of infection after challenge of BALB/c mice with a superantigen-producing *Y. pseudotuberculosis* or with the corresponding superantigen-deficient mutant. Cytokines were quantified in the spleen by mRNA detection and in the blood by ELISA, 12 hours, 1, 3 and 5 days after infection. In the spleen, the main differences in cytokine profiles were observed on the fifth day of infection:

The Genus Yersinia, Edited by Skurnik *et al.*
Kluwer Academic/Plenum Publishers, New York, 2003

IL-6 and IFNγ were produced mainly by mice infected by the superantigen-deficient mutant whereas IL-2 was only detected in wild type strain-infected animals. Interestingly, infection of mice with wild type *Y. pseudotuberculosis* was associated with a significant IL-4 production whereas this cytokine was not detected following challenge by the mutant or in non-infected animals. IL-10 and TNFα productions were never observed. The absence of production of TNFα during *Y. pseudotuberculosis* infection, especially in the presence of YPM, was attributed to the anti-inflammatory property of the YopJ effector (Palmer *et al.*, 1998). Results obtained *in situ* in the spleen were confirmed by the assay of the cytokines in the blood of infected animal: high level of IL-4 (35 to 768 pg/ml) was found at day 5 in the serum of mice infected with the wild type strain but not in the mutant-infected mice. In conclusion, this cytokine response analysis in *Y. pseudotuberculosis*-infected mice revealed an association of YPM with a high production IL-4, a cytokine seldom associated with a superantigenic activity.

2. YPM EXPANDS Vβ8 AND Vβ7 T CELLS *IN VIVO*

During the course of infection we also observed an important splenomegaly in animals infected by the wild type strain, indicating a cellular expansion in the spleen in the presence of YPM. Characterisation of the different cell populations on day 5 of infection was performed by flow cytometry. Using cell-specific antibodies, we showed that the splenomegaly found in the presence of YPM was due to B and T cell expansion but not to granulocyte recruitment. Further analysis revealed that, when mice were infected with the YPM^+ strain, 64.2 ± 8.5 % of the splenic $CD3^+$ T cells expressed Vβ7, Vβ8 on their TcR. When the challenge was performed with the superantigen-deficient mutant the Vβ7, Vβ8 T cells represent only 37.3 ± 4.2 % of the splenocytes. This ratio was identical in non-infected mice (38 ± 2.5%). To confirm that the expansion of Vβ7 and Vβ8 T cells was specific of YPM, we analysed the Vβ6 T cell population and found that this population was not expanded in the presence of YPM. This demonstrates that, *in vivo,* YPM specifically stimulates T cells expressing the Vβ7 and Vβ8 variable regions.

The phenotype of the Vβ7, Vβ8 T cells was further characterised. We found that most of the cells expanded in the presence of YPM were $CD4^+$: in wild type infected mice, Vβ7/Vβ8 $CD4^+$ cells represent 68.6 ± 11 % of the $CD3^+$ T cells whereas in mutant-infected mice these cells represent 34.7 ± 9.7 %. We also observed an increase of $CD8^+$ T cells expressing the Vβ7, Vβ8 variable regions. The expansion of these $CD8^+$ T cells in the presence of YPM was attributed to a direct effect of YPM since the Vβ6 $CD8^+$ T cells were not expanded in wild type-infected animals when compared to the ratio of these cells in control or in mutant-infected mice. This indicates that proliferation of Vβ7/Vβ8 $CD8^+$ T cells is due to a direct interaction with YPM rather than a consequence of the activation of the $CD4^+$ cells.

Altogether, this preliminary results indicate that, *in vivo*, YPM induces high expression of IL-4 and stimulates T cells with Vβ7 and Vβ8 variable regions. This demonstrates the impact of the superantigenic toxin of *Y. pseudotuberculosis* on the murine immune system. So far, the relationship between the exacerbation of virulence in the presence of YPM (Carnoy *et al.,* 2000), the skewed Vβ population, and the high production of IL-4, is still speculative.

REFERENCES

Abe, J., Takeda, T., Watanabe, Y., Nakao, H., Kobayashi, N., Leung, D.Y.M., and Kohsaka, T., 1993, Evidence for superantigen production by *Yersinia pseudotuberculosis. J. Immunol.* **151**:4183-4188.

Baba, K., Takeda, N., and Tanaka, M., 1991, Cases of *Yersinia pseudotuberculosis* infection having diagnostic criteria of Kawasaki disease. *Contrib. Microbiol. Immunol.* **12**:292-296.

Carnoy, C., Mullet, C. , Müller-Alouf, H., Leteurtre, E., and Simonet, M., 2000. Superantigen YPMa exacerbates the virulence of *Yersinia pseudotuberculosis* in mice. *Infect. Immun.* **68**:2553-2559.

Konishi, N., Baba, K., Abe, J., Maruko, T., Waki, K., Takeda, N., and Tanaka, M., 1997, A case of Kawasaki disease with coronary artery aneurysms documenting *Yersinia pseudotuberculosis* infection. *Acta Paediatr.* **86**:661-664.

Miyoshi-Akiyama, T., W. Fujimaki, X. J. Yan, J. Yagi, K. Imanishi, H. Kato, K. Tomonari, and T. Uchiyama. 1997. Identification of murine T cells reactive with the bacterial superantigen *Yersinia pseudotuberculosis*-derived mitogen (YPM) and factors involved in YPM-induced toxicity in mice. *Microbiol. Immunol.* **41**:345-352.

Palmer, L.E., Hobbie, S., Galan, J.E., Bliska, J.B., 1998, YopJ of *Yersinia pseudotuberculosis* is required for the inhibition of macrophage TNFα production and down regulation of the MAP kinase p38 and JNK. Mol. Microbiol. **27**:953-965.

Chapter 27

Role of T Cells and Gamma Interferon in *Yersinia pseudotuberculosis* -Derived Mitogen (YPM)-Induced Toxicity in Mice

Hirotsugu KANO[1,2], Yasuhiko ITO[1], Kentaro MATSUOKA[3], Tutomu IWATA[2], Hirohisa SAITO[1], Takao KOHSAKA[4] and Jun ABE[1]

[1]Dep. of Allergy and Immunology, National Research Institute for Child Health and Development, Tokyo, [2]Dep. of Pediatrics, Faculty of Medicine, The University of Tokyo, Deps. of [3]Clinical Pathology and [4]Gastroenterology, National Center for Child Health and Development, Tokyo, Japan

1. INTRODUCTION

Bacterial superantigens have been implicated in the pathogenesis of several human diseases. In *Yersinia pseudotuberculosis* infection, it is reported that more than 95% of the clinical strains have a superantigen gene, *ypm*, and IgG anti-YPM antibodies were detected in more than half of the infected patients in Japan. Moreover, the patients with systemic symptoms such as coronary aneurysms and renal failure had significantly higher titers of anti-YPM antibodies than the patients with gastrointestinal symptoms alone. The mechanisms by which the organism mediates these systemic symptoms are poorly understood. However, the above findings may suggest that a superantigen, YPM could be responsible for these symptoms. In this study, we generated an experimental murine model of YPM-mediated acute toxic shock utilizing the point mutants of YPM to investigate whether the superantigenicity of YPM lead to an acute toxic shock like syndrome *in vivo* as other pyrogenic staphylococcal toxins does and to which degree that effect depends on the Th1 cytokines generated *in vivo*.

The Genus Yersinia, Edited by Skurnik *et al.*
'uwer Academic/Plenum Publishers, New York, 2003

2. MATERIALS AND METHODS

Male and female BALB/c mice (8-12 wk) and C.B-17/lcr.$^{scid/scid}$ mice (6-10 wk) were used. The production of single amino acid substituted mutants of wild-type YPM, L7Q and D88G was described previously. Purification of wYPM, L7Q and D88G were done using pMAL-p2-HisTag as an expression vector and NovaBlue (Novagen) as host bacteria. The purified protein was free of significant endotoxin contamination. Two kinds of hybridoma cell lines (YSA8E3, and YSA12B8) were generously provided from Drs. H. Kato and T. Nakano (Yamasa Corporation, Japan).

BALB/c mice were pretreated i.p. with 30 mg of D-galactosamine and thirty minutes after injection, mice were given i.v. with 100 μg of wYPM, D88G or L7Q. Control mice were injected with 0.1 ml of PBS instead of YPM. Six C.B-17/lcr.$^{scid/scid}$ mice and 4 T cell-reconstituted C.B-17/lcr.$^{scid/scid}$ mice were treated as the same way as BALB/c mice. One milligram of monoclonal antibody against YPM (YSA8E3 or YSA12B8) was injected i.p. three times, i.e., 2 hours before, at the same time of, and 2 hours after wYPM injection. Mice were bled by tail vein puncture at several time points after YPM injection.

3. RESULTS

Survival rates of BALB/c mice injected with wYPM, L7Q, D88G, and control PBS were 1/7 (14%), 2/7 (28%), 6/7 (86%), and 7/7 (100%), respectively. The mitogenic activity of each antigen on mouse splenocytes was positively correlated to its lethal effectiveness. Serum IFN-γ level was significantly increased at 4h after wYPM challenge compared with the D88G or PBS (1141.5 vs. 42.7, 0 pg/ml, $p< 0.0001$, 0.0001). Serum IFN-γ level of L7Q treated mice (754.4 pg/ml) was also higher than that of D88G or PBS treated mice ($p=0.0133$, 0.0129). Serum TNF-α level at 2h after wYPM challenge was increased compared with D88G or PBS challenge, although the differences were not statistically significant. There was no difference in serum IL-10 levels among the four groups. When a monoclonal antibody against wYPM (YSA8E3) was administered at the time of YPM injection, 6 out of 8 (75%) mice were protected from death. Six out of 6 (100%) CB17/*scid* mice survived after injection of the same amount of the wYPM, whereas 4 out of 4 CB17/*scid* mice reconstituted with the T cells from normal BALB/c spleens died within 2 days. Histologically, massive necrosis and congestion were seen in the liver and spleen of the wYPM injected mice. Moreover, focal necrosis with calcification was found in the epicardium of the heart. These findings were alleviated by the pre-administration of the YSA8E3 antibody.

4. CONCLUSION

YPM had an acute toxic effect on BALB/c mice, which was dependent on the presence of T cells and correlated with the elevation of serum IFN-gamma. One monoclonal antibody (YSA8E3) rescued BALB/c mice from development of toxic shock after YPM injection.

ACKNOWLEDGEMENTS

This work was supported in part by grants from the Japan Human Science Foundation and from the Ministry of Health, Labour, and Welfare.

Chapter 28

Yersinia pestis Pla Has Multiple Virulence-Associated Functions

Kaarina LÄHTEENMÄKI, Maini KUKKONEN, Silja JAATINEN, Marjo SUOMALAINEN, Hanna SORANUMMI, Ritva VIRKOLA, Hannu LÅNG and Timo K. KORHONEN
Division of General Microbiology, Department of Biosciences, FIN-00014 University of Helsinki, Helsinki, Finland

1. INTRODUCTION

The surface protein Pla of *Yersinia pestis* is an important virulence factor in plague infection. In a murine infection model, Pla is required in bacterial migration from the subcutaneous infection site into circulation (Sodeinde *et al.*, 1992), and its function has been linked to the presence of plasminogen in the host (Goguen *et al.*, 2000). Pla activates plasminogen to the broad-spectrum serine protease plasmin, which facilitates degradation of fibrin (Beesley *et al.*, 1967) and extracellular matrices (Lähteenmäki *et al.*, 1998). Pla has homologs among other enterobacterial species: PgtE in *Salmonella typhimurium* and *Salmonella typhi*, OmpT and OmpP in *Escherichia coli*, and SopA in *Shigella flexneri*. Together, these proteins form the omptin family of bacterial surface proteases. The three-dimensional structure of OmpT has been resolved (Vandeputte-Rutten *et al.*, 2001), and the predicted structures of Pla and PgtE are highly similar with it. The omptins are β-barrel structures composed of 10 transmembrane β-strands and five surface-exposed loops (L1-L5). Omptins are aspartate proteases with conserved catalytic amino acid residues in the L2 and L4 (Kukkonen *et al.*, 2001; Vandeputte-Rutten *et al.*, 2001; Kramer *et al.*, 2001). Alltogether, Pla and PgtE share 72% and Pla and OmpT 48% amino acid sequence identity.

The Genus Yersinia, Edited by Skurnik *et al.*
'uwer Academic/Plenum Publishers, New York, 2003

2. FUNCTIONS OF OMPTINS

Pla, PgtE, and OmpT have differences in their functions (Lähteenmäki *et al.*, 1998; 2001; Kukkonen *et al.*, 2001). The identified functions of the three omptins are summarized in Table 1. We have found that in addition to generating plasmin, Pla also proteolytically inactivates α_2-antiplasmin, the main inhibitor of plasmin in mammalian plasma (Kukkonen *et al.*, 2001). This way, Pla generates uncontrolled proteolytic activity in the vicinity of the bacteria, which probably promotes spread of *Y. pestis* through tissue barriers. PgtE also converts plasminogen to plasmin (Sodeinde and Goguen, 1989), whereas OmpT is a poor plasminogen activator (Kukkonen *et al.*, 2001). Unlike Pla and PgtE, OmpT cleaves the Val-Leu-Lys-p-nitroaniline substrate of plasmin. We have identified amino acid residues in Pla and OmpT that are responsible for their differing substrate specificities, and by substitution mutagenesis modified OmpT into a Pla-like protease (Kukkonen *et al.*, 2001).

Table 1. Identified functions of *Y. pestis* Pla, *Salmonella* PgtE and *E. coli* OmpT. + = positive reaction, – = no reaction.

Function	Pla	PgtE	OmpT	References
Proteolysis:				
Plasminogen activation	+	+	–	Beesley *et al.*, 1967; Sodeinde & Goguen, 1989; Kukkonen *et al.*, 2001
Cleavage of α_2-antiplasmin	+		–	Kukkonen *et al.*, 2001
Cleavage of antimicrobial peptides		+	+	Guina *et al.*, 2000; Stumpe *et al.*, 1998
Cleavage of complement C3	+			Sodeinde *et al.*, 1992
Adhesion to:				
Human extracellular matrix	+	+	(–)	Lähteenmäki *et al.*, 1998; 2001
Murine laminin	+	+	(–)	Lähteenmäki *et al.*, 1998
Murine type IV collagen	+			Kienle *et al.*, 1992
Invasion into:				
Human HeLa cells	+			Cowan *et al.*, 2000
Human ECV304 cells	+	–	–	Lähteenmäki *et al.*, 2001

Pla also promotes bacterial adherence to laminin, which is an abundant component of extracellular matrices (Lähteenmäki *et al.*, 1998), and mediates bacterial invasion into cultured human cells (Lähteenmäki *et al.*, 2001; Figure 1). In a gentamicin protection assay, 1-8% of the recombinant *E. coli* expressing Pla are internalized into human ECV304 cells (Lähteenmäki *et al.*, 2001). *Y. pestis* has been reported to invade HeLa epithelial cells with 30-50% invasion efficiency; the invasion is dependent on the plasmid pPCP1 which encodes for Pla as well as the pesticin and pesticin immunity protein (Cowan *et al.*, 2000). Although recombinant *E. coli* expressing Pla are less efficient in invasion than wild-type *Y. pestis*, our results indicate that Pla is at least partly responsible for the invasiveness of *Y. pestis*. Thus, Pla appears to be a multifunctional protein that can promote adherence and generate localized proteolysis on extracellular matrices and may also enhance transcellular migration of *Y. pestis*.

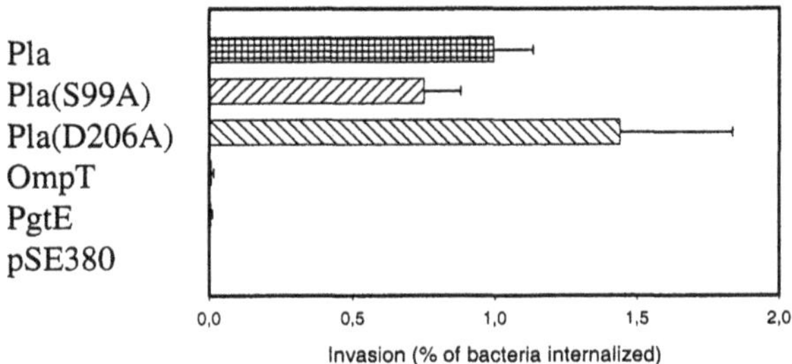

Figure 1. Invasion into human ECV304 cells of *E. coli* expressing *Y. pestis* Pla or its proteolytically inactive derivatives (S99A and D206A), *E. coli* OmpT, or *Salmonella* PgtE. Invasion of the host strain with the expression vector pSE380 only is shown as a control. Invasion is measured as the percentage of inoculated bacteria surviving gentamicin treatment.

Pla-mediated invasion seems not to be due to proteolysis, as *E. coli* expressing proteolytically inactive derivatives of Pla are as invasive as bacteria expressing wild-type Pla (Figure 1). Also, addition of plasminogen has no effect on invasion. Pla-expressing bacteria adhere efficiently to the extracellular matrix of ECV304 cells (Lähteenmäki *et al.*, 2001), and this adherence probably is involved in invasion. However, other factors also seem to play a role, as PgtE-expressing bacteria are not invasive although PgtE adheres to ECV304 extracellular matrix nearly as efficiently as Pla and its derivatives (Figure 2). OmpT is weak in both adhesive and invasive properties (Lähteenmäki *et al.*, 2001).

We have constructed recombinant bacteria expressing Pla-OmpT and Pla-PgtE hybrid proteins (Kukkonen *et al.*, 2001; unpublished), and are using these constructs to locate regions critical for Pla-mediated invasion and other functions of omptins. The corresponding surface loops can be changed between the omptins; the resulting chimeric proteins are functional and expressed on bacterial surface (Kukkonen *et al.*, 2001). Our ongoing work suggests that Pla-mediated invasion is due to a discontinous epitope situated within several surface loops.

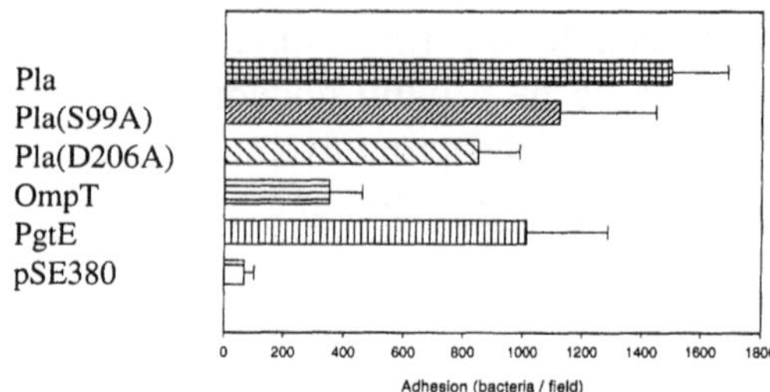

Figure 2. Adhesion to the extracellular matrix derived from human ECV304 cells of *E. coli* expressing *Y. pestis* Pla or its proteolytically inactive derivatives (S99A and D206A), *E. coli* OmpT, or *Salmonella* PgtE. Adhesion of the host strain with pSE380 plasmid only is also shown.

3. CONCLUSIONS

Our work indicates that *Y. pestis* Pla is a multifunctional surface protein. It is an efficient protease and also an adhesin and an invasin. The *Salmonella* homolog of Pla, PgtE, also has proteolytic and adhesive properties, whereas *E. coli* OmpT is only weakly adhesive, and its substrate specificity differs from that of Pla and PgtE. PgtE and OmpT do not mediate bacterial invasion into any of the human epithelial and endothelial cell lines tested. We are currently characterizing the mechanism and relevance of Pla-mediated invasion as well as the structural requirements of the other functions of Pla.

ACKNOWLEDGEMENTS

This work has been supported by the Academy of Finland (project numbers 42103, 45162 and 180666) and by the University of Helsinki.

REFERENCES

Beesley, E.D., Brubaker, R.R., Janssen, W.A., and Surgalla, M.J., 1967, Pesticins. III. Expression of coagulase and mechanism of fibrinolysis. *J. Bacteriol.* **94**:19-26.

Cowan, C., Jones, H.A., Kaya, Y.H., Perry, R.D., and Straley, S.C., 2000, Invasion of epithelial cells by *Yersinia pestis*: evidence for a *Y. pestis*-specific invasin. *Infect. Immun.* **68**: 4523-4530.

Goguen, J.D., Bugge, T., and Degen, J.L., 2000, Role of the pleiotropic effects of plasminogen deficiency in infection experiments with plasminogen-deficient mice. *Methods* **21**: 179-183.

Guina, T., Yi, E.C., Wang, H., Hackett, M., and Miller, S.I., 2000, A PhoP-regulated outer membrane protease of *Salmonella enterica* serovar typhimurium promotes resistance to alpha-helical antimicrobial peptides. *J. Bacteriol.* **182**, 4077-4086.

Kienle, Z., Emödy, L., Svanborg, C., and O'Toole P.W., 1992, Adhesive properties conferred by the plasminogen activator of *Yersinia pestis*. *J. Gen. Microbiol.* **138**: 1679-1687.

Kramer, R.A., Vandeputte-Rutten, L., de Roon, G.J., Gros, P., Dekker, N., and Egmond, M.R., 2001, Identification of essential acidic residues of outer membrane protease OmpT supports a novel active site. *FEBS Lett.* **505**: 426-430.

Kukkonen M., Lähteenmäki, K., Suomalainen, M., Kalkkinen, N., Emödy, L., Lång, H. and Korhonen, T.K., 2001, Protein regions important for plasminogen activation and inactivation of α_2-antiplasmin in the surface protease Pla of *Yersinia pestis*. *Mol. Microbiol.* **40**: 1097-1111.

Lähteenmäki, K., Virkola, R., Sarèn, A., Emödy, L., and Korhonen, T. K., 1998, Expression of plasminogen activator Pla of *Yersinia pestis* enhances bacterial attachment to the mammalian extracellular matrix. *Infect. Immun.* **66**: 5755-5762.

Lähteenmäki, K., Kukkonen, M., and Korhonen, T.K., 2001, The Pla surface protease/adhesin of *Yersinia pestis* mediates bacterial invasion into human endothelial cells. *FEBS Lett.* **504**: 69-72.

Sodeinde, O.A., Subrahmanyam, Y.V.B.K., Stark, K., Quan, T., Bao, Y., and Goguen, J. D., 1992, A surface protease and the invasive character of plague. *Science* **258**: 1004-1007.

Sodeinde, O.A., and Goguen, J.D., 1989, Nucleotide sequence of the plasminogen activator gene of *Yersinia pestis*: relationship to *ompT* of *Escherichia coli* and gene *E* of *Salmonella typhimurium*. *Infect. Immun.* **57**: 1517-1523.

Stumpe, S., Schmid, R., Stephens, D.L., Georgiou, G., and Bakker, E.P., 1998, Identification of OmpT as the protease that hydrolyzes the antimicrobial peptide protamine before it enters growing cells of *Escherichia coli*. *J. Bacteriol.* **180**: 4002-4006.

Vandeputte-Rutten, L., Kramer, R.A., Kroon, J., Dekker, N., Egmond, M.R., and Gros, P., 2001, Crystal structure of the outer membrane protease OmpT from *Escherichia coli* suggests a novel catalytic site. *EMBO J.* **20**: 5033-5039.

Chapter 29

Polyclonal B-Cell Activation in Mice Infected by Intragastric Route with *Yersinia enterocolitica* O:8

Beatriz M.M. MEDEIROS, Orivaldo P. RAMOS, Eloisa E.E. SILVA and Deise P. FALCÃO
Department of Biological Sciences, School of Pharmaceutical Sciences, UNESP, Araraquara, SP, Brazil

1. INTRODUCTION

Mice infected by oral route with *Yersinia enterocolitica* O:8 develop systemic infections involving the Peyer's patches, mesenteric lymphonodes, liver and spleen. The Peyer's patches constitute the primary site of cell-pathogen interaction. In this study we investigated the B lymphocytes activation induced by bacteria in Peyer's patches and spleen.

2. MATERIALS AND METHODS

SPF Swiss mice were infected by intragastric route with a strain of *Y. enterocolitica* (Ye 2707, B2, O:8, X_z) and with its plasmidless isogenic pair using an inoculum of 10^8 CFU/mL. Groups of infected animals and controls were sacrificed weekly for 6 weeks and the number of Ig-secreting cells of the different immunoglobulin isotypes (IgG, IgA and IgM) present in the spleens and Peyer's patches was determined by ELISPOT (Czerkinsky *et al.*, 1983). The presence of specific anti-*Yersinia* antibodies was determined in mouse sera by ELISA (Medeiros *et al.*, 1991). The presence of autoantibodies was determined by the Dot Blot technique (Heinicke *et al.*, 1992). The autoantigens used were: myosin, myelin, actin, type I and type II collagen, phosphorylcholine , thyroglobulin and cardiolipin.

The Genus Yersinia, Edited by Skurnik *et al.*
Kluwer Academic/Plenum Publishers, New York, 2003

3. RESULTS

A stronger activation of non-specific Ig-secreting cells was found in the Peyer's patches than that observed in the spleen, both in the animals infected with the virulent strain and in those with the plasmidless strain. In the animals infected with the plasmidless strain, the strongest activation occurred in the IgA isotype, on the 21st day after infection, which exhibited a 16.1-fold increase compared to the controls (Figure 1).

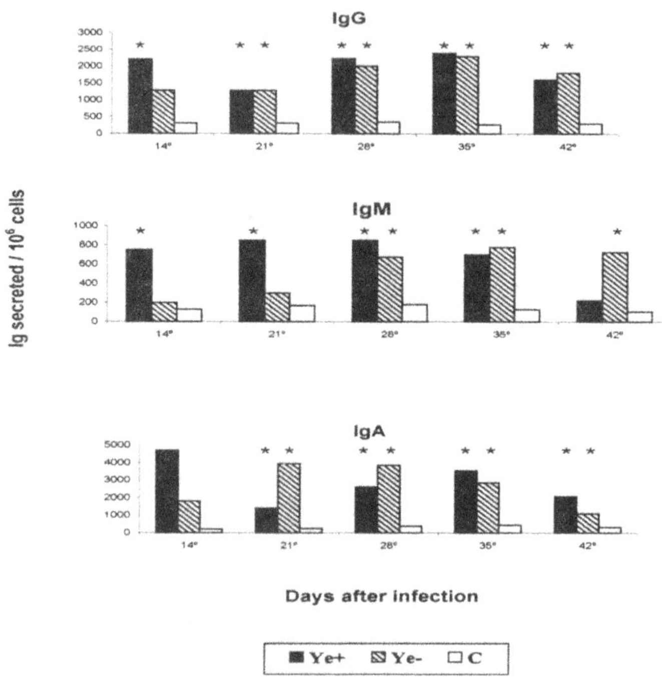

Figure 1. Number of Ig-secreting cells in Peyer's patches (*$p<0.05$ compared to control).

In the spleens, the largest activation occurred in the animals infected with the virulent sample, in the IgM isotype, which increased 9.3-fold with respect to the controls on the 35th day after infection. It was not possible to detect specific Ig-secreting cells by the technique of ELISPOT, during 42-days observation period. Further experiments were done using the ELISA technique, in the attempt to detect specific anti-*Yersinia* antibodies in the serum of the infected animals. Those infected with the virulent strain produced IgG-specific antibodies, from the 28th day after infection. No specific antibodies were produced in the animals infected with the avirulent strain. Concerning the tests for autoantibodies, the sera of animals infected with both the virulent and avirulent strains showed reactivity to several autologous constituents, which coincided with the period during which the greatest non-specific B lymphocyte activation was observed. Clinical and pathological alterations were more intense in the animals infected with the plasmid-bearing strain. Neither arthritic nor inflammatory signs were observed in the joints of the infected animals.

4. CONCLUSION

The infection of mice with virulent and avirulent strains of *Y. enterocolitica* O:8 resulted in polyclonal activation of the repertory of spleen and Peyer's patches B lymphocytes, including some autoreactives clones.

ACKNOWLEDGEMENTS

The work from our laboratory was supported by grants from FAPESP and PADC/UNESP.

REFERENCES

Czerkinsky, C.C., Nilsson, L. A., Nygren, H., Ouchterlony, O., and Tarkowsi, A., 1983, A solid-phase enzyme-linked immunospot (ELISPOT) assay for enumeration of specific antibody-secreting cells. *J. Immunol. Methods*, **65**:109-121.

Heinicke, E., Kumar, U., and Munoz, D. G., 1992, Quantitative dot-blot assay for proteins using enhanced chemiluminescence. *J. Immunol. Methods.*, **152**:227-236.

Medeiros, B.M.M., Mendes-Giannini, M.J.S. and Falcão, D.P., 1991, Immunoglobulin isotypes produced by mice experimentally infected with *Yersinia* sp. *Contrib Microbiol Immunol,* **21**:117-122 specific antibodies were produced in the animals infected with the [illegible] strain. Concerning the tests for autoantibodies, the [illegible] were both [illegible] thymocyte [illegible] autoantigen [illegible] components, which coincided with the [illegible] specific [illegible] B lymphocyte [illegible] and [illegible] after [illegible] the [illegible] strain. [illegible] in the [illegible] animals.

Chapter 30

Polyclonal Activation as a Consequence of Infection of Mice with *Yersinia enterocolitica* O:3 Isolated from Patients with or without Arthritis

Beatriz M. M. MEDEIROS, Eloisa E. E. SILVA, Orivaldo P. RAMOS and Deise P. FALCÃO
Department of Biological Sciences, School of Pharmaceutical Sciences, UNESP, Araraquara, SP, Brazil

1. INTRODUCTION

Yersinia enterocolitica causes acute gastroenteritis. At 1-3 weeks after infection, a subset of patients develop reactive arthritis that varies from mild arthralgia to severe polyarthritis and usually resolves within a few weeks or months in most patients. A subset of these patients (10%) will develop a chronic inflammatory arthritis (Bottone, 1997). Clinical *Y. enterocolitica* infections are followed by the development of autoantibodies including antibodies against smooth muscle, connective tissue, renal tubular epithelium, and the basement membrane of thyroid epithelial cells (Gripenberg *et al.*, 1978, Montplaisir *et al.*, 1979, Shenkman and Bottone, 1981) The production of these autoantibodies may result from polyclonal B-lymphocyte activation. Resting B cells with receptors directed against autologous structures may be stimulated as a result of the microorganism-host relationship, which can contribute for the development of reactive arthritis (Medeiros *et al.*, 1997). The objective of this work was to evaluate the extent of the polyclonal activation in the spleen and the production of autoantibodies in sera of infected Swiss mice induced by an arthritogenic strain of *Y. enterocolitica* O:3, and by a nonarthritogenic strain and its plasmidless pair.

The Genus Yersinia, Edited by Skurnik *et al.*
ıwer Academic/Plenum Publishers, New York, 2003

2. MATERIALS AND METHODS

SPF Swiss mice were infected intravenously with a suspension (10^8 cells/ml) of a arthritogenic strain of *Y. enterocolitica* O:3 (FCF 526), a virulent but non-arthritogenic strain (FCF 397 [+]) or an avirulent strain (FCF 397 [-]). Groups of 25 non-inoculated animals were used as controls. Four mice from each group were bled by heart puncture and their spleens were removed for cell collection on the 7th, 14th, 21st and 28th day after infection. The number of Ig-secreting spleen cells (IgG1, IgG2a, IgG2b, IgG3, IgM and IgA), specific and nonspecific, was determined by ELISPOT (Czerkinsky *et al.,* 1983). Production of autoantibodies directed against different autologous constituents (actin, collagens I and II, phosphorylcholine, myosin, cardiolipin, myelin and thyroglobulin) was determined by Dot Blot (Heinicke *et al.,* 1992).

3. RESULTS

The arthritogenic strain (FCF 526) activated mostly nonspecific Ig-secreting cells of the IgG2a and IgG3 isotypes (a 10,4- and 17-fold increase compared to control) on the 7th day after infection. The major activation observed with strain 397 [+] was found with the IgG2a and IgG3 isotypes (a 10,6- and 12-fold increase) on the 14th day after infection. The strain FCF 397 [-] activated the isotypes IgG3 on the 14th day (a 17,5-fold increase compared control) and IgA (a 4,9-fold increase compared control) on the 21th day after infection. The arthritogenic strain FCF 526 gave the greatest activation of specific antibody-secreting cells of the IgG2b and IgM isotypes (a 16- and 120-fold increase compared control) at day 7 after infection. The number of specific antibody-secreting cells was relatively low for both the nonarthritogenic strain FCF 397 [+] and the plasmidless pair FCF 397 [-]. The greatest reactivity to FCF 526 occurred 7 days post infection; autoantibodies were found for all constituents, except for type II collagen and myelin. Strains FCF 397 [+] and [-] provoked the greatest reactivity 28 days post infection; strain FCF 397 [+] did not react with actin, while FCF 397 [-] reacted to everything it was tested against. The greatest reactivity for cardiolipin was observed 21 days post infection.

4. CONCLUSIONS

It was observed polyclonal activation of B-lymphocytes with all the bacterial strains tested. The IgG2a and IgG3 isotypes were that which present the major activation. The three bacterial strains stimulated the production of autoantibodies against all the autologous constituents tested.

ACKNOWLEDGEMENTS

The work from our laboratory was supported by grants from FAPESP and PADC/UNESP.

REFERENCES

Bottone, E. J., 1997, *Yersinia enterocolitica*: the charisma continues. *Clin. Microbiol. Rev.*, **10**: 257-276.

Czerkinsky, C.C., Nilsson, L. A., Nygren, H., Ouchterlony, O., and Tarkowski, A.,1983, A solid-phase enzyme-linked immunospot (ELISPOT) assay for enumeration of specific antibody-secreting cells. *J.Immunol. Methods.*, **65**: 109-121.

Gripenberg, M.; Miettinen, A.; Kurki, P.; Linder, E., 1978, Humoral immune stimulation and antiepithelial antibodies in *Yersinia* infection. *Arthritis Rheum.,* **21**: 904-908.

Heinicke, E., Kumar, U., and Munoz, D. G., 1992, Quantitative dot-blot assay for proteins using enhanced chemiluminescence. *J. Immunol. Methods.*, **152**:227-236.

Medeiros, B. M. M.; Costa, A M.; Araújo, P. M. F.; Falcão, D. P., 1997, Association between polyclonal B cell activation and the presence of autoantibodies in mice infected with *Yersinia enterocolitica* O:3. *Braz. J. Med. Biol.* ***Res.***, **30**: 401-405.

Montplaisir, S. ; Gagné, M.; Dubois, R. ; Gengoux, P.; Pelletier, M., 1979, Brush border membrane antibodies in *Yersinia* infection and other diseases. *Contrib. Microbiol. Immunol.* , **5**: 351-358, 1979.

Shenkman, L.; Bottone, E. J., 1981, *The occurrence of antibodies to* Yersinia enterocolitica *in thyroid diseases*. In *Yersinia enterocolitica* (Bottone, E. J., Ed.), CRC Press, Boca Raton:, pp.135-144.

Chapter 31

The Response of Murine Macrophages to Infection with *Yersinia pestis* as Revealed by DNA Microarray Analysis

Lee-Ching NG[1], Ola FORSLUND[2], Susie KOH[1], Kerstin KUOPPA[3] and Anders SJÖSTEDT[4]
[1]Centre for Chemical Defence, DSO National Laboratories, 20 Science Park Drive, Singapore; [2]Div. of Microbioloy, Malmoe University Hospital. Sweden. [3]FOI, NBC Defence. 20 Cementsvagen, SE-901-82, Umea, Sweden; [4]Department of Clinical Microbiology, Clinical Bacteriology, Umea University, SE-901 85 Umea. Sweden

1. ABSTRACT

Macrophages play a crucial role in recognition and phagocytosis of pathogens and in the induction of response, immunity and immunopathology. A key strategy employed by numerous pathogens such as *Yersinia pestis* is to circumvent the immune response of the host via actively down-regulating the activation of macrophages. The study on host-pathogen interaction and gene expression is imperative for the development of alternative therapeutics. We have combined Suppression Subtractive Hybridisation (SSH), Microarray techniques, Northern blot analysis and quantitative reverse transcription coupled PCR (RT-PCR) to gain a view of differential host gene expression in response to *Y. pestis*-26°C infection. In our study, a total of 22 different genes were identified as up-regulated in response to the *Y. pestis* infection. These genes include unknown EST's, cytokines, enzyme of cytokine, receptors, ligands, transcriptional factors, inhibitor of transcriptional factor, and proteins involved with cytoskeleton. More interestingly, among them are 7 genes that encode for factors known to be associated with cell cycling and cell proliferation, with 3 of them playing a role in apoptosis. Our data also indicate that macrophage cells undergo apoptosis during an infection with *Y. pestis*-37°C, however an infection with

The Genus Yersinia, Edited by Skurnik *et al.*
˜uwer Academic/Plenum Publishers, New York, 2003

26°C cultures results in a delayed apoptosis. The correlation between the delayed apoptosis and the up-regulation of anti-apoptotic gene is currently being studied.

2. INTRODUCTION

Macrophages play a central role in host defense to infection by killing microbes but also have a major role in initiating, maintaining, and resolving host inflammatory responses by releasing cytokines and chemokines. They are also a primary habitat for a group of pathogens that utilize the intracellular environment as a sanctuary to avoid extracellular defense mechanisms. To survive within or escape from the hostile environment of an activated macrophage, intracellular microorganisms have developed a multitude of survival strategies. *Y. pestis* is thought to be predominantly extracellularly located during infection. However, some other studies had demonstrated that the pathogen is highly disseminative, capable of surviving and replicating within nonactivated macrophages (Straley and Harmon, 1984), the first line defence of our immune system. It is believed that *Y. pestis* introduced into a host via fleas leads an intial phase of intracellular lifestyle. The aim of the present study was to characterize the macrophage response to the invasive form of the bacterium by use of a large-scale DNA microarray. Changes in specific host gene mRNA populations following infections with intracellular pathogens could yield important clues to characterize and understand the events that occur inside infected macrophage

3. RESULTS AND DISCUSSION

3.1 Survival and multiplication of bacteria in J774 macrophages

Monolayers of cells were infected with various pathogens for 15 min, washed and further incubated for 60 min. before bacteria were recovered. The total number of surviving bacteria was obtained from the infected macrophage cells J774 incubated in medium in the absence of gentamicin, while the number of intracellular bacteria was recovered from infected cells incubated in the presence of 50μg/ml gentamicin. Figure 1 shows that *Y. pestis* grown at 26°C was taken by the macrophages while negligible number of bacteria grown at 37°C were internalized. The level of uptake of *Y. pestis* grown at 26°C was equivalent to infections by the other well-known intracellular bacteria, *Francisella tularensis*, *Salmonella typhimurium* and *Listeria monocytogenes*. The experiment confirmed the dual lifestyle of *Y. pestis* and that the lifestyle opted for *Y. pestis* is dependent on the temperature to which the pathogen is subjected to prior to infection.

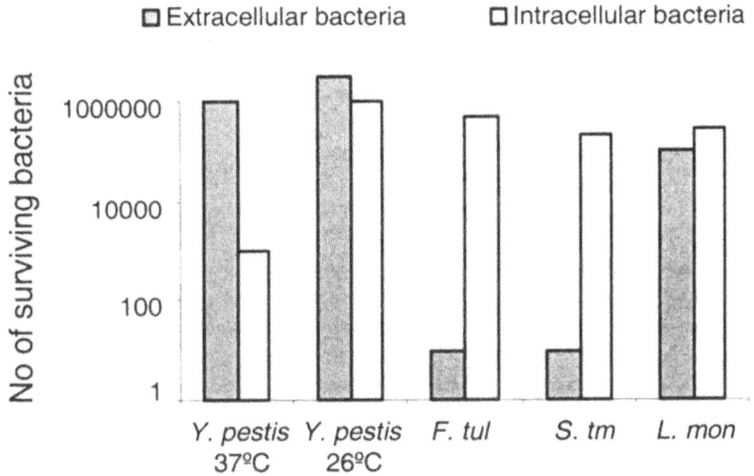

Figure 1. Number of surviving bacteria after 15 min infection followed by 60 min incubation. (*F. tul, F. tularensis; S. tm, S. typhimurium; L. mon, L. monocytogenes*).

3.2 Microarray for Identification of Host Genes Associated with *Y. pestis* Infection.

Differential; macrophage host expression effected by infection with *Y. pestis*-26°C was analysed with microarrays, which were fabricated with in-house generated subtractive suppression hybridisation.mouse EST libraries. In our study, a total of 22 different genes were identified as upregulated in response to the *Y. pestis* infection. These genes include unknown EST's, cytokines, enzyme of cytokine, receptors, ligands, transcriptional factors, inhibitor of transcriptional factor, and proteins involved with cytoskeleton. More interestingly, among them are 7 genes that encode for factors known to be associated with cell cycling and cell proliferation, with 3 of them playing a role in apoptosis (Table 1).

Table 1. Genes up-regulated by *Y. pestis*-26°C infection

Gene id	Function	Category
Nuclear factor of kappa light chain gene enhancer in B-cells inhibitor, alpha	Inhibitor of transcriptional factor NFkappaB	Inhibitor of transcriptional factor
ELK3, Ets-domain protein	Transcription factor, associated with ras-signalling pathway and plays a role in both B-cell development and IgH gene regulation.	Transcriptional factor associated with ras signaling pathway
Sequestosome 1, alias ubiquitin-binding protein p62	Ligand of the p56-lck SH2 domain (T-cell-specific src family tyrosine kinase) which is involved in T-cell signal transduction. Also binds to ras GTP-ase-activating protein, a negative regulator of the ras signaling pathway	Ligand associated with ras signaling pathway
Interferon regulatory factor 1	Functions as a transcriptional activator for the type I IFN genes,. Also found to be a tumor suppressor gene	Transcriptional factor associated with cell proliferation
Peroxiredoxin 1	Regulates NK activity and help cope with oxidative stress	Associated with cell proliferation
Cytokine inducible SH2-containing protein 3	Immediate-early cytokine-responsive genes, growth inhibitory function	Associated with cell proliferation
SNO proto-oncogene protein FL isoform	Overexpression rendered cells resistant to the growth-inhibitory effects of TGFB1, component of the SMAD pathway.	Associated with cell proliferation and apoptosis
Platelet-derived growth factor, B chain precursor (PDGF)	Promotes cellular proliferation and inhibits apoptosis, activates the RAS/PIK3/AKT1/Ikk/NFkB1 pathway	Associated with cell proliferation and apoptosis
Immediate early response 3 (IER3)	Protect cells from TNF induced apoptosis, contains binding site for NF-kappa-B, CEBP and SP1	Associated with cell proliferation and apoptosis
Potassium channel, inwardly rectifying,	Plays an important role in developmental signaling	Ion channel

Gene id	Function	Category
subfamily J, member 2		
Regulator of G-protein signaling 1	Immediate early gene, expression in B lymphocytes able to down-regulate chemotaxis to lymphoid chemokines	Associated with chemotaxis
Actin cross-linking protein 7	A hybrid of dystonin and dystrophin that can interact with the actin and microtubule cytoskeletons	Associated with cytoskeleton
Small inducible cytokine A4, MIP1B	Mediators of inflammation from T cell, B cell and monocyte	Cytokine
Tumor necrosis factor (TNF)	Proinflammatory cytokines	Cytokine
Sequestosome 1	A key cytokine for efficient type 1 immune responses through differential regulation of macrophage IL12 and IL10 cytokine expression	Cytokine
Interleukin 1B	Implicated in inflammation, septic shock	Cytokine
C-type lectin, Superfamily member 9	Inducible by Lps, TNF-alpha, IL-6, and IFN-gamma, target of NF-IL6 in macrophages	Receptor

A number of studies have shown that apoptosis is triggered in host cells in response to infection by a variety of extra- and intra-cellular bacterial pathogens, including extracellular *Y. pseudotuberculosis*. However, in our study of intracellular *Y. pestis*, two activated genes, namely SNO, PDGF and IER3, are known to possess anti-apoptotic function. The result may shed some light on the bacterial strategy in maintaining the two lifestyles, probably to its great advantage.

4. CONCLUSION

Using an in-vitro system, this study has confirmed the two lifestyles, intracellular and extracellular, adopted by *Y. pestis* in its infection of macrophage cells. The activation of as many as 7 genes that play significant roles in cell proliferation and anti-apoptosis implies the importance of such pathways to the intracellular infection by *Y. pestis*. In the case of an extracellular infection by *Y. pestis*, macrophage cell death may be the goal of the pathogen as the elimination of macrophage would lead to the abolishment of host's first-line defence. However, in the case of an intracellular infection, the bacteria lack the toxins required for the killing of host. The activation of host anti-apoptotic genes may thus be a bacterial strategy that prevents the infected cell from undergoing apoptosis so that the host can function as a sanctuary for the pathogen.

REFERENCE

Straley and Harmon. 1984. *Yersinia pestis* grows within phagolysosomes in mouse peritoneal macrophages. Infect Immun. 45:655-9.

Chapter 32

Defensive Function of Phagocytes in Pseudotuberculosis

Natalya G. PLEKHOVA, Larisa M. SOMOVA-ISACHKOVA and Felix N. SHUBIN
Research Institute of Epidemiology and Microbiology, Siberian Branch of RAMS, Selskay 1, Vladivostok, Russia

1. INTRODUCTION

Yersinia pseudotuberculosis resistance to phagocytosis is based on the ability of virulent bacteria to survive and multiply within polymorphonuclear leukocyte and macrophages. The virulence of *Yersinia* is associated with the presence of plasmids ranging 42 - 48 and 82 megadaltons (MDa) in molecular size. We demonstrate here that there is an inverse relationship between the degree of *Y. pseudotuberculosis* virulence and the functional activity of phagocytes.

The ability to kill bacteria is one of characteristics of "professional" phagocyte. In recent years the bactericidal and cytotoxic actions of neutrophils have been attributed mainly to H_2O_2 -(myelo)peroxidase-halide systems (Levy, 2000). The progenitor cells of the macrophages, the promonocytes of bone marrow are endowed with myeloperoxidase in their granules. This enzyme, however, is lost as the monocyte reaches the tissue and becomes a mature macrophage. Recent studies (Leung and Goren, 1989) suggest that macrophages could acquire peroxidase activity from inflammatory neutrophils, i.e. macrophages might obtain functional myeloperoxidase from granulocyte organelles in inflammatory foci.

In this study, we tested at cytochemical (*in vivo*) and ultrastructural (*in vitro)* level various functions of neutrophils and macrophages infected by virulent strains of *Y. pseudotuberculosis*.

The Genus Yersinia, Edited by Skurnik *et al.*
ˈluwer Academic/Plenum Publishers, New York, 2003

2. ACTIVATED NEUTROPHILS

Figure 1 summarises the cytochemical studies on the neutrophil enzymes (cationic proteins) of infected guinea pigs. The results demonstrated that the bactericidal activity of the blood cells of guinea pigs infected with either the avirulent plasmid-free strain or a strain carrying only the 82 MDa plasmid was considerably enhanced by the 28 days of infection. On the contrary, interaction of neutrophils with virulent *Y. pseudotuberculosis* strains carrying either two plasmids (48 and 82 MDa) or only the 45 MDa plasmid caused a decrease in phagocytic activity. Interestingly, there was a reduction in the intracellular amounts of enzymes 3-7 days post-infection. These data correlated with the enzyme (myeloperoxidase and lysozyme) activity results.

In vitro the phagocytosis of bacteria by neutrophils was weakly affected. The same effect of the virulence characteristics of *Yersinia* on phagocytosis was discovered when the functional activity of peritoneal exudate neutrophils (obtained from inflammation foci) was studied in vitro. The most clear effect was seen with *Y. pseudotuberculosis* strains carrying both plasmids. This data indicates that reduction of the antimicrobial activity of neutrophils results from degranulation.

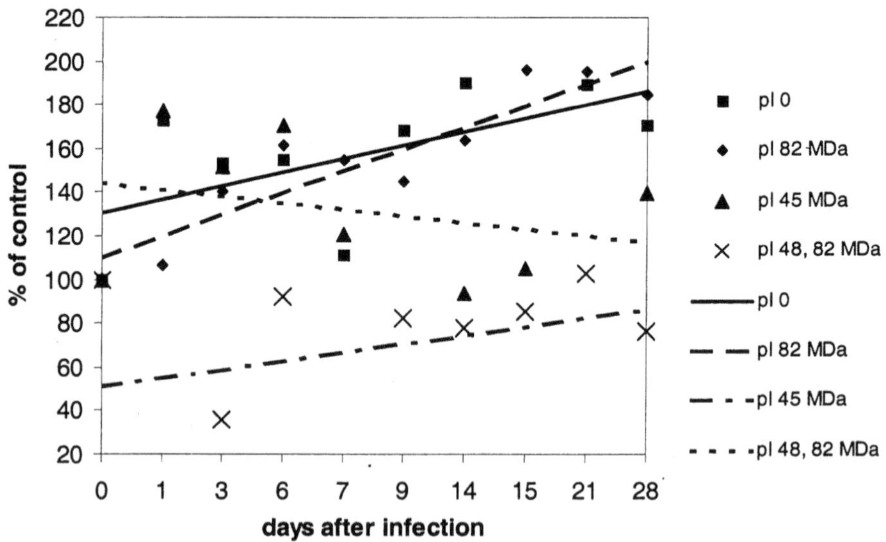

Figure 1. Intracellular contents of cationic proteins in neutrophils of infected guinea pigs. Control for neutrophils of intact animals is set to 100 % and was 1,41± 0,075 units for 100 cells.

3. IN VITRO-ACTIVATED MACROPHAGES

Similar results were obtained when the functional activity of macrophages was studied in vitro. The phagocytic activity of macrophages was inversely proportional to the degree of virulence of the *Y. pseudotuberculosis* strains. The macrophages more actively ingested avirulent plasmid-free bacteria. The bacterial numbers in phagocytes increased significantly (*$p<0.01$*) between 30 and 120 minutes after which no further increase was seen. The phagocytes were able to ingest more the avirulent bacteria that carried only the 82 MDa plasmid than the virulent bacteria carrying either two plasmids (48 and 82 MDa) or only the 45 MDa plasmid. These results demonstrate that the presence of plasmids influence the bactericidal activity of phagocytes.

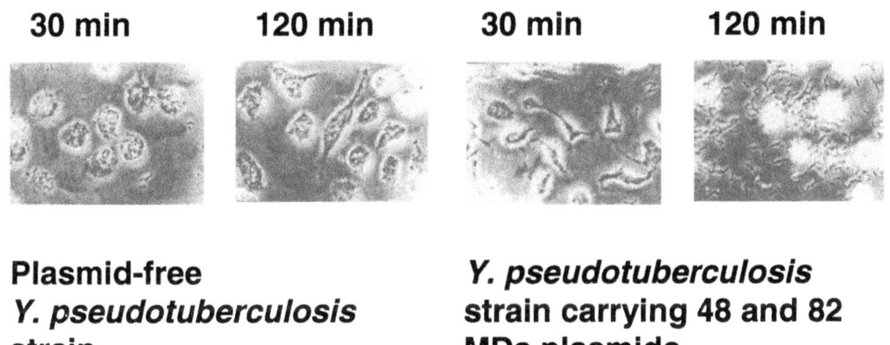

Figure 2. Phase contrast micrographs of mouse resident macrophages cultivated for 72 h in medium with *Y. pseudotuberculosis*. X 800

Phase contrast microscopy revealed clear differences in the effect of virulent and avirulent bacteria on the cultured macrophages (Figure 2). Activation of macrophages was detected after their contact with plasmid-free strains of bacteria. It is worth noting that the amount of intracellular vacuoles present in these macrophages was increased strongly suggesting that bacteria had been ingested. The findings with phagocytes infected with the *Y. pseudotuberculosis* strains carrying only one plasmid (45 or 82 MDa) were similar although the effects were not so dramatic. Moreover, the number of ingested bacteria was smaller and chains of extracellularly replicating bacteria were seen. In contrast, most bacteria carrying the 48 and 82 MDa plasmids survived due to their strong cytophatic effect on macrophages. This was manifested by an increase in the amount of damaged phagocytes and presence of extracellularly multiplying bacteria.

4. CONCLUSIONS

The results of our investigations showed that plasmid-harbouring *Y. pseudotuberculosis* actively suppress the phagocytic activity of neutrophils and macrophages. It is likely that deactivation of phagocytes is caused by secretion of Yops. Moreover, bacteriological studies showed that only a few of the ingested bacteria were destroyed and the remaining killed the macrophages.

However, a collaborative action of neutrophils and macrophages on the virulent extra- and intracellular *Y. pseudotuberculosis* takes place. For the bactericidal activity macrophages require exposure to extracellular biologically-active products (in particular, myeloperoxidase) of neutrophils already activated by bacteria. Neutrophils are able to cause L-transformation of *Y. pseudotuberculosis* (Plekhova and Isachkova, 1998). Myeloperoxidase together with other oxygen-dependent bactericidal systems breaks the permeability of the cell wall of *Yersinia* and thus enables the macrophages to ingest and digest the bacteria with the help hydrolytic enzymes.

REFERENCES

Leung K.- P., and Goren M.B., 1989, Uptake and utilization of human polymorphonuclear leukocyte granule myeloperoxidase by mouse peritoneal. *Cell and Tiss. Res.* **257** (3): 653-656.

Levy O., 2000, Antimicrobial proteins and peptides of blood: templates for novel antimicrobial agents. *Blood* **96** (8): 2664-2672.

Plekhova N.G., and Isachkova L.M., 1998, Electronic-cytochemical characteristic of process interaction neutrophils and macrophages with Yersinia pseudotuberculosis *Med. J. Russia* **1-2**: 61-65.

Chapter 33

Mechanisms of *Yersinia enterocolitica* Evasion of the Host Innate Immune Response by V Antigen

Andreas SING, Dagmar ROST, Natalia TVARDOVSKAIA, Andreas ROGGENKAMP, Anna M. GEIGER, Carsten J. KIRSCHNING[1], Agnès WIEDEMANN, Martin AEPFELBACHER and Jürgen HEESEMANN
Max von Pettenkofer-Institut für Hygiene und Medizinische Mikrobiologie, Ludwig Maximilians-Universität München, Pettenkoferstrasse 9a, 80336 München, Germany; [1]Institut für Medizinische Mikrobiologie, Immunologie und Hygiene, Technische Universität München, Trogerstrasse 32, 81675 München, Germany

1. INTRODUCTION

V antigen (LcrV) belongs to the common virulence-associated, pYV-encoded antigens of the three human pathogenic *Yersinia* spp. and is a secreted, multifunctional protein. Besides participating in regulation of Yop production, translocation of Yops into host cells and membrane channel formation LcrV exhibits immunoregulatory features such as TNFα and IFNγ suppression (Motin *et al.*, 1994) and IL-10 amplification (Nedialkov *et al.*, 1997). Recently, we could assign TNFα suppression to LcrV-mediated IL-10 production (Sing *et al.*, 2002a). The possible *in vivo* relevance became evident by showing that IL-10 $^{-/-}$ mice were highly resistant against i.p. *Y. enterocolitica* infection. We conclude that LcrV helps to evade the innate immunity effector TNFα by stimulating IL-10 production, such exploiting the TNF-α down-regulating capacity of endogenous host IL-10. In the present study, we analyzed innate immunity receptors involved in LcrV-dependent signaling.

The Genus Yersinia, Edited by Skurnik *et al.*
Kluwer Academic/Plenum Publishers, New York, 2003

2. CD14 IS NEEDED FOR LCRV-INDUCED IMMUNOMODULATION

In contrast to wild-type proteose-peptone elicited peritoneal macrophages (PPMs), IL-10 induction by rLcrV is virtually absent in CD14 $^{-/-}$ PPMs (Sing *et al.*, 2002b). CD14-dependency of LcrV-induced immunomodulation was confirmed with human MonoMac-6 cells: anti-CD14-blocking, but not anti-CD14-non-blocking mAbs abrogate rLcrV-mediated TNFα suppression.

3. LCRV AND LCRV-DERIVED PEPTIDES SIGNAL IN A CD14/TLR2-DEPENDENT MANNER

Toll-like receptor (TLR) 2 - in combination with CD14- was identified as the responsible TLR for LcrV-signaling (Sing *et al.*, 2002b). Moreover, rLcrV constructs lacking different amino acid (aa) regions in the C-terminus of the protein (aa190-278) and *r*LcrV constructs comprising the first 130 and 100 N-terminal aa (rV130 and rV100) as well as peptides derived from the N-terminus of LcrV were also able to induce NFκB activation in a TLR2/CD14-dependent manner (Figure 1).

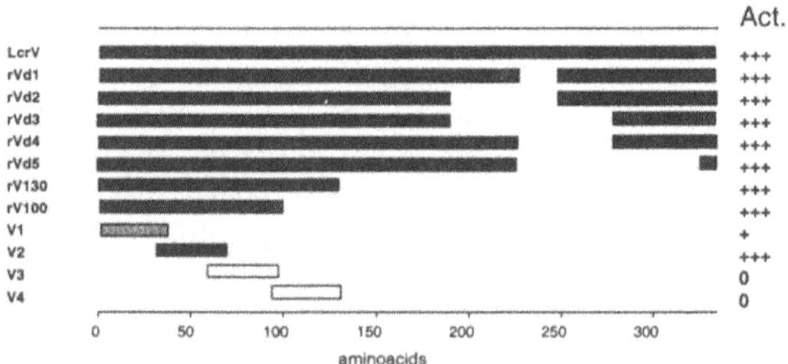

Figure 1. Structure-function analysis of LcrV. Recombinant constructs and peptides and their respective activity in CD14/TLR2-transfected 293 HEK cells.

4. TLR2 IS INVOLVED IN IL-10 PRODUCTION BY LCRV AND IN *Yersinia* SUSCEPTIBILITY

Further evidence for recognition of LcrV by TLR2 leading to IL-10 induction was obtained by comparison of PPMs from $TLR2^{-/-}$ and congenic $TLR2^{+/+}$ mice of SV129xB57BL/6 or C57/BL6 background, respectively. TLR2-deficient PPMs did not adequately produce IL-10 upon rLcrV-stimulus when compared to wild-type PPMs. Moreover, $TLR2^{-/-}$ mice of SV129 and C57/BL6 background were less susceptible to oral *Y. enterocolitica* infection (Sing *et al.,* 2002b).

5. CONCLUSIONS

Our data demonstrate a new ligand specificity of TLR2, since LcrV is the first known secreted and non-lipidated virulence-associated protein of a Gram-negative bacterium utilizing TLR2 for cell activation. We conclude that yersiniae might exploit host innate pattern recognition molecules and defense mechanisms to evade the host immune response.

ACKNOWLEDGEMENTS

This work was supported in part by the Deutsche Forschungsgemeinschaft (GRK 303 and SPP 1089).

REFERENCES

Motin, V.L., Nakajima, R., Smirnov, G.B., and Brubaker, R.R, 1994, Passive immunity to yersiniae mediated by anti-recombinant V antigen and protein A-V antigen fusion peptide. *Infect. Immun.* **62**:4192-4201.

Nedialkov, Y.A., Motin, V.L., and Brubaker, R.R., 1997, Resistance to lipopolysaccharide mediated by the *Yersinia pestis* V antigen-polyhistidine fusion peptide: amplification of interleukin-10. *Infect. Immun.* **65**:1196-1203.

Sing, A., Roggenkamp, A., Geiger, A.M., and Heesemann, J., 2002a, *Yersinia enterocolitica* evasion of the host innate immune response by V antigen-induced IL-10 production of macrophages is abrogated in IL-10-deficient mice. *J. Immunol.* **168**: 1315-1321.

Sing, A., Rost, D., Tvardovskaia, N., Roggenkamp, A., Wiedemann A., Kirschning, C.J., Aepfelbacher, M., and Heesemann, J., 2002b, Yersinia V-antigen exploits toll-like receptor 2 and CD14 for interleukin-10 mediated immunosuppression. J. Exp. Med. **196**: 1017-1024.

Chapter 34

Yersinia pseudotuberculosis, Its Toxins and Plant Cells

Nelly TIMCHENKO[1], Marina ELISEIKINA[3], Viktor BULGAKOV[2], Elena BULAKH[1], Elena YASNETSKAYA[2], Elena NEDASHKOVSKAYA[1] and Yury ZHURAVLEV[2]

[1]Institute of Epidemiology and Microbiology, Academy of Medical Science; [2]Institute of Biology and Soil Sciences; [3]Institute of Marine Biology, Vladivostok, Russia

1. INTRODUCTION

Pathogenic strategy of *Y. pseudotuberculosis* interacting with ectoterm organisms is not yet clear. The presence of *Y. pseudotuberculosis* has been demonstrated on vegetables, vegetable juice, and edible roots at low temperature (Somov, 1979). However, very little information of the *Yersinia*-plant interaction is available (Venediktov *et al.*, 1989). Therefore, in this work we have studied the interactions between *Y. pseudotuberculosis* and plant cells. We used callus cultures of *Panax ginseng*, *Aristolochia manshuriensis*, *Lithospermum erythrorhizon* and Brassica, *Y. pseudotuberculosis* strains 512, 282, 1179, 2517 (both pYV+ and pYV-) and the 45-kDa *Yersinia* thermostable toxin. Plant-microbe systems were incubated at 23°C and bacteriological, cytological and electron-microscopic data were recorded.

2. RESULTS

A complex picture of plant-microbe interaction was observed in different experiments. All the cultures of tested plant cells were damaged by *Y. pseudotuberculosis* and no difference was found between the pYV+ and pYV- bacteria regarding to their damaging potential. Bacteria multiplied for

The Genus Yersinia, Edited by Skurnik *et al.*
'uwer Academic/Plenum Publishers, New York, 2003

a long time (50-60 days) in association with live and dying *L. erythrorhizon* and Brassica cells. On the contrary, they rapidly died in association with *P. ginseng* and *A. manshuriensis* cells, killing simultaneously the plant cells. It is interesting, that the dynamics of bacterial death strongly suggested the involvement of the plant cells in this process. When the number of injured plant cells reached 50% and more, the massive death of bacteria could be easily detected. Extracts of *P. ginzeng*, depending on their concentration, caused significant stimulative or supressive effect on the bacterial growth.

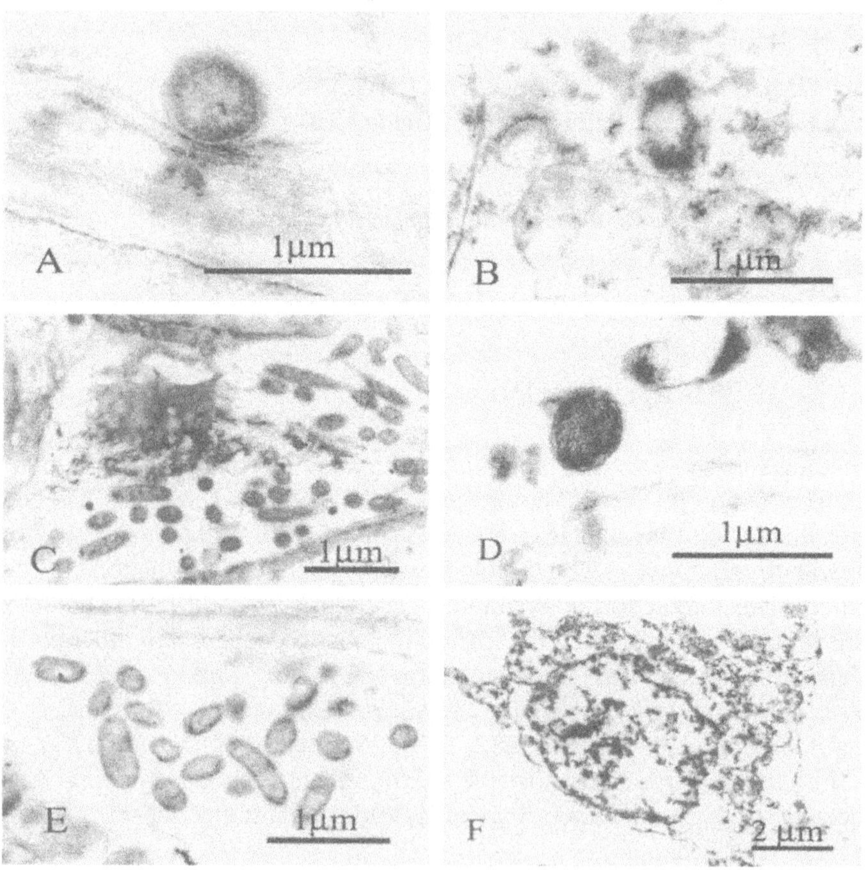

Figure 1. Interaction of *Y. pseudotuberculosis* and cultured plant cells: *Brassica*.A. Penetration of bacteria through plant cells wall. B. *Y. pseudotuberculosis* among the organoids of plant cells. C. Destruction and lysis of the plant cell; *P. ginseng*. D. Adsorption of the bacteria by the wall of plant cells. E. The bacteria in the intercellular space. F. Necrosis of the plant cell cytoplasm.

The interaction of *Y. pseudotuberculosis* and tissue culture cells of *Brassica* and *P. ginseng* was studied after incubation for 6h, 1d and 4d.

During the first hours after infection the bacteria surrounded the *Brassica* cells and penetrated deep into the intercellular space. It is likely that bacteria produced some lytic factors which destroyed the plant cell wall structures. Some bacteria were able to penetrate through the plant cell wall (Figure 1A). There were also bacteria present between the organoids of plant cell cytoplasm (Figure 1B). By the influence of bacterial factors both the walls and the cytoplasm of cabbage cells were damaged. On the later stage of the experiment most of the cells were devastated and filled with bacteria (Figure 1C).

During the interaction of *Y. pseudotuberculosus* and the cells of *P. ginseng* adhesion of bacteria to the tissue culture cells was rarely seen during the first stages of experiment (Figure 1D). Most of *Y. pseudotuberculosis* bacteria were free and had no contact with the plant cell wall components. However, by means of some lytic factors bacteria were able to destroy components of intercellular structures and entered to those free spaces (Figure 1E). The cytoplasm of plant cells was necrotic and the plant cell walls were also damaged. On the later stages most of tissue culture cells were devastated (Figure 1F). The thermostable toxin of *Y. pseudotuberculosis* produced necrosis and death of plant cells (data not shown).

3. CONCLUSIONS

Y. pseudotuberculosis was able to grow in the presence of plant cells at low temperature. The bacteria were apparently able to produce factors that promoted adhesion and invasion to plant cells as well as toxins which caused pathogenic effects while interacting with plant cells. Taking into consideration the major differences between plant and mammalian cells it is very likely that *Y. pseudotuberculosis* uses different strategies of interaction with endo- and ectoterm organisms. However, the mechanisms of these processes are unknown.

REFERENCES

Venediktov V.C.,Timchenko N.F.,Bulgakov V.P., Zhuravlev Yu.N.,1989, *Yersinia* and plant. Conf. Yersiniosis,Vladivostok, 12-13.

Somov G.P.,1979, Scarlet-like fiver. Medicine, Moscow.

James, E., 1988, Plants and Fungi. In *Botanical Research Techniques* (J. Willis and P. Peterson, eds.), Academic Press, London, pp.7-15.

Chapter 35

Influence of *Yersinia pseudotuberculosis* on the Immunity of Echinoderms

Marina ELISEIKINA[1], Nelly TIMCHENKO[2], Aleksandr BULGAKOV[3], Timyr MAGARLAMOV[1] and Irina PETROVA[1]
[1]*Institute of Marine Biology Far East Branch Academy of Sciences;*[2]*Institute of Epidemiology and Microbiology, Academy of Medical Sciences;* [3]*Pacific Institute of Bioorganic Chemistry, Far East Branch Academy of Sciences, Russia, Vladivostok*

1. INTRODUCTION

There is little information about the influence of *Yersinia spp.* on the immunity of aquatic animals. However, we have reliable data concerning the finding of these bacteria in seawater and sea animals in the places of anthropogenic pollution. In this connection, *Yersinia* is a potential pathogen to marine animals and echinoderms, in particular. As is known, the immunity of Echinoderms can be schematically divided into cellular and humoral responses (Chia and Xing, 1996). Cellular response is mediated by the coelomocytes which are the cells that circulate in the coelomic cavity, while humoral response depends upon molecules that are present in the coelomic fluid. The most essential types of coelomocytes are the amoebocytes (that accomplish phagocytosis) and the morular cells that participate in biosynthesis of the humoral components of immunity including agglutinins or lectins. Lectins are sugar-binding proteins other than enzymes and antibodies (Barondes, 1988). Lectins are involved in opsonization and lytic functions in addition to functioning in clot formation and wound repair (Gros *et al.*, 1999). The purpose of this research was to study participation of the main types of coelomocytes and mannan-binding lectins (MBL) in the defense reactions after introduction of *Yersinia* bacteria in the coelomic cavity of echinoderms. At the same time, it was interesting to investigate the effect of these defense components on the bacterial cells.

The Genus Yersinia, Edited by Skurnik *et al.*
'uwer Academic/Plenum Publishers, New York, 2003

2. DEFENCE REACTIONS OF ECHINODERMS

Two species of echinoderms were used in this study: *Strongylocentrotus nudus* and *Apostichopus japonicus. Y. pseudotuberculosis* 512 (pYV+) and 2517 (pYV+) were introduced in the coelomic cavity. In both echinoderm species, after 1 hour upon inoculation, numerous bacterial cells were found in the coelomic cavity. The structure of the bacterial cells changed. They had electron-dense capsules and an electron-light nucleoid.

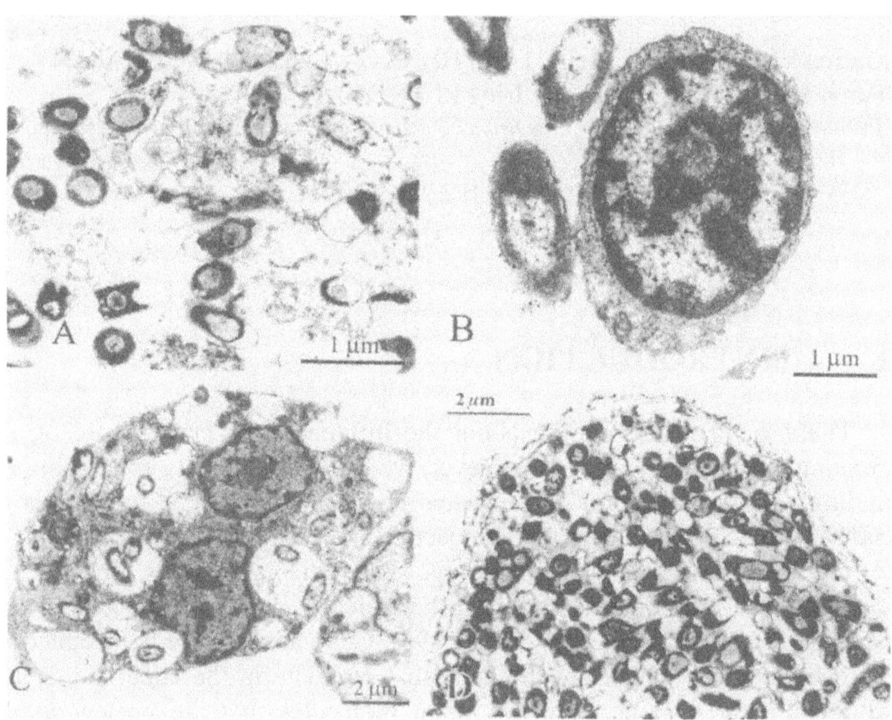

Figure 1. Interaction of *Y. pseudotuberculosis* with coelomocytes of echinoderms. A. Bacteria in the fibrillar substance inside of the coelomic cavity. B. Adhesion of bacterial cells to the amoebocytes. C. Fusion of bacteria containing amoebocytes. D. Group of bacterial cells in the coelomic cavity.

Some of the bacterial cells were free and others were enclosed by amorphous substance produced by morular cells (Figure 1A). Likely, that amorphous substance includes the MBL and ensures the agglutination of the bacterial cells. Part of bacteria was aggregated to the surface of the amoebocytes (Figure 1B) and was inside the phagosomes. After 1 day, most of the amoebocytes contained numerous bacterial cells in their vacuoles. Lysosomal enzymes damaged some bacteria, but the greater part of them had the usual structure. Sometimes, the amoebocytes formed a syncytium (Figure 1C). At the same time, there were bacteria in the coelomic cavity. Sometimes, we observed dividing *Yersinia*. Later in the experiment, we noted an increase of phagocytosis and a decrease in the number of free bacterial cells. The aggregates of coelomocytes became larger. There were large clusters of bacterial cells that were bordered by long pseudopodia of the amoebocytes in the coelomic cavity (Figure 1D). But throughout the experiments there were always free bacteria in the coelomic fluid.

The number of differentiated cells decreased as determined by calculating the different types of coelomocyte types at the end of experiment. Using the hemagglutinatin test and ELISA, we showed the changes in the quantity of MBL in the coelomic liquid of echinoderms after inoculation. These fluctuations coincided with the variations of the number of morular coelomocytes in the coelomic cavity which participated in biosynthesis of MBL. As a rule, infected animals died within a short time after introduction of bacteria. So, the components of the defence system of echinoderms were ineffective against *Y. pseudotuberculosis*.

3. CONCLUSIONS

The inoculation *Y. pseudotuberculosis* induced a defence reaction in two species of echinoderms. The main mechanisms of defence were the agglutination of bacterial cells by MBL, the phagocytosis of bacteria by the amoeboebocytes and lysis. At the same time, the bacteria suppressed the defence system of echinoderms by toxins.

REFERENCES

Barondes, S. 1988. Bifunctional properties of lectins: lectins redefined. *Trends in Biochemical Sciences*. 13: 480-483.

Chia, F.-S. and Xing, J. 1996. Echinoderm coelomocytes. *Zoological Studies*.35: 231-254.

Gross, S.P., Walid, Z.-A., Lori A.C., and Smith C.L. 1999, Echinoderm immunity and evolution of the complement system. *Dev. & Comp. Immunol.*23: 429-442.

Chapter 36

Acute and Chronic Experimental Infection Processes Caused by *Yersinia pseudotuberculosis* and Evaluation of Interleukin Action on Their Development and Outcome

Galina Y. TSENEVA[1], Ekaterina A. VOSKRESSENSKAYA[1], Andrey S. SIMBIRTSEV[2], Irina V.SHESTAKOVA[3], Tatiana E. DEMAKOVA[4] and Irina A. CHMYR[5]

1St.Petersburg Pasteur Institute; 2State Research Institute of Higly Pure Biopreparations, St.Petersburg; 3Semashko medical stomatological institute, Moscow; 4State Sanitary Epidemiological Inspection Centre in the St.Petersburg region; 5North-West Plague Station, St.Petersburg

A characteristic feature of the diseases caused by *Y. pseudotuberculosis* (*Y. ps*) is a wide diversity of the forms and variants of the disease courses. Up to 30% of cases present chronic development or relapses indicating that there is a need for new methods of treatment for this disease.

The aim of this study was to evaluate by microbiological methods the efficacy of IL-1 treatment in an experimental infection caused by *Y.ps.*

1. MATERIALS AND METHODS

The acute and chronic experimental processes were modeled using 250 white mice (weight 14-16 g). The animals were infected parentherally with *Y. ps* serotype I virulent strain N 2160 (LD_{50}, 10^6 cells/ml) by injection of 1 ml of suspension with a concentration 10^6 and 10^7 cells/ml. Duration of the experiment was 10 and 30 days for acute and chronic infections, respectively. IL-1 was applied parentherally in doses 200; 20; 2; 0,2; 0,02 ng per mouse after 30 min of infection. The mice were autopsied after 3, 6, 10, 17, 23 days, and total numbers of *Y. ps* in blood and organ (liver, spleen, intestine, mesenteric node) homogenates were quantitatively determined. Colony forming units (CFU) per 1 g of organ were analyzed for a total of

1250 samples. A 50% lethal dose (LD_{50}) and the levels of dissemination and accumulation of *Yersinia* bacteria in the organs were determined.

2. RESULTS

We assessed the differential protective action of IL-1 preparation on the course of acute and chronic infections for all the doses tested. A maximal protective effect was achieved for the dose of 2 ng/mouse. In the experiments with acute infection the concentration of *Yersinia* in the livers and spleens of IL-1 infected animals was 4-6 times lower than in the control group (Figure 1). No bacteria were found in the other organs. Lethality was 30% lower than in the control group.

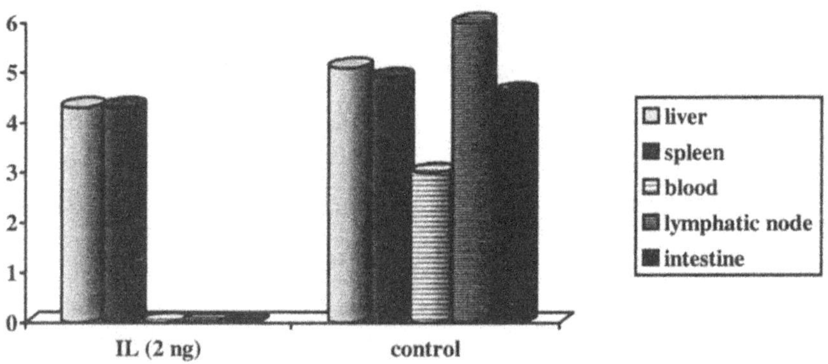

Figure 1. The concentration (log of CFU/g) of *Y. ps* (strain 2160, dose 10^6) bacteria in organs and tissues of surviving animals (acute infection).

Duration of chronic infection process was 17 days in the IL-1 protected animals versus 23 days in control animals (the duration of the experiment). The concentrations of *Y. ps* bacteria in the IL-1 treated infected mice was lower than those in the control group and no bacteria were grown from the spleens and livers. The *Yersinia* concentrations were lower in the experimental group than in the control group in all the samples recovered in the autopsy.

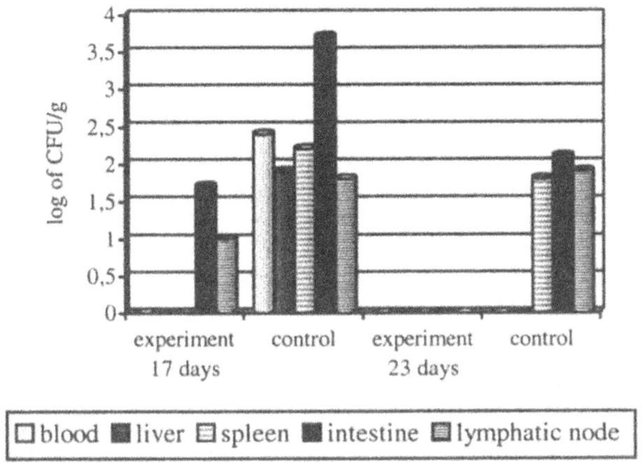

Figure 2. Protective activity of IL-1 (0.2 ng) during chronic infection.

3. CONCLUSIONS

Our data demonstrate that the administration of IL-1 results in decrease of the severity of the experimental disease course caused by *Y. ps* infection in mice. Specifically IL-1 administration decreased the number of lethal infections and also accumulation of the pathogen in organs and tissues.

Chapter 37

Role of Apoptosis of Phagocytic Cells in the Development of Immunodeficiency in Plague

Galina I. VASILIEVA, Victor N. KOZLOVSKY, Alla K. KISELEVA, Michail B. MISHANKIN and Boris N. MISHANKIN
Research Institute for Plague Control, Department of Immunology, Rostov-on-Don, Russia

1. INTRODUCTION

Apoptosis represents the physiological process of active programmed cell death. This process is subject to regulation and it is genetically encoded. At present, apoptosis is considered to have similar fundamental importance as mitosis has on the biological processes. The basic difference between necrosis and apoptosis is that the latter is a "conscious" choice of destruction of one cell to rescue the other cellular population (Kerr *et al.,* 1972). At present, there is enough data to confirm this concept and to indicate a key role of apoptosis in the destiny of all nuclear cells of a multicellular organisms (Nagata, 1997). Violation of the normal balance between the cell proliferation and apoptotic cell death plays an essential role in the formation of different immunopathological processes in humans including secondary immunodeficiencies, autoimmune and oncological diseases, chronic inflammatory processes, viral and other infections.

Taking into account that the apoptotic destruction of phagocytes plays an essential role in formation of immunopathological reactions caused by some superantigens (Moulding *et al.,* 1999), we have suggested that damage of macrophages and neutrophils under the action of "murine" toxin (MT), which belongs to superantigens, occurs as a result of this process (Vasilieva *et al.,* 2000). The objective of this research was to analyse the ability of MT to induce the apoptotic program in phagocytes.

The Genus Yersinia, Edited by Skurnik *et al.*
·uwer Academic/Plenum Publishers, New York, 2003

2. MATERIALS AND METHODS

The ability of MT to cause apoptotic cell death was evaluated with the help of several methods including histological preparations and analysis of DNA degradation.

In histological studies apoptotic cells are identified based on the presence of the following apoptosis-specific morphological changes: cell shrinkage, membrane blebbing, presence of condensed unstructured chromatin and membrane-bound fragments known as apoptotic bodies. Acridine orange stains normal cellular DNA green and RNA present in the cell nucleolus, red. Apoptotic cells were distinguished from normal ones based on the presence of green color in the cytoplasm and by the presence of green apoptotic bodies in the cytoplasm that have escaped from the fragmented nucleus. These methods gave comparative results in the detection of apoptotic cells. Degradation of DNA in apoptosis was determined with a colorimetric method with diphenylamine and with polyacrylamide gel electrophoresis. Also these two methods produced comparative results.

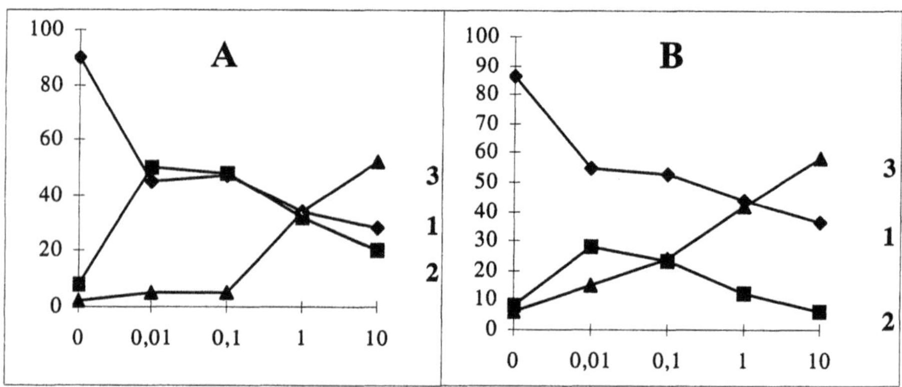

Figure 1. Apoptogenic effect of "murine" toxin (MT) on mouse macrophages (A) and neutrophils (B) (stained with acridine orange) X-axis – dose of MT, mkg/ml; Y-axis – percentage of normal cells (1), cells with morphological features of apoptosis (2) and necrosis (3).

3. RESULTS AND DISCUSSION

Experiments performed with the above mentioned techniques demonstrated the ability of MT to cause apoptosis in guinea pig and mouse macrophages and neutrophils. It is necessary to note that decrease in the working concentration of MT resulted in increase in the percentage of phagocytes with apoptotic changes. On the contrary, increase in the MT concentration started to cause the necrotic type of the cell destruction (Figure 1).

Our observations confirm those of others (Cynthia *et al.,* 1998) that a substance can cause either apoptosis or necrosis depending on the diverse circumstances. In this respect the superantigenic properties of MT are similar to those of other superantigens when MT is used at a very low concentration. At higher concentrations the apoptotic effect is masked by the cytotoxicity of "murine" toxin. Based on our results we conclude that the death of immunocompetent cells by apoptosis was caused by the superantigenic properties of MT.

The features of morphological alterations of macrophages and neutrophils characteristic for apoptosis were confirmed with the colorimetric method for the estimation of DNA degradation using the smallest doses of MT (0.01 – 0.1 mkg/ml). These results correlated with the data obtained with the polyacrylamide gel electrophoresis and morphological methods. Thus, our results demonstrate that the mechanism for the active programmed cell death is part of the general cytotoxic effect of MT.

REFERENCES

Cynthia A., Grain T., Streetz K., Trautwein C., Brenner D.A., and Lemasters J.J., 1998, The mitochondrial permeability transition is required for tumor necrosis factor alpha-mediated apoptosis and cytochrome C release. *Mol. Cell. Biol.* **18**:6353-6364.

Kerr I.F., Wyllie A.H., and Curric A.R., 1972, Apoptosis: a basic biological phenomenon with wide ranging implication in tissue kinetics. *Brit. J. Cancer.* **26**:239-257.

Moulding DA., Walter C., Hart C.A., and Edwards S.W., 1999, Effects of staphylococcal enterotoxins on human neutrophil functions and apoptosis. *Infect. Immun.* **67**: 2312-2318.

Nagata S., 1997, Apoptosis by death factor. *Cell.* **88**:355-365.

Vasilieva G.I., Mishankin M.B., Kiseleva A.K., Kozlovsky V.N., and Mishankin B.N., 2000, Interaction of recombinant *Yersinia pestis* "murine" toxin with cell lines and poly- and mononuclear phagocytes of experimental animals and man. *Biotechnology.* **4**:28-34.

PART III

LIPOPOLYSACCHARIDE, BACTERIOPHAGES AND CELL WALL FUNCTION

Picture 17. Joanna Radziejewska-Lebrecht entering the Castle of Turku for the Symposium dinner

Chapter 38

Molecular Genetics, Biochemistry and Biological Role of *Yersinia* Lipopolysaccharide

Mikael SKURNIK
Department of Bacteriology and Immunology, the Haartman Institute, University of Helsinki and Department of Medical Biochemistry and Molecular Biology, University of Turku, Finland

1. SUMMARY

Lipopolysaccharide (LPS) is the major component of the outer leaflet of the outer membrane of Gram-negative bacteria. The LPS molecule is composed of two biosynthetic entities: the lipid A – core and the O-polysaccharide (O-antigen). Most biological effects of LPS are due to the lipid A part, however, there is an increasing body of evidence also with *Yersinia* indicating that O-antigen plays an important role in effective colonization of host tissues, resistance to complement-mediated killing and in the resistance to cationic antimicrobial peptides that are key elements of the innate immune system.

The biosynthesis of O-antigen requires numerous enzymatic activities and includes the biosynthesis of individual NDP-activated precursor sugars in the cytoplasm, linkage and sugar-specific transferases, O-unit flippase, O-antigen polymerase and O-chain length determinant. Based on this enzymatic mode of O-antigen biosynthesis LPS isolated from bacteria is a heterologous population of molecules; some do not carry any O-antigen while others that do have variation in the O-antigen chain lengths.

The genes required for the O-antigen biosynthesis are located in O-antigen gene clusters that in genus *Yersinia* is located between the *hemH* and *gsk* genes.

Temperature regulates the O-antigen expression in *Y. enterocolitica* and *Y. pseudotuberculosis*; bacteria grown at room temperature (RT, 22-25°C)

The Genus Yersinia, Edited by Skurnik *et al.*
Kluwer Academic/Plenum Publishers, New York, 2003

produce in abundance O-antigen while only trace amounts are present in bacteria grown at 37°C. Even though the amount of O-antigen is known to fluctuate under different growth conditions in many bacteria very little detailed information is available on the control of the O-antigen biosynthetic machinery.

2. INTRODUCTION

Genus *Yersinia* contains 11 recognized species of which *Y. pestis* does not have any O-serotypes, *Y. pseudotuberculosis* has 21 O-serotypes, *Y. enterocolitica* and closely related species have more than 70 O-serotypes and the fish pathogen *Y. ruckeri* has 5 O-serotypes. O-serotypes are specified by the antigenically variable LPS O side chains or O-antigens (O-ag). *Y. pestis* is long known to express a rough type LPS that lacks the O-ag.

2.1 'Evolution' of LPS studies

LPS research in my laboratory started by chance due to availability of a monoclonal antibody specific for the *Y. enterocolitica* serotype O:3 (*Ye* O:3) O-ag (Pekkola-Heino *et al.,* 1987). That monoclonal antibody was used as a screening tool when cloning the *Ye* O:3 O-ag gene cluster (Al-Hendy *et al.,* 1991). The gene cluster was later completely sequenced and characterized (Zhang *et al.,* 1993). Next we started to study the temperature regulation of the *Ye* O:3 O-ag and by using a bacteriophage as a selection tool found out that the phage receptor is the outer core (OC) hexasaccharide of LPS, and we cloned and characterized the gene cluster directing the OC biosynthesis of *Ye* O:3 (Skurnik *et al.,* 1999; Skurnik *et al.,* 1995). We expanded the studies to *Ye* serotype O:8 (Bengoechea *et al.,* 2002; Bengoechea *et al.,* 2002; Zhang *et al.,* 1997; Zhang *et al.,* 1996; Zhang *et al.,* 1995), to serotype O:9 (Lübeck *et al.,* 2003) and to *Y. pseudotuberculosis* and *Y. pestis* (Skurnik *et al.,* 2000). We have also isolated and characterized some bacteriophages that use LPS as a receptor (Pajunen *et al.,* 2000; Pajunen *et al.,* 2002).

2.2 LPS structure and biosynthesis

LPS that bacteria carry on their outer membrane is a heterologous population of molecules. This is clearly seen when isolated LPS is analysed in polyacrylamide gel electrophoresis; a wide range of LPS molecules of different sizes are present. The smallest LPS molecules are composed of lipid A and (inner) core moieties, a little larger molecules have lipid A and complete core structure and the largest ones contain lipid A, complete core and variable numbers of O-ag repeat units. Therefore these LPS populations appear as a ladder in the stained gel. If the repeating unit is a single sugar, the O-ag is called homopolymeric, if it is composed of 2 or more different sugar residues, the O-ag is heteropolymeric. Biosynthesis of LPS takes place as two biosynthetic entities:

(i) lipid A is synthesized on the cytoplasmic leaflet of the inner membrane, the core sugar residues are sequentially transferred on it by specific glycosyltransferases, and the completed lipid A core is somehow translocated to face the periplasmic side of the inner membrane, and

(ii) the heteropolymeric O-units are synthetized onto a lipid carrier molecule, undecaprenyl phosphate (Und-P), a complete O-unit is flipped to the periplasmic face of the inner membrane where Wzy (the O-ag polymerase) and Wzz (the O-ag chain length determinator) act in concert to produce O-side chains with the desired number of repeat units (between 7 and 11 in the case of *Ye* O:8). The homopolymeric polysaccharide chain synthetized onto Und-P is completed to full length in the cytoplasm after which it is translocated to the periplasmic space by Wzm and Wzt, an ATP binding cassette transporter system.

Thus formed O-side chain is transferred from the Und-P by WaaL (the O-ag ligase) to the preformed lipid A core molecule. The complete LPS molecule is then translocated onto the outer membrane by a still unknown mechanism.

3. LPS GENETICS IN *Yersiniae* AND EVOLUTION OF *Y. pestis*

Nucleotide sequences of the O-ag and the OC gene clusters indicated at that point that a locus between the *hemH* and *gsk* genes is universally used for the O-ag and OC gene clusters in genus *Yersinia.* We generated primers specific for the *hemH* (forward primer) and the *gsk* genes (reverse primer) and used them to amplify the fragment between the *hemH* and *gsk* genes from various *Yersinia* strains. *Ye* O:3, O:5,27, O:9 and some other serotypes produced a ca. 13 kb fragment, *Ye* O:8, a 20 kb fragment as did all the studied *Y. pseudotuberculosis* serotypes and, surprisingly, also *Y. pestis.* The latter result indicated that *Y. pestis* carries also the O-ag gene cluster even though it does not express any O-ag. At that time (1996) the genome of *Y. pestis* had not been sequenced, however, it was known that many inactive genes are present in the *Y. pestis* genome. One of the first identified pseudogenes was the *yadA* gene that carries a one bp deletion (Rosqvist *et al.,* 1988; Skurnik and Wolf-Watz, 1989). Also the presence of a large number of IS elements in the *Y. pestis* genome was known thus making it very likely that the O-ag gene cluster of *Y. pestis* would also be inactivated. This and the fact that *Y. pestis* is very closely related to *Y. pseudotuberculosis* prompted us to compare the *hemH* - *gsk* intergenic regions of *Y. pestis* and of all known *Y. pseudotuberculosis* serotypes (Skurnik *et al.*, 2000). We showed that the *Y. pestis* O-ag gene cluster is almost 100% identical to that of *Y. pseudotuberculosis* O:1b and that five of the 17 genes of *Y. pestis* were inactivated either by frame shift mutations (4 genes) or a small deletion of 62 bp (one gene) (Figure 1, top).

3.1 Genetic setup of O-antigen gene clusters of *Y. pseudotuberculosis*

We have recently used PCR and sequencing to study the genetic set up of the O-ag gene clusters of different *Y. pseudotuberculosis* serotypes (Figure 1).

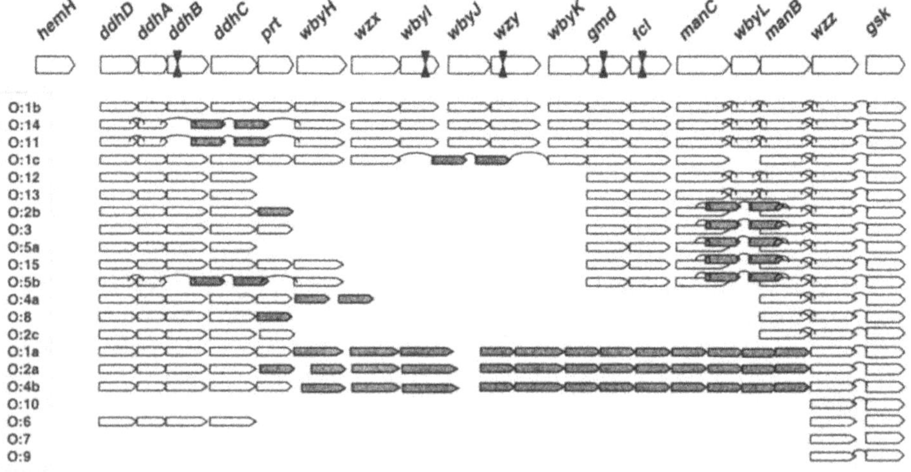

Figure 1. Overview of genetic composition of O-antigen gene clusters of *Yersinia pseudotuberculosis* serogroups O:1 - O:15. On top, the O-antigen gene cluster of *Y. pestis* EV76 along with the inactivated genes indicated by inverted arrowheads. The genetic setup of the gene clusters of different *Y. pseudotuberculosis* serotypes is presented below. Open arrows indicate that a positive gene specific PCR result was obtained, grey arrows indicate that nonhomologous genes are present in these locations in these serotypes. The grey arrow data for serotypes O:1a, O:2a, O:4a and O:4b is from published sequences, and for serotypes O:1c, O:2b, O:3, O:5a O:5b, O:11, O:14, and O:15 from our unpublished results. Empty areas indicate that the genetic setup has not been studied (however, see Reeves *et al.*, 2003).

The results indicate that the gene clusters carry biosynthetic modules and that the O-ag variability is based on the module make up. Sequencing revealed traces of very active horizontal gene transfer that most likely has involved homologous recombination events and gene transfer by bacteriophages and/or conjugative plasmids carrying IS-elements. For a more detailed treatment of this subject, see (Reeves *et al.*, 2003).

4. GENETICS AND BIOSYNTHESIS OF *Ye* O:8 O-ANTIGEN

The O-antigen of *Y. enterocolitica* O:8 (*Ye* O:8) is composed of a polysaccharide having pentasaccharide repeat units (Figure 2). The O-antigen gene cluster contains 18 genes which are sufficient to express the O-antigen in a heterologous host like *E. coli*.

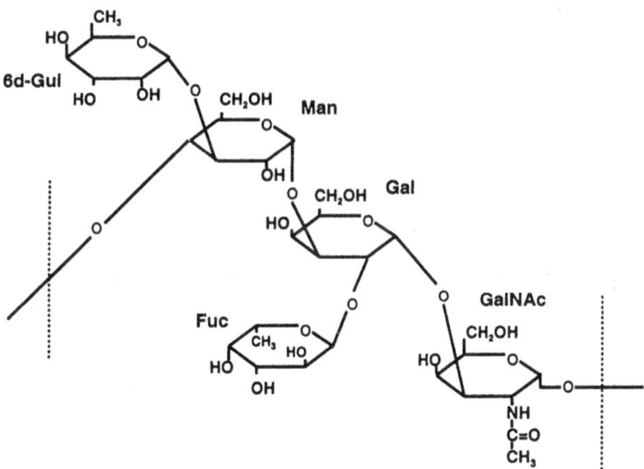

Figure 2. Y. enterocolitica O:8 O-antigen structure. 6d-Gul, 6-deoxy-gulose; Man, mannose; Fuc, fucose; Gal, galactose; GalNAc, N-acetylgalactosamine.

The NDP-activated sugar precursors needed for the *Ye* O:8 O-unit biosynthesis are:

1. **CDP-6-deoxy-gulose,** synthetized enzymatically from glucose-1-phosphate by the action of DdhA, DdhB, WbcA, WbcB and WbcC.
2. **GDP-mannose,** synthetized from fructose-1-phosphate by ManA, ManB and ManC.
3. **GDP-fucose,** synthetized from GDP-mannose by Gmd and Fcl.

4. **UDP-N-acetylgalactosamine,** synthetized from UDP-N-acetylglucosamine by Gne.
5. **UDP-galactose,** synthetized from UDP-glucose by GalE.

The glucosyltransferases needed to assemble the O-unit on Und-P are WecA, WbcF, WbcG, WbcH and WbcI. The assembled O-unit is translocated by Wzx to face the periplasmic space, the flippase, where Wzy and Wzz build up the heteropolymeric O-antigen. The O-antigen is finally ligated to the lipid A core oligosaccharide and the completed LPS molecule is then transported to the outer membrane.

All the genes except *manA, galE* and *wecA* are located in the O-antigen gene cluster. These three other genes are involved also in other cellular processes and their enzymatic activity is also used in the O-antigen biosynthesis.

4.1 Properly expressed *Ye* O:8 O-antigen is necessary for virulence

We have earlier shown that *Ye* O:8 strain missing completely the O-antigen (rough strain) is about 100-fold attenuated in orally infected mice (Zhang *et al.,* 1997). We engineered mutants that were missing either the Wzy, the O-antigen polymerase, or the Wzz, the O-antigen chain length determinant, activities. The *wzy* strain expresses LPS with a single O-unit ligated to lipidA core oligosaccharide while the *wzz* strain expresses O-antigens with a random distribution of the chain lengths (Bengoechea *et al.,* 2002). Virulence potential of the wild type, rough, *wzy* and *wzz* strains was tested in intragastrically infected rabbits (Najdenski *et al.,* 2003). Bacterial loads in different tissues and organs were analysed and the results clearly indicated that compared to the wild type strain the three mutants were attenuated in virulence. Very surprising to us was the finding that the most attenuated strain was the *wzz* strain.

At present, we can only speculate of the specific role the O-antigen and its properly regulated expression plays in virulence. The *in vitro* experiments have indicated that O-antigen plays a role in the resistance of bacteria to cationic antimicrobial peptides and to complement. In addition, especially when considering the results obtained with the *wzz* mutant, the proper expression of O-antigen chain lengths may be synchronized with the expression of other virulence factors such as invasin and the Yop-translocating type III secretion system. Clearly, further studies are needed to address these questions.

5. THE LPS OF *Y. enterocolitica* O:3

The structure of the LPS of *Ye* O:3 has some peculiarities rarely seen in other enterobacteria. Its O-antigen is a homopolymer of 1-2 linked 6-deoxy-L-altrose that is attached to the inner core region of the LPS. In addition to O-antigen also the outer core hexasaccharide (OC) is attached to the inner core thus forming a short branch in the LPS molecule. This peculiar structure has allowed us to construct mutants that are missing either the O-antigen, the OC or both. We have demonstrated that mutants lacking either the O-antigen or the OC are less virulent in mouse experiments and that the O-antigen appears to be required in the colonization of the Peyer's patches while the OC, in prolonged survival of the bacteria in Peyer's patches and in invasion to deeper tissues like liver and spleen (Skurnik *et al.*, 1999). It has been long known that the expression of O-antigen is regulated by temperature so that at 37°C less O-antigen and shorter chain lengths are expressed than at room temperature (RT) (Acker *et al.*, 1981; Al-Hendy *et al.*, 1991; Kawaoka *et al.*, 1983; Wartenberg *et al.*, 1983).

5.1 Serum resistance of *Ye* O:3

We have long been interested in the mechanisms that contribute to the survival of *Ye* O:3 in serum. Several factors have been implicated in the resistance including YadA, Ail and O-antigen (Balligand *et al.*, 1985; Pierson and Falkow, 1993; Pilz *et al.*, 1992; Wachter and Brade, 1989), in addition we showed that the OC is perhaps indirectly involved (Skurnik *et al.*, 1999).

In order to study the roles of these four factors in serum resistance we have constructed a set of 24 strains that express all the possible combinations of these 4 factors. The survival of these strains grown either at RT or at 37°C was tested in 66.7% normal and EGTA-Mg treated human sera, to assess the classical (CP) and the alternative (AP) complement activation pathway mediated killing, respectively. Our results (Biedzka, Venho and Skurnik, unpublished results) demonstrated that based on the different combinations of the 4 factors the strains could be divided into several complement (serum) resistance phenotypes (Figure 3).

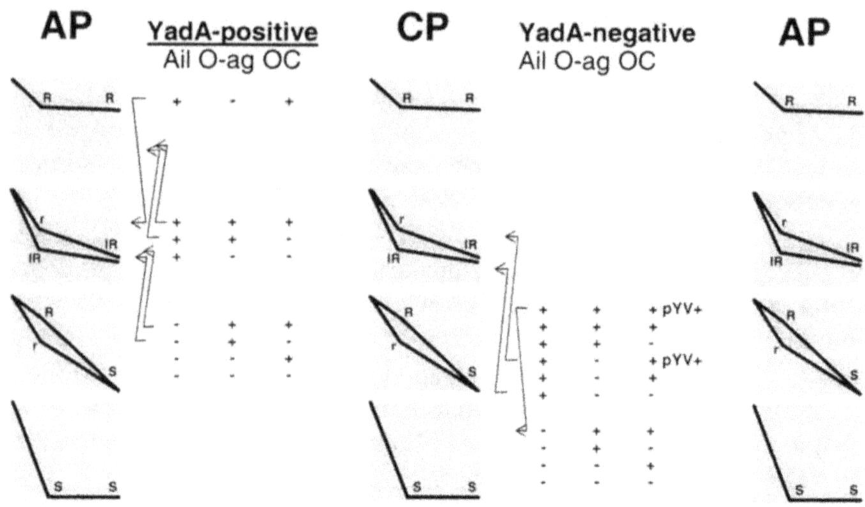

Figure 3. Serum resistance phenotypes (the curves illustrate surviving bacteria after 30 and 120 min incubation in sera) of *Ye* O:3 strains expressing different combinations of YadA, Ail, O-ag and OC (horizontal plus and minus signs, for some combinations presence of pYV is indicated). Different serum resistance phenotypes in AP and CP are indicated by kinked arrows.

In general, YadA appeared to be the most important single serum resistance factor, although all the Ail-negative YadA-positive strains also were killed efficiently by CP. The O-antigen or the OC appeared to play a minor role. With a few strains the classical and the alternative pathway-mediated killing was different. The most resistant strain in CP was O-antigen negative, however, in the AP killing this strain was more susceptible to killing. O-antigen appears to be important in the AP resistance.

In YadA-negative strains Ail was more potent in strains lacking either the O-antigen or the outer core or both. This suggests that Ail function requires direct interaction with some complement component(s) to prevent complement activation and its killing potential and that in these strains Ail would be properly exposed to carry out its function. Further work is required to elucidate the detailed roles of the serum resistance factors .

5.2 Temperature-regulation of O-antigen expression

An old dogma with *Ye* O3 is that it expresses less O-antigen when grown at 37C. This has been demonstrated by bacteriophage susceptibility tests (Kawaoka *et al.,* 1983), exposure of antigenic determinants to antibodies (Acker *et al.,* 1981), and by chemical analysis of isolated LPS (Wartenberg *et al.,* 1983). We have also used Northern blotting to verify this (Al-Hendy *et al.,* 1991). All the time, however, there has been a controversy between the *in vitro* analyses and the *in vivo* data, where patients with *Ye* O:3 infection have always demonstrated very high levels of specific antibodies to O-antigen in their sera. This has indirectly indicated that *Ye* O3 bacteria do express the O-antigen *in vivo* during infection. We have constructed reporter strains to study more closely the regulation of O-antigen expression (see Lahtinen *et al.,* 2003). Our main finding from these reporter studies is that during the exponential phase of growth the O-antigen expression is not under temperature regulation whereas immediately when the bacteria enter into the stationary phase of growth the expression becomes temperature-regulated. This finding explains nicely the *in vivo* data discussed above indicating that O-antigen is expressed during infection in bacteria that are actively growing.

6. CONCLUSIONS

Our work has elucidated the genetics and biochemistry of the *Yersinia* LPS, also we have strong indications of the central role LPS and its components play in pathogenesis. More work is still needed to completely understand the biological role of LPS in *Yersiniae.* We also need to take into consideration the fact that LPS may play different roles in different species, therefore, we cannot transfer the data obtained with *Salmonella* to *Yersinia,* and we need to be cautious even when different serotype of the same species are discussed. In future, it will be interesting to uncover the genetics of the regulation of LPS and also the roles LPS components have in the bacterium-host interactions.

ACKNOWLEDGEMENTS

I want to thank all the former and present members of my laboratory who have contributed to the LPS work: Ayman Al-Hendy, José Antonio Bengoechea, Marta Biedzka, Agniezka Brzezinska, Tatiana Bogdanovich, Pia Lahtinen, Anne Peippo, Elise Pinta, Dassanayake Sirisena, Reija Venho and Lijuan Zhang. I am also very grateful for the contributions made by the visiting scientists Clemens Oertelt and Peter Stephensen Lübeck, and my international collaborators Elisabeth Carniel, Hiroshi Fukushima, Otto Holst, Jeffrey Hoorfar, Hubert Meyer, Hristo Najdenski, Joanna Radziejewska-Lebrecht and Richard Titball.

REFERENCES

Acker, G., Knapp, W., Wartenberg, K., and Mayer, H. (1981) Localization of enterobacterial common antigen in *Yersinia enterocolitica* by the immunoferritin technique. *J Bacteriol* 147: 602-611.

Al-Hendy, A., Toivanen, P., and Skurnik, M. (1991) The effect of growth temperature on the biosynthesis of *Yersinia enterocolitica* O:3 lipopolysaccharide: temperature regulates the transcription of the *rfb* but not of the *rfa* region. *Microb Pathog* 10: 81-86.

Al-Hendy, A., Toivanen, P., and Skurnik, M. (1991) Expression cloning of *Yersinia enterocolitica* O:3 *rfb* gene cluster in *Escherichia coli* K12. *Microb Pathog* 10: 47-59.

Balligand, G., Laroche, Y., and Cornelis, G. (1985) Genetic analysis of a virulence plasmid from a serogroup 9 *Yersinia enterocolitica* strain: role of outer membrane protein P1 in resistance to human serum and autoagglutination. *Infect Immun* 48: 782-786.

Bengoechea, J. A., Pinta, E., Salminen, T., Oertelt, C., Holst, O., Radziejewska-Lebrecht, J., Piotrowska-Seget, Z., Venho, R., and Skurnik, M. (2002) Functional characterization of Gne (UDP-N-acetylglucosamine-4- epimerase), Wzz (chain length determinant), and Wzy (O-antigen polymerase) of *Yersinia enterocolitica* serotype O:8. *J Bacteriol* 184: 4277-4287.

Bengoechea, J. A., Zhang, L., Toivanen, P., and Skurnik, M. (2002) Regulatory network of lipopolysaccharide O-antigen biosynthesis in *Yersinia enterocolitica* includes cell envelope-dependent signals. *Mol Microbiol* 44: 1045-1062.

Kawaoka, Y., Otsuki, K., and Tsubokura, M. (1983) Growth temperature-dependent variation in the bacteriophage-inactivating capacity and antigenicity of *Yersinia enterocolitica* lipopolysaccharide. *J Gen Microbiol* 129: 2739-2747.

Kawaoka, Y., Otsuki, K., and Tsubokura, M. (1983) Serological evidence that *Yersinia enterocolitica* lipopolysaccharide produced during growth *in vivo* resembles that produced during growth *in vitro* at 25°C. *J Gen Microbiol* 129: 2749-2751.

Lahtinen, P., Brzezinska, A., and Skurnik, M. (2003) Temperature and growth phase regulate the transcription of the O-antigen gene cluster of *Yersinia enterocolitica* O:3. In *The Genus Yersinia: entering the functional genomic era*, M. Skurnik, K. Granfors and J. A. Bengoechea, eds.: Kluwer Academic/Plenum Publishers, pp. 289-292.

Lübeck, P. S., Hoorfar, J., Ahrens, P., and Skurnik, M. (2003) Cloning and characterization of the *Yersinia enterocolitica* serotype O:9 lipopolysaccharide O-antigen gene cluster. In *The Genus Yersinia: entering the functional genomic era*, M. Skurnik, K. Granfors and J. A. Bengoechea, eds.: Kluwer Academic/Plenum Publishers, pp. 207-209.

Najdenski, H., Golkocheva, E., Vesselinova, A., Bengoechea, J. A., and Skurnik, M. (2003) Proper expression of the O-antigen of lipopolysaccharide is essential for the virulence of *Yersinia enterocolitica* O:8 in experimental oral infection of rabbits. *FEMS Immunol Med Microbiol*: In press.

Pajunen, M., Kiljunen, S., and Skurnik, M. (2000) Bacteriophage φYeO3-12, specific for *Yersinia enterocolitica* serotype O:3, is related to coliphages T3 and T7. *J Bacteriol* 182: 5114-5120.

Pajunen, M. I., Elizondo, M. R., Skurnik, M., Kieleczawa, J., and Molineux, I. J. (2002) Complete nucleotide sequence and likely recombinatorial origin of bacteriophage T3. *J Mol Biol* 319: 1115-1132.

Pekkola-Heino, K., Viljanen, M. K., Ståhlberg, T. H., Granfors, K., and Toivanen, A. (1987) Monoclonal antibodies reacting selectively with core and O-polysaccharide of *Yersinia enterocolitica* O:3 lipopolysaccharide. *APMIS* 95: 27-34.

Pierson, D. E., and Falkow, S. (1993) The *ail* gene of *Yersinia enterocolitica* has a role in the ability of the organism to survive serum killing. *Infect Immun* 61: 1846-1852.

Pilz, D., Vocke, T., Heesemann, J., and Brade, V. (1992) Mechanism of YadA-mediated serum resistance of *Yersinia enterocolitica* serotype O3. *Infect Immun* 60: 189-195.

Reeves, P. R., Pacinelli, E., and Wang, L. (2003) O antigen Gene Clusters of *Yersinia pseudotuberculosis*. In *The Genus Yersinia: entering the functional genomic era*, M. Skurnik, K. Granfors and J. A. Bengoechea, eds.: Kluwer Academic/Plenum Publishers, pp. 199-206.

Rosqvist, R., Skurnik, M., and Wolf-Watz, H. (1988) Increased virulence of *Yersinia pseudotuberculosis* by two independent mutations. *Nature* 334: 522-525.

Skurnik, M., Peippo, A., and Ervelä, E. (2000) Characterization of the O-antigen gene clusters of *Yersinia pseudotuberculosis* and the cryptic O-antigen gene cluster of *Yersinia pestis* shows that the plague bacillus is most closely related to and has evolved from *Y. pseudotuberculosis* serotype O:1b. *Mol Microbiol* 37: 316-330.

Skurnik, M., Venho, R., Bengoechea, J.-A., and Moriyón, I. (1999) The lipopolysaccharide outer core of *Yersinia enterocolitica* serotype O:3 is required for virulence and plays a role in outer membrane integrity. *Mol Microbiol* 31: 1443-1462.

Skurnik, M., Venho, R., Toivanen, P., and Al-Hendy, A. (1995) A novel locus of *Yersinia enterocolitica* serotype O:3 involved in lipopolysaccharide outer core biosynthesis. *Mol Microbiol* 17: 575-594.

Skurnik, M., and Wolf-Watz, H. (1989) Analysis of the *yopA* gene encoding the Yop1 virulence determinants of *Yersinia* spp. *Mol Microbiol* 3: 517-529.

Wachter, E., and Brade, V. (1989) Influence of surface modulations by enzymes and monoclonal antibodies on alternative complement pathway activation by *Yersinia enterocolitica. Infect Immun* 57: 1984-1989.

Wartenberg, K., Knapp, W., Ahamed, N. M., Widemann, C., and Mayer, H. (1983) Temperature-dependent changes in the sugar and fatty acid composition of lipopolysaccharide from *Yersinia enterocolitica* strains. *Zentralblatt fur Bacteriologie - International Journal of Medical Microbiology* 253: 523-530.

Zhang, L., Al-Hendy, A., Toivanen, P., and Skurnik, M. (1993) Genetic organization and sequence of the *rfb* gene cluster of *Yersinia enterocolitica* serotype O:3: Similarities to the dTDP-L-rhamnose biosynthesis pathway of *Salmonella* and to the bacterial polysaccharide transport systems. *Mol Microbiol* 9: 309-321.

Zhang, L., Radziejewska-Lebrecht, J., Krajewska-Pietrasik, D., Toivanen, P., and Skurnik, M. (1997) Molecular and chemical characterization of the lipopolysaccharide O-antigen and its role in the virulence of *Yersinia enterocolitica* serotype O:8. *Mol Microbiol* 23: 63-76.

Zhang, L., Toivanen, P., and Skurnik, M. (1996) The gene cluster directing O-antigen biosynthesis in *Yersinia enterocolitica* serotype O:8: Identification of the genes for mannose and galactose biosynthesis and the gene for the O-antigen polymerase. *Microbiology* 142: 277-288.

Zhang, L. J., Toivanen, P., and Skurnik, M. (1995) Genetic characterization of a novel locus of *Yersinia enterocolitica* serotype O:8 for down-regulation of the lipopolysaccharide O side chain at 37 degrees C. In *Yersiniosis: Present and Future*, G. Ravagnan and C. Chiesa, eds. Basel, Switzerland: Karger, pp. 310-313.

Chapter 39

O Antigen Gene Clusters of *Yersinia pseudotuberculosis*

Peter R. REEVES, Elvia PACINELLI and Lei WANG[1]
School of Molecular and Microbial Biosciences, University of Sydney, NSW2006, Australia;
[1]Current address: College of Life Science, Nankai University, Tianjin 300071, China.

Yersinia pseudotuberculosis is typical member of *Enterobacteriaceae* in having a range of O antigen forms, in this case 21, with structures reported for 12 of them (Figure 1). Some of the structures were determined many years ago and those reanalysed have generally been refined. The structures in Figure 1 include additional refinements based on the genetic data discussed below, but these need to be confirmed or rejected by structural analysis. As for other species the genes required specifically for synthesis of the O repeat unit are clustered, in this case of *Yersinia* between the *hemH* and *gsk* genes (see Skurnik, 2003). We now have O antigen gene cluster sequences for 11 of the reported structures (Figure 2), and the relationship between the genes and chemical structure is the focus of this paper.

The reader is referred to reviews for general principles of O antigen synthesis and assembly (Reeves, 1994; Whitfield, 1995), but see also the BPGD website referred to in Figure 3 for current gene names. The *Y. pseudotuberculosis* structures and sequences fall into reasonably well defined groups (Figures 1 and 2). Some of the genes in the O antigen gene clusters are present in more than one group and we look first at these. Each gene cluster had *wzx*, *wzy* and *wzz* genes for O-unit processing, showing that all fall into the same class for synthesis. Nine of the O antigens contain a dideoxyhexose (DDH) residue with CDP-sugar precursor (Figure 3). The *ddhA, ddhB, ddhC,* and *ddhD* genes for synthesis of intermediate CDP-4-keto 3,6 dideoxyglucose were at the 5′ end in all cases. The *abe*, *prt*, *tyv*, *ascE* and *ascF* genes for later steps (Figure 3) were located immediately downstream of the *ddh* genes. The 4 *ddh* genes show little variation. The 5 *prt* genes in the paratose and tyvelose containing strains are also very similar as are the 2 functional *tyv* genes and 3 *abe* genes. Pairwise comparisons all show over 97% identity and many are over 99%. We now look at the relationships of the genes and chemical structure for each group.

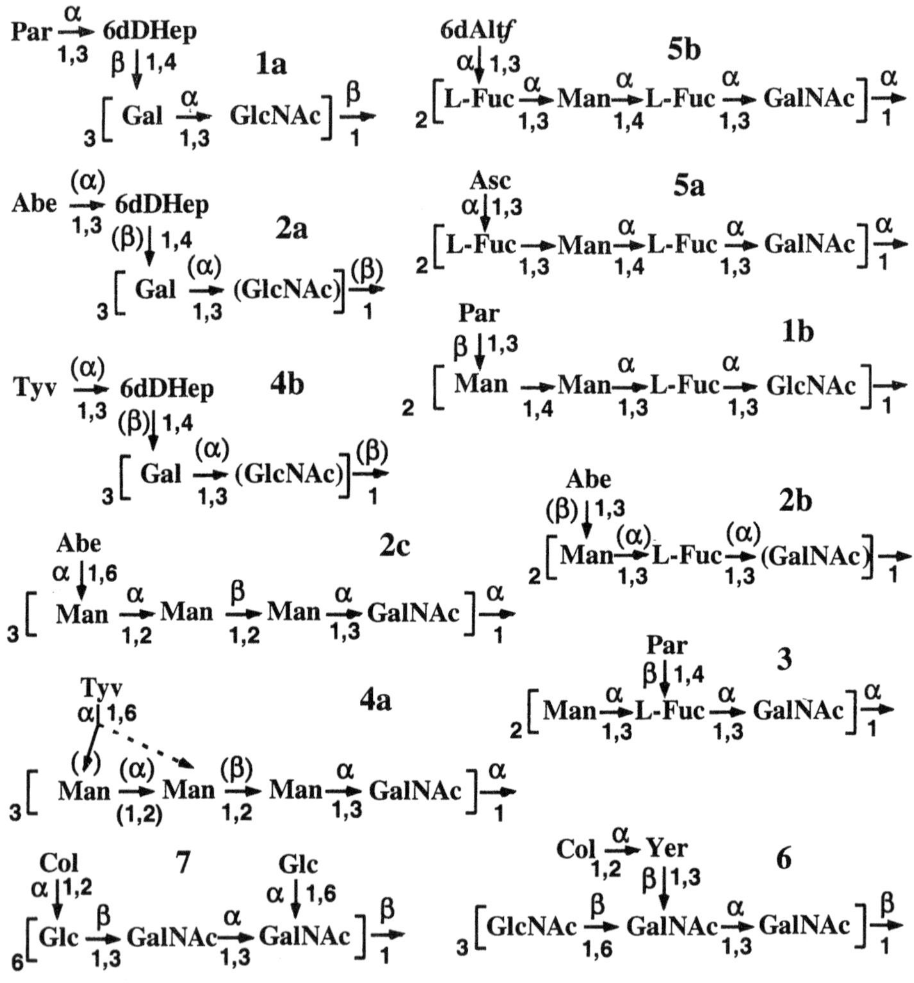

1a
Par →(α, 1,3) 6dDHep
β ↓ 1,4
3[Gal →(α, 1,3) GlcNAc] →(β, 1)

2a
Abe →((α), 1,3) 6dDHep
(β) ↓ 1,4
3[Gal →((α), 1,3) (GlcNAc)] →((β), 1)

4b
Tyv →((α), 1,3) 6dDHep
(β) ↓ 1,4
3[Gal →((α), 1,3) (GlcNAc)] →((β), 1)

2c
Abe
α ↓ 1,6
3[Man →(α, 1,2) Man →(β, 1,2) Man →(α, 1,3) GalNAc] →(α, 1)

4a
Tyv
α ↓ 1,6 (β)
3[Man →((α), (1,2)) Man →((β), 1,2) Man →(α, 1,3) GalNAc] →(α, 1)

7
Col α ↓ 1,2 ; Glc α ↓ 1,6
6[Glc →(β, 1,3) GalNAc →(α, 1,3) GalNAc] →(β, 1)

5b
6dAlt*f*
α ↓ 1,3
2[L-Fuc →(α, 1,3) Man →(α, 1,4) L-Fuc →(α, 1,3) GalNAc] →(α, 1)

5a
Asc
α ↓ 1,3
2[L-Fuc →(1,3) Man →(α, 1,4) L-Fuc →(α, 1,3) GalNAc] →(α, 1)

1b
Par
β ↓ 1,3
2[Man →(1,4) Man →(α, 1,3) L-Fuc →(α, 1,3) GlcNAc] →(1)

2b
Abe
(β) ↓ 1,3
2[Man →((α), 1,3) L-Fuc →((α), 1,3) (GalNAc)] →(1)

3
Par
β ↓ 1,4
2[Man →(α, 1,3) L-Fuc →(α, 1,3) GalNAc] →(α, 1)

6
Col →(α, 1,2) Yer
β ↓ 1,3
3[GlcNAc →(β, 1,6) GalNAc →(α, 1,3) GalNAc] →(β, 1)

Figure 1. Structures of *Y. pseudotuberculosis* O antigens. Abe, abequose; 6dAlt*f*, 6-deoxyaltrofuranose; Asc, ascarylose; Col, colitose; Fuc, fucose; Gal, galactose; GalNAc, N-acetylgalactosamine; Glc, glucose; GlcNAc, N-acetylglucosamine; 6dDHep, 6-deoxy-D-mannoheptose; Man, mannose; Par, paratose; Tyv, tyvelose; Yer, yersiniose. Bonds and residues in parentheses were absent or different in the published structure, but included here on the basis of genetic data (see text). See Knirel and Kochetkov (1994) for original references.

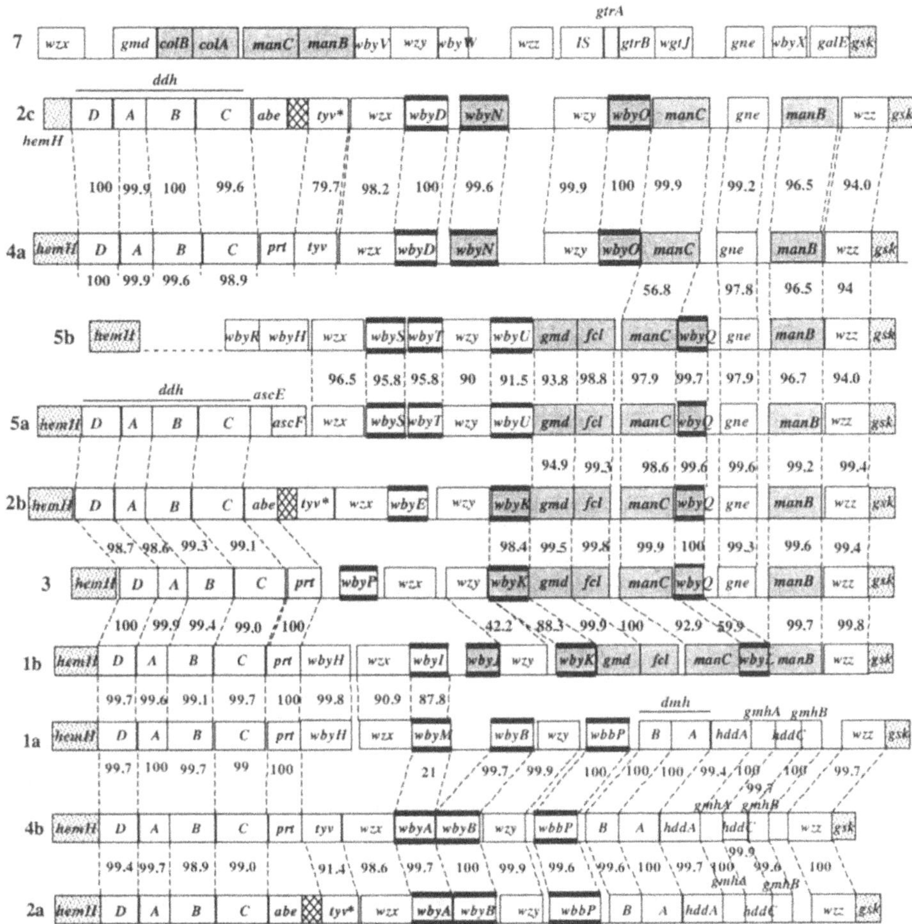

Figure 2. Y. pseudotuberculosis O antigens gene clusters. Data from Skurnik *et al.,* (2000) (1b), Pacinelli *et al.,* (2002) (1a, 2a, 4b) and unpublished or published in part only (2b, 2c, 3, 4a, 5a, 5b, 7). For gene functions see Figure 3. *tyv** indicates a non-functional *tyv* gene. The new sequences are of PCR products from primers in the JumpStart sequence in the promoter region and *gsk.* 5b has JumpStart sequences downstream of *hemH* and upstream of *wbyR*, The presence of genes between *hemH* and the second JumpStart sequence was observed by Skurnik (2003). The location of the *hemH* genes shown in relation to the first O antigen gene of each cluster was confirmed by PCR. Light shading, pathway and transferase genes for dideoxyhexoses . Dark shading, pathway and transferase genes for GDP-mannose, GDP-fucose and GDP-colitose. Stippled indicates flanking non-O-antigen genes. Bars on top and bottom of box indicates transferase gene. Numbers indicate % nucleotide base identity. See text for basis for allocation of function.

1. THE O ANTIGEN GENE CLUSTERS

1.1 O antigens 1a, 2a and 4b

The 3 O units with 6-deoxy-D-mannoheptose (6dDHep) have similar structures (Figure 1) and gene clusters (Figure 2). 6dDHep is a rare sugar and genes for synthesis of precursor GDP-6dDHep (see Figure 3) were identified (Pacinelli *et al.,* 2002). Precursor UDP-Gal is synthesised by housekeeping genes.

The first step in O unit synthesis is thought to be formation of UndPP-GlcNAc from GlcNAc-P and UndP by WecA, which is generally present in the *Enterobacteriaceae.* The galactose transferase is *wbbP*, previously identified in other species. Based on homologies *wbyB* is proposed to encode the 6dDHep transferase, and *wbyA* and *wbyM* the two dideoxyhexose transferases. As *wbyB* and *wbyC* are so similar in all cases we propose that the linkages are the same and so have refined the O:2a and O:4b structures to match the more recent NMR based O:1a structure. Except for *wbyH*, which is thought to have no function in O:1a (see below), all genes are accounted for. The differences, in the middle of the gene clusters, all relate to the different DDH side branch sugars and are discussed later.

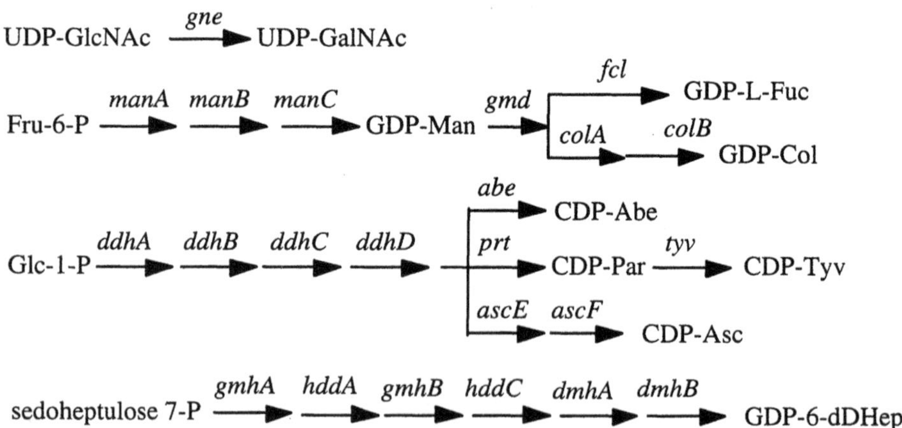

Figure 3. Nucleotide sugar biosynthesis genes. Fru, Fructose. See Figure 1 for other abbreviations and for more detail consult the Bacterial Polysaccharide Genes Database (BPGD): http://www.microbio.usyd.edu.au/BPGD/sugarpathways.htm

1.2 O antigens 2c and 4a

The 2c and 4a O units each have one GalNAc and 3 mannose residues, with abequose or tyvelose side branches respectively (Figure 1). The gene clusters (Figure 2) have DDH genes at the 5′ end, and at the 3′ end the genes for synthesis of UDP-GalNAc and GDP-mannose, plus three sugar transferase genes, all being near identical. WbyD is identified as the DDH transferase by homology, leaving WbyN and WbyO as mannose transferases, one presumably transferring 2 residues. Similarity of WbyN and WbyO led us to refine the O:4a structure to match the more recent O:2c structure. Again only a segment in the middle is significantly different as discussed later.

1.3 O antigens 1b, 2b, 3, 5a and 5b

These 5 O antigens have related structures that include fucose and mannose. The gene clusters have the expected pathway genes. The 4 genes for synthesis of GDP-mannose and GDP-fucose (Figure 3) occur as a block of *gmd*, *fcl* and *man*C, separated from *manB* by *wbyL* (1b) or by *wbyQ* plus *gne* (2b, 3, 5a 5b). Most of the remaining genes are transferase genes based on homologies, but *wbyH* in O:1b is the exception. The 1b O unit is unusual in having paratose in the furanose form and WbyH is proposed to be a mutase that converts CDP-Par*p* to CDP-Par*f* on the basis of similarity to Glf, which carries out a similar reaction to form UDP-Gal*f* from UDP-Gal*p*.

There are good structures for O:3, O:5a and O:5b, all including GalNAc. We expect a GlcNAc-4-epimerase gene for UDP-GalNAc synthesis (see Figure 3), and the one candidate gene was named *gne*. The gene cluster had no predicted protein with the transmembrane segments found in those transferases that add a sugar phosphate to UndP, and we assume that, as in other cases, GalNAc-P like GlcNAcP is transferred by WecA. The only putative transferase gene common to the three is *wbyQ*, thus identified as the gene for the fucosyl α(1-3) GalNAc linkage. O:1b has no *gne* gene and has GlcNAc in place of GalNAc. It also has *wbyL* in place of *wbyQ*, the proteins being 60% identical. WbyL was identified as the transferase for the Fucosyl α(1-3) GlcNAc linkage. The O:1b and O:3 O units have a mannose α(1-3) fucose linkage and WbyK is proposed to be the transferase based on its distribution. O:1b has an additional unique transferase gene *wbyJ*, which must encode the unique mannose (1-4) mannose transferase. There is no recent structure information for O:2b but as the genes for synthesis of the main chain are the same as in O:3, we can assume that it has the same mannose-fucose-GalNAc chain, although this was not reported in the early structural studies.

The 1b, 2b, 3 and 5a O units all have a DDH side branch. WbyI and WbyP are identified as the paratose transferases for O:1b and O:3 respectively on the basis of similarity to other dideoxyhexose transferase genes, and WbyE and WbyS as abequose and ascarylose transferases for O:2b and O:5a respectively.

The main chain of O:5a differs from those of 1b, 2b and 3 in having a mannose α(1-4) Fucose linkage in place of the α(1-3) linkage, and a final fucose residue on the mannose. It has the *gmd* to *manB* genes but *wbyU* in place of *wbyK*, and also the *wbyT* gene, both of which appear to be transferase genes and are proposed to encode these 2 steps. The expected *ddh* genes are present and also *ascE* and *ascF* genes (Thorson *et al.*, 1994) for synthesis of CDP-ascarylose.

O:5b has 6dAlt*f* as a side branch but otherwise is very similar to O:5a. A close relative of *wbyH* is present and indeed it may be a form of *wbyH*. It seems clear that this gene encodes the mutase that forms the furanose ring, probably as a final step converting NDP-6dAlt*p* to NDP-6dAlt*f*. Allocation of genes for synthesis of 6dAlt*p* is not yet possible as the 5´end is yet to be sequenced (see Figure 2).

Thus in this group of O units with related main chains we have again been able to identify most genes, but 2 transferase genes for O:5a and O:5b were not allocated to a specific reaction, and the pathway for the 6dAlt*f* precursor is yet to be resolved. The transferase genes were in general allocated by correlation of their distribution with that of specific linkages and experimental confirmation is needed.

1.4 O antigen 7

The O:7 repeat unit includes colitose, the only DDH that is transferred from a GDP-sugar precursor. The genes for GDP-colitose synthesis are near the 5' end of the gene cluster, in contrast to the 3´ location of genes for synthesis of GDP sugars in other gene clusters. The O:7 gene cluster is also unusual in that it starts with *wzx*, present in other *Y. pseudotuberculosis* O-units in the central region, and in having *wzz* in the central region in contrast to its location as the last gene in all other reported *Yersinia* gene clusters.

O:7 has a side branch glucose and has a set of 3 genes (*gtrA*, *gtrB* and *wgtJ*) homologous to those involved in addition of side-branch glucose residues in several bacterial species (Allison and Verma, 2000). These gene sets involve translocation across the membrane of Und-P-Glc, and then transfer of the glucose residue to the O unit. They are generally phage encoded and it as unusual to find them in an O antigen gene cluster. We expect 3 transferases in addition to WecA and WgtJ, and these may be WbyV, WbyW and WbyX. We have the further complication of finding 2 *galE*/*gne* homologues. We expect a *gne* gene but even with 2 GalNAc residues, there is no need for 2 *gne* genes. Perhaps the gene cluster was assembled with the GalNAc transferases coming from 2 sources, each with a *gne* gene, and both have been retained. We await the sequence for the O:6 gene cluster as O:6 and O:7 have similar structures and it may cast light on the matter, but there is more work to be done on the O:7 genes.

1.5 Relationships and origins of the gene clusters

For the 1a, 1b, 2a, 2b, 2c, 3, 4a, 4b and 5a gene clusters we can identify all sugar nucleotide pathway genes and processing genes, and with reasonable confidence all but 2 transferase genes. In each case we find the genes needed and, except for 1a (see below), no additional functional genes. Where we have related structures, the gene clusters are related in gene order and have high levels of similarity for common genes. The genes common to groups of related structures are generally at the ends and the genes that differentiate clusters in the middle.

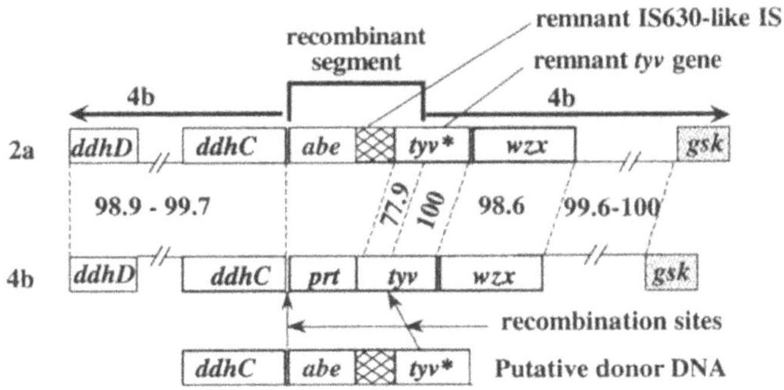

Figure 4. Close up of segment proposed to have been transferred from the O:4b gene cluster to give the O:2a gene cluster. Numbers are level of nucleotide identity. Level of shading indicates level of amino acid identity. Although shown to illustrate recombination for formation of O:2a from O:4b, the event could have involved O:4a to give O:2c. See text.

We can see examples of recombination. Genes related to the main chains of O:1a, O:2a and O:4b are near identical, and must have a recent common ancestor. The differences lie in the genes between *ddhC* and *wbyB* that determine which DDH is present. In O:2a, downstream of the *abe* gene there is a defective *tyv* gene, that would have had no role in this cluster before loss of function (see Figure 3). Alignment of the sequences (Figure 4) suggests that the O:2a cluster arose from the O:4b cluster by recombination, with replacement of *prt* by *abe*, the tyv gene becoming redundant. The alignment shows that the 3′ end of *tyv* is identical in O:4b and O:2a, suggesting recombination at the junction with more divergent DNA. The *abe* gene, an IS element and the remnant *tyv* gene of O:2a must have come from an unidentified source. The relationship of the near identical 4a and 2c gene clusters parallels that of O:4b and O:2a. the *abe* containing segment seen in O:2a and O:2c presumably arose but once, converting either O:4a or O:4b from a tyvelose to abequose containing structure. Once formed the new *abe* containing segment could readily transfer to the other by homologous recombination to give a second abequose containing structure.

A second example is the apparent origin of the O:1a gene cluster by recombination between those of O:1b and O:4b, as can be seen from the alignment shown in Figure 2. Such an origin would account for the presence of *wbyH* in O:1a, although the paratose is not in furanose form.

Finally one can compare the *Y. pseudotuberculosis* situation with that in the DDH containing A, B, C2 and D O units of *Salmonella enterica*. The organisation of DDH genes is similar, but in *S. enterica* all DDH containing O antigens have related main chains (Reeves, 1994) whereas there are several basic main chains structures in the *Y. pseudotuberculosis* forms. Also in *S. enterica* the paratose form is a recent mutational derivative of a tyvelose form, but clearly a primary form in *Y. pseudotuberculosis*, where the abequose containing forms seem to be of relatively recent origin.

REFERENCES

Allison, G.E. and Verma, N.K., 2000. Serotype-converting bacteriophages and O-antigen modification in *Shigella flexneri*. Trends Microbiol. 8, 17-23

Knirel, Y.A. and Kochetkov, N.K., 1994. The structure of lipopolysaccharides of gram-negative bacteria .3. The structure of O-antigens: A review. Biochemistry (Moscow). 59, 1325-1383.

Pacinelli, E., Wang, L. and Reeves, P.R., 2002. Relationship of *Yersinia pseudotuberculosis* O antigens IA, IIA, and IVB: the IIA gene cluster was derived from that of IVB. Infect. Immun. 70, 3271-3276.

Reeves, P.R., 1994. Biosynthesis and assembly of lipopolysaccharide. In: Neuberger, A. and van Deenen, L.L.M. (Eds.), Bacterial cell wall. Elsevier . pp. 281-314.

Skurnik, M., 2003. Molecular genetics, biochemistry and biological role of *Yersinia* lipopolysaccharide. In: Skurnik, M., Granfors, K. and Bengoechea, J.A. (Eds.), The Genus Yersinia: Entering the functional genomics era. Kluwer Plenum, pp. 187-197.

Skurnik, M., Peippo, A. and Ervelä, E., 2000. Characterization of the O-antigen gene clusters of *Yersinia pseudotuberculosis* and the cryptic O-antigen gene cluster of *Yersinia pestis* shows that the plague bacillus is most closely related to and has evolved from *Y. pseudotuberculosis* serotype O:1b. Mol. Microbiol. 37, 316-330.

Thorson, J.S., Lo, S.F., Ploux, O., He, X. and Liu, H.W., 1994. Studies of the biosynthesis of 3,6-dideoxyhexoses: Molecular cloning and characterization of the *asc* (Ascarylose) region from *Yersinia pseudotuberculosis* serogroup VA. J. Bacteriol. 176, 5483-5493.

Whitfield, C. 1995. Biosynthesis of lipopolysaccharide O antigens.Trends Microbiol. 3, 178-185.

Chapter 40

Cloning and Characterization of the *Yersinia enterocolitica* Serotype O:9 Lipopolysaccharide O-Antigen Gene Cluster

Peter S. LÜBECK[1], Jeffrey HOORFAR[1], Peter AHRENS[1] and Mikael SKURNIK[2]
[1]*Department of Bacteriology, Danish Veterinary Institute, Copenhagen, Denmark;*
[2]*Department of Bacteriology and Immunology, Haartman Institute, University of Helsinki and Department of Medical Biochemistry and Molecular Biology, University of Turku, Finland.*

1. INTRODUCTION

Several serotypes of *Yersinia enterocolitica* are pathogenic for animals and they are phenotypically very similar to each other. The serological typing of the strains is based on the variability of the lipopolysaccharide (LPS) O-antigen. The O-antigen of *Y. enterocolitica* serotype O:9 (YeO9) is a homopolymer of N-formyl-perosamine (Nf-Per) (Caroff *et al.*, 1984).

We have previously cloned and characterized the O-antigen gene clusters of *Y. enterocolitica* serotypes O:3 and O:8 (Skurnik & Zhang 1996, Zhang *et al.*, 1993, 1997). A similar work is presented here for YeO9 O-antigen gene cluster. The results are important in order to study the molecular genetics and biological role of YeO9 O-antigen cluster and to establish the basis for future applications in typing and differential diagnostics.

2. METHODS AND RESULTS

Genomic libraries of YeO9 strain Ruokola/71-c were constructed in *E. coli* strain C600 using pUC18 as vector. The perosamine synthetase (*per*) gene sequences of *Vibrio cholerae, V. anguillarum, E. coli* O157:H7 and *Brucella melitensis* (50-60 % identical to each other) were used to design

The Genus Yersinia, Edited by Skurnik *et al.*
'uwer Academic/Plenum Publishers, New York, 2003

degenerate primers for PCR amplification of a *per* gene fragment of YeO9, from which a probe was prepared to screen the libraries by colony hybridisation.

We estimated that the YeO9 O-antigen gene cluster would contain about 10 genes (biosynthesis of GDP-Nf-Per needs five enzymes, polymerisation of the Nf-Per residues into the homopolymer, two, and translocation of the homopolymer through the inner membrane, two). From one genomic library a clone containing a ca. 10 kb *Pst*I fragment was identified and sequenced.

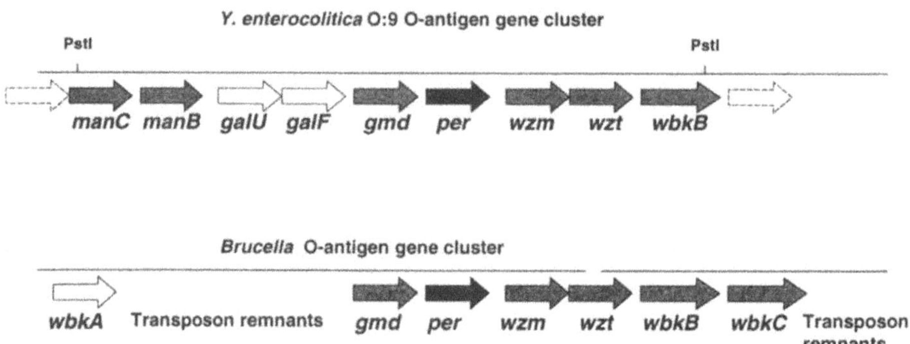

Figure 1. Comparison of the O-antigen gene clusters of *Y. enterocolitica* O:9 and *Brucella melitensis* (Godfroid *et al.*, 1998). The genes needed for the synthesis of the GDP-Nf-Per are *gmd, per* and *wbkC* that use GDP-mannose as substrate. GDP-mannose biosynthesis starts from fructose-1-phosphate which ManA, ManB and ManC convert to GDP-mannose. The *wzm* and *wzt* genes are needed for the translocation of the O-antigen to periplasm. WbkC may be a Nf-Per-transferase, but WbkA is apparently not involved in O-antigen biosynthesis. The *galU* and *galF* gene products are involved in the biosynthesis of UDP-glucose, therefore, there appears to be no direc link to biosynthesis of the O-antigen

Sequencing data of the cloned 10 kb *Pst*I fragment and database comparisons revealed the presence of nine genes in the fragment (Figure 1). The genes were named based on their sequence similarities to known genes (*manC, manB, galU, galF, gmd, per, wzm, wzt* and *wbkB*). Parts of the *manC* and *wbkB* genes were outside the *Pst*I-fragment. The *manB, manC, gmd, per* genes are involved in the biosynthesis of GDP-perosamine and the *wzm* and *wzt* genes in translocation. The *wbkB* gene apparently codes for a Nf-Per transferase. The *galU* and *galF* genes do not have any clear function in the O-antigen biosynthesis.

Organisation of the YeO9 gene cluster is identical over the 5 kb region carrying the *gmd, per, wzm, wzt* and *wbkB* genes to that of *B. melitensis* (Figure 1). The genes are ca. 60 % identical to each other. Upstream the *gmd* gene there is no similarity between the two organisms.

The data indicated that the *Pst*I fragment does not carry all the genes required for the YeO9 O-antigen biosynthesis; apparently missing are the genes for the second Nf-Per transferase and the formyltransferase that in *Brucella* is the *wbkC* gene located downstream of the *wbkB* gene. Cloning of the 5' and 3' regions is under way to characterize the complete O-antigen gene cluster.

ACKNOWLEDGMENTS

This work was supported by the Danish Council of Science (SJVF) and by the Academy of Finland.

REFERENCES

Caroff, M., D. R. Bundle, and M. B. Perry. 1984. Structure of the O-chain of the phenol-phase soluble cellular lipopolysaccharide of *Yersinia enterocolitica* serotype O:9. Eur. J. Biochem. 139:195-200.

Godfroid, F., Taminiau, B., Danese, I., Denoel, P., Tibor, A., Weynants, V., Cloeckaert, A., Godfroid, J. & Letesson, J. J.: Identification of the perosamine synthetase gene of *Brucella melitensis* 16M and involvement of lipopolysaccharide O side chain in *Brucella* survival in mice and in macrophages. Infect. Immun. 66: 5485-5493, 1998.

Skurnik M, Zhang L. 1996. Molecular genetics and biochemistry of *Yersinia* lipopolysaccharide. APMIS 104:849-72.

Zhang L, al-Hendy A, Toivanen P, Skurnik M. 1993. Genetic organization and sequence of the *rfb* gene cluster of *Yersinia enterocolitica* serotype O:3: similarities to the dTDP-L-rhamnose biosynthesis pathway of *Salmonella* and to the bacterial polysaccharide transport systems. Mol. Microbiol. 9:309-21.

Zhang L, Radziejewska-Lebrecht J, Krajewska-Pietrasik D, Toivanen P, Skurnik M. 1997. Molecular and chemical characterization of the lipopolysaccharide O-antigen and its role in the virulence of *Yersinia enterocolitica* serotype O:8. Mol. Microbiol. 23:63-76.

Chapter 41

Characterization of the Lipopolysaccharide Outer Core Biosynthesis of *Yersinia enterocolitica* Serotype O:3

Elise PINTA[1], Reija VENHO[1], José Antonio BENGOECHEA[3] and Mikael SKURNIK[1,2]
[1]*Department of Medical Biochemistry and Molecular Biology, University of Turku and* [2]*Department of Bacteriology and Immunology, Haartman Institute, University of Helsinki, Finland;* [3]*Unidad de Investigacion, Hospital Son Dureta, Andrea Doria 55, 07014 Palma Mallorca, Spain*

1. INTRODUCTION

Lipopolysaccharide (LPS) is an essential component of the cell wall of Gram-negative bacteria. LPS of *Yersinia enterocolitica* is a virulent factor (Skurnik, 1999) and therefore detailed information of its structure, biosynthesis and genetics is needed to understand its role in the disease process.

In *Y. enterocolitica* O:3 (YeO3) the outer core comprises of a hexasaccharide of two glucoses, two N-acetyl-D-galactosamines, one galactose and one N-acetyl-D-fucosamine (Radziejewska-Lebrecht *et al.*, 1998; Skurnik, 1999; Skurnik *et al.*, 1999).

The YeO3 outer core gene cluster contains 9 genes (Skurnik *et al.*, 1995). The functions of the gene products were predicted based on amino acid sequence similarities to known proteins, however these functions are still putative. Six of the genes were predicted to encode for glycosyltransferases (one for each sugar residue in the hexasaccharide), two for enzymes for NDP-sugar precursor biosynthesis and one for flippase.

The Genus Yersinia, Edited by Skurnik *et al.*
Kluwer Academic/Plenum Publishers, New York, 2003

2. AIMS AND METHODS

In order to definitely assign the specific functions of the gene products biochemical and genetic analyses were needed. To facilitate the analysis we constructed a mobilizable plasmid pRV16NP in which the YeO3 outer core gene cluster is expressed under the vector tetracycline promoter, and an outer core mutant host strain YeO3-c-OC that has the outer core gene cluster completely deleted.

Mobilization of pRV16NP into YeO3-c-OC complemented fully the outer core deficient phenotype (Figure 1). This allows us to easily construct single and double mutants into the outer core cluster genes on pRV16NP and analyze their functions in the YeO3-c-OC background.

Mutations to different genes were constructed as follows: suitable restriction enzymes were used to digest pRV16NP and the so created single-stranded overhangs were blunt-ended by either mung bean nuclease or T4 DNA polymerase and then ligated. These plasmids were then introduced to YeO3-c-OC by triparental conjugation and the LPS-phenotypes of the resulting strains were analyzed by gel electrophoresis (Figure 1).

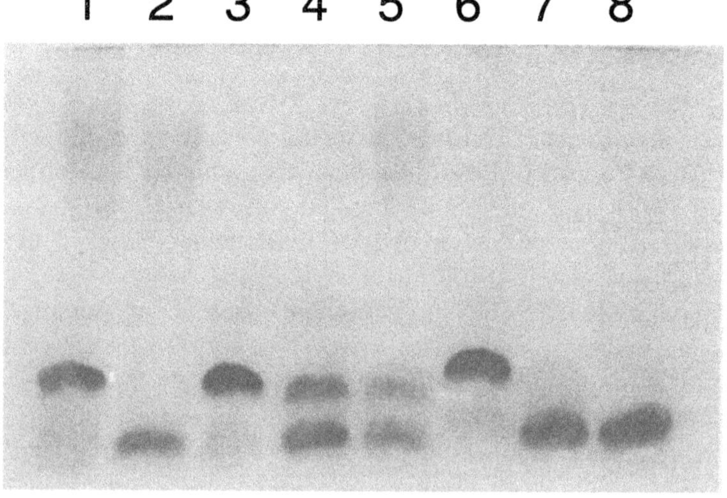

Figure 1. Silver stained deoxycholate polyacrylamide gel electrophoresis analysis of lipopolysaccharides. Lanes: 1, YeO3-c; 2, YeO3-c-OC; 3, YeO3-c-OC-R / pRV16NP; 4, YeO3-c-OC-R / pRV16NPwbcL; 5, YeO3-c-OC / pRV16NPwbcL; 6, YeO3-c-OC-R / pRV16NPwbcK; 7, YeO3-c-OC-R / pRV16Npgne; 8, YeO3-c-OC-R / pRV16NPwzx.

3. RESULTS AND FUTURE

So far we have been able to create single mutations to the genes *gne*, *wbcL*, *wbcK* and *wzx*. It seems that both *gne* (codes for UDP-*N*-acetylglucosamine-4-epimerase) and *wzx* are needed for full expression of outer core, since mutating one of them was already enough to prevent the synthesis of outer core.

In the case of both *wbcL* and *wbcK* the situation was different. Mutating *wbcL* did not fully prevent outer core synthesis and there were LPSs both with and without outer core present. When *wbcK* was mutated it seemed that outer core was synthesized but the size was a bit bigger than the wild type outer core. This result is still very preliminary and needs further studies.

In conclusion, we have shown that the approach we have chosen is functional however more mutations need to be constructed and the mutant LPSs need to be chemically analyzed to exactly reveal the defects generated by the mutations.

REFERENCES

Radziejewska-Lebrecht, J., Skurnik, M., Shashkov, A. S., Brade, L., Rozalski, A., Bartodziejska, B., and Mayer, H. (1998) Immunochemical studies on R mutants of *Yersinia enterocolitica* O:3. *Acta Biochimica Polonica* 45: 1011-1019.

Skurnik, M. (1999) Molecular genetics of *Yersinia* lipopolysaccharide. In *Genetics of Bacterial Polysaccharides*, J. Goldberg, ed. Boca Raton, FL: CRC Press, pp. 23-51.

Skurnik, M., Venho, R., Bengoechea, J.-A., and Moriyón, I. (1999) The lipopolysaccharide outer core of *Yersinia enterocolitica* serotype O:3 is required for virulence and plays a role in outer membrane integrity. *Mol Microbiol* 31: 1443-1462.

Skurnik, M., Venho, R., Toivanen, P., and Al-Hendy, A. (1995) A novel locus of *Yersinia enterocolitica* serotype O:3 involved in lipopolysaccharide outer core biosynthesis. *Mol Microbiol* 17: 575-594.

Chapter 42

ECA-Antibodies in Antisera Against R Mutants of *Yersinia enterocolitica* O:3

Joanna RADZIEJEWSKA-LEBRECHT[1], Katarzyna KASPERKIEWICZ[1], Mikael SKURNIK[2], Lore BRADE[3], Ivo STEINMETZ[4], Anna S. ŚWIERZKO[5] and Artur MUSZYŃSKI[1]

[1]Department of Microbiology, University of Silesia, Katowice, Poland; [2]Department of Bacteriology and Immunology, Haartman Institute, University of Helsinki and Department of Medical Biochemistry and Molecular Biology, University of Turku, Finland; [3]Research Center Borstel, Borstel, Germany; [4]Institute of Medical Microbiology, Medical School Hannover, Hannover, Germany; [5]Center of Microbiology and Virology, Polish Academy of Sciences, Łódź, Poland.

1. INTRODUCTION

Enterobacterial Common Antigen (ECA), a high molecular weight polymer built of trisaccharide repeating units, is anchored to the outer membrane of the cell wall either via its glycerophospholipid (ECA_{PG}) or via lipid A in some R-type lipopolysaccharides (ECA_{LPS}). Only the latter form is immunogenic in rabbits. Previously it was reported that mainly enterobacterial R forms having the *E. coli* R1, R4 or K-12 full core type are ECA-immunogenic (Kuhn *et al.*, 1988). We recently showed that the *Yersinia enterocolitica* serotype O:3 mutant strain Ye75R that expresses Rc type LPS (lipid A plus inner core) was ECA-immunogenic (Radziejewska *et al.*, 1998).

2. METHODS, RESULTS AND DISCUSSION

In this study we have examined the ECA immunogenicity of five other R mutants of *Y. enterocolitica* O:3 (Skurnik *et al.*, 1999) that express LPS with

The Genus Yersinia, Edited by Skurnik *et al.*
'uwer Academic/Plenum Publishers, New York, 2003

either complete (YeO3-c-R1), truncated (YeO3-c-trs22-R) or no outer core (OC) oligosaccharide (YeO3-c-RfbR7, YeO3-c-trs8-R, YeO3-c-trs24-R). The ECA-immunogenic strain Ye75R was used as a control. White New Zealand rabbits were immunized with repeated intravenous injections of heat killed bacterial suspensions over a period of 51 days. In serological studies monoclonal antibody Mab 899 specific for ECA (gift of Prof. D. Bitter Suermann, Medical School Hannover, Germany) as well as ECA_{PG} of *S. montevideo* SH94 and polyclonal antiserum against *E. coli* O14, rich in antibodies against ECA (gifts of Drs H. Mayer, H.-M. Kuhn, MPI, Freiburg, Germany), were used as controls. ECA and LPS were extracted from the dry bacterial mass of the *Yersinia* strains by a combination of hot phenol/water and phenol/chloroform/petroleum ether (PCP) method of Männel and Mayer described in Kuhn *et al.,* (1988), modified such that products after phenol/water extraction were purified by ultracentrifugation prior to subjecting them to the PCP procedure. In SDS-PAGE analysis LPS/PCP of *S. montevideo* SH 94 (MPI, Freiburg, Germany) was used as a control for S form LPS.

All anti-*Yersinia* antisera reacted with ECA_{PG} in an ELISA assay and gave the following titers: Ye75R, 6400; YeO3-c-RfbR7 and YeO3-c-trs22-R, 3200; YeO3-c-trs8-R, 1600; YeO3-c-R1, 400; and YO3-c-trs24-R, <200. In Western blot analysis all the antisera produced the typical ladder-like banding pattern characteristic for ECA (Figure 1 and Figure 2B, lane S), however, that of anti-YeO3-c-trs24-R was very weak (Figure 1, lane F, the bracket).

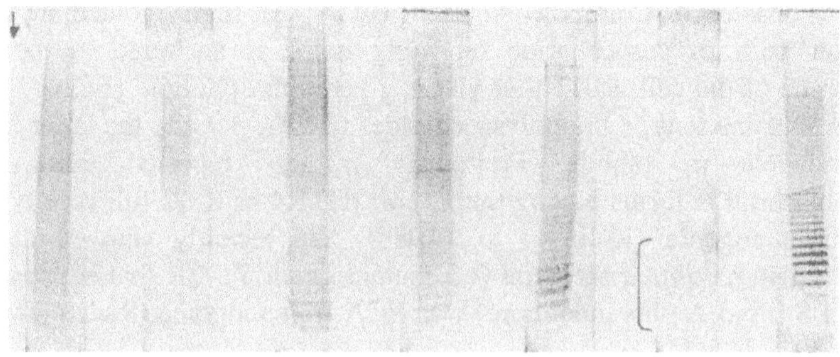

Figure 1. Western Blot analysis of anti-*Yersinia* antisera using as antigen ECA_{PG} of *S. montevideo* SH 94 (7 μg). Antisera: A, Ye75R (control); B, YeO3-c-R1; C, YeO3-c-RfbR7; D, YeO3-c-trs8-R; E, YeO3-c-trs22-R; F, YeO3-c-trs24-R; G, *E.coli* O14 (control).

Some variablity in the reaction pattern between the antisera was evident such that anti-ECA antibodies raised by some strains recognized better the longer (e.g., YeO3-c-R1) and by other strains, the shorter (YeO3-c-trs24-R), chains of ECA_{PG}. In summary, these results provided clear evidence that the five examined R mutants as well as Ye75R are ECA-immunogenic strains.

In order to study whether ECA is ligated to the full R core region of the LPS in the ECA immunogenic strains, LPS/PCP was isolated from these strains. PCP extraction produces LPS free from ECA_{PG}. With all the *Yersinia* strains SDS-PAGE/silver-staining analysis of LPS/PCP showed the presence of fast migrating molecules (Figure 2A).

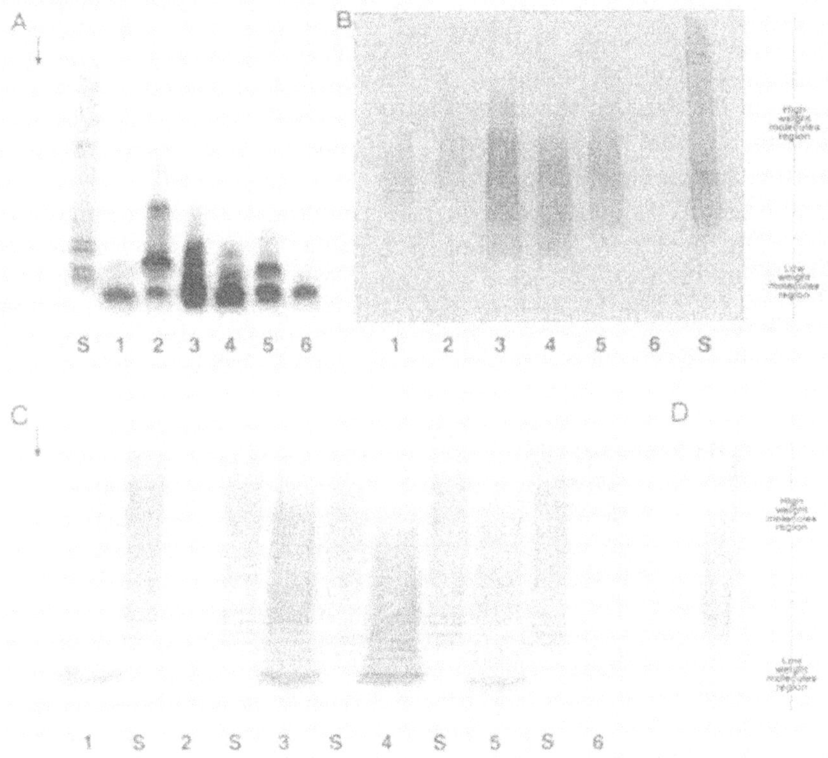

Figure 2. Analysis of LPS isolated from ECA-immunogenic *Yersinia* strains.

Panel A. SDS-PAGE analysis. Lane S, LPS/PCP of *S. montevideo*; lanes 1-6, LPS/PCP samples (2 μg) of *Yersinia* strains: lane 1, Ye75R; lane 2, YeO3-c-R1; lane 3, YeO3-c-Rfb-R7; lane 4, YeO3-c-trs8-R; lane 5, YeO3-c-trs22-R; lane 6, YeO3-c-trs24-R.

Panel B. Immunoblot analysis of the same samples with ECA-spefific Mab ECA 898. Lanes 1-6 as in panel A (5 μg of LPS/PCP samples per lane); lane S, ECA_{PG} of *S. montevideo.*

Panel C. Immunoblot analysis of the same samples with anti *E. coli* O14 antiseum. Lanes 1-6 as in panel A (15 μg of LPS/PCP samples per lane); lane S ECA_{PG} of *S. montevideo.*

Panel D. Immunoblot analysis of L1 phase material (15 μg) of YeO3-c-trs24-R with ECA-specific Mab ECA 898.

LPS/PCP of all mutants, besides YeO3-c-trs24-R, gave a ladder-like profile in an immunoblot with ECA-specific Mab 898 (Figure 2B) and reacted strongly (titers from 8000 to 256000) in ELISA with the same antibody. Similar results were obtained for LPS/PCP of all mutants in both tests when anti *E. coli* O14 serum, rich in antibodies against ECA, was used instead of Mab 898. Results of immunoblot analysis are shown at Figure 2C. A ladder-like ECA profile was again observed in a high weight molecules migration region for all *Yersinia* lipopolysaccharides, besides of LPS/PCP YeO3-c-trs24-R mutant, when tested with *E. coli* O14 antiserum. In addition, a reaction in the region of low weight (core plus lipid A) molecules was observed. This occurred most probably due to the presence of common epitopes in the R-cores shared by the five *Yersinia* lipopolysaccharides and LPS with the core type coli R4 of *E. coli* O14 strain against which the antiserum was produced.

These results suggested that in five *Yersinia* R mutant strains ECA is ligated to the inner core structure of LPS. Exception to this was strain YeO3-c-trs24-R. Its ECA was found in a supernatant L_1 (Figure 2D) obtained after ultracentrifugation of crude LPS. Interestingly, in this L_1 fraction also LPS was present. Further studies are needed to reveal whether it was ECA_{LPS}.

REFERENCES

Kuhn H.-M., Meier-Dieter, U., Mayer, H., 1988, ECA, the enterobacterial common antigen. FEMS Microbiol. Rev. 4: 195-222.

Radziejewska-Lebrecht, J., Skurnik, M., Shashkov, AS., Brade, L., Rozalski, A., Bartodziejska, B., Mayer, H., 1998, Immunochemical studies on R mutants of *Yersinia enterocolitica* O:3. Acta Biochim. Pol. 45: 1011-1019.

Skurnik, M., Venho, R., Bengoechea, JA., Moriyon. I. 1999, The lipopolysaccharide outer core of *Yersinia enterocolitica* serotype O:3 is required for virulence and plays a role in outer membrane integrity. Mol. Microbiol. 31: 1443-62.

Chapter 43

Lipopolysaccharides of *Yersinia*

An overview

Otto HOLST
Division of Structural Biochemistry, Research Center Borstel, Parkallee 22, D-23845 Borstel, Germany

1. INTRODUCTION

Lipopolysaccharides (LPSs) are the endotoxins of Gram-negative bacteria and well known for their immunological, pharmacological and pathophysiological effects displayed in eucaryotic cells and organisms. To date, much emphasis has been put on the elucidation of the chemical structures of LPSs and on their relation, or that of substructures, to the various biological effects. Lipopolysaccharides (Alexander and Rietschel, 2001) can be classified into two types which are characterized by the size of the saccharide portion, i.e. smooth- and rough-form (S- and R-form). Both types consist of lipid A and, covalently linked to it, a saccharide portion composed of up to fifteen sugars, the core region (Holst, 1999; Holst, 2002). In S-form LPS, this core region is substituted by the O-specific polysaccharide (Knirel and Kochetkov, 1994; Jansson, 1999). Both LPS types are present in wild-type Gram-negative bacteria, S-form for example in *Escherichia coli* or *Vibrio cholerae*, and R-form in *Neisseria meningitidis or Bordetella pertussis*. Since mutants that are not able to synthesize a minimal core structure are not viable, the core region and lipid A represent a common structural unit occurring in all LPSs and important for viability and membrane function of Gram-negative bacteria. The lipid part of LPS, the lipid A, was proven to represent the toxic principle of endotoxin. However, lipid A toxicity depends strongly on its structure, and is influenced by the core region.

The Genus Yersinia, Edited by Skurnik *et al.*
Kluwer Academic/Plenum Publishers, New York, 2003

2. LPS OF YERSINIA

This short overview deals mainly with structures and biological functions of LPSs from the three major human pathogenic species, i.e. *Yersinia enterocolitica*, *Y. pestis* and *Y. pseudotuberculosis*. Structures of lipid A and the core regions of LPSs from the first two species have been elucidated, as were some biological properties of the lipid A moieties. Of *Y. pseudotuberculosis*, only a partial core structure was reported, and no structure of lipid A. With regard to O-specific polysaccharide structures, quite a number of LPSs from *Y. enterocolitica* and *Y. pseudotuberculosis* are known. *Y. pestis* LPSs are of the rough type and, thus, do not possess any O-specific polysaccharides.

2.1 Lipid A structures

In *Y. enterocolitica*, the structures of the lipid A from LPSs of the serotypes O:3, O:8, O:9, O:11,23 and O:11,24 (Figs. 1 and 2) have been elucidated (Aussel *et al.*, 2000; Oertelt *et al.*, unpublished.). The structures of the major lipid A molecules are all very similar, i.e. they consist of a β-(1→6)-linked 2-amino-2-deoxy-D-glucopyranose (Glc*p*N) disaccharide that is bisphosphorylated at positions C-1 and C-4' and that is hexa-acylated. Two amide- and two ester-linked 3-hydroxymyristic acid [14:0(3-OH)] residues are present. Whereas in serotypes O:3, O:8 and O:9 the acyloxyacyl pattern at the non-reducing Glc*p*N is identical, that of serotypes O:11,23 and O:11,24 differs by the presence of an 14:0[3-O(12:0)] that is amide-linked to the non-reducing Glc*p*N. In serotype O:8, the structure of the lipid A was found to be temperature-dependent (Figure 2) (Oertelt *et al.*, unpublished). The above described hexa-acylated lipid A was the major component in LPSs of bacteria grown at 21°C. In LPSs of bacteria grown at 37°C, the major components were a tri- and a tetra-acylated lipid A. This preparation contained also small amounts of hexa-acylated lipid A, as contained the preparation from the bacteria grown at 37°C small amounts of the tri- and a tetra-acylated lipid A. Unfortunately, it was not possible to isolate pure tri- and a tetra-acylated lipid A fractions, thus, their biological properties could not be established.

The lipid A structures of LPSs from serotypes O:11,23, O:11,24, O:3, O:8 and O:9 are very similar to the structure of lipid A in LPSs of *Escherichia coli* (Zähringer *et al.*, 1994).

Similar to the lipid A of *Y. enterocolitica*, the structures of lipid A of LPSs from *Y. pestis* vary with the growth temperature (Figure 3). Again, the hexa-acylated carbohydrate backbone is produced as major component at lower temperature (27°C), whereas at 37°C two less acylated lipid A molecules are dominant, i.e. a tri- and a tetra-acylated compound (Kawahara *et al.*, 2002; Aussel *et al.*, 2000).

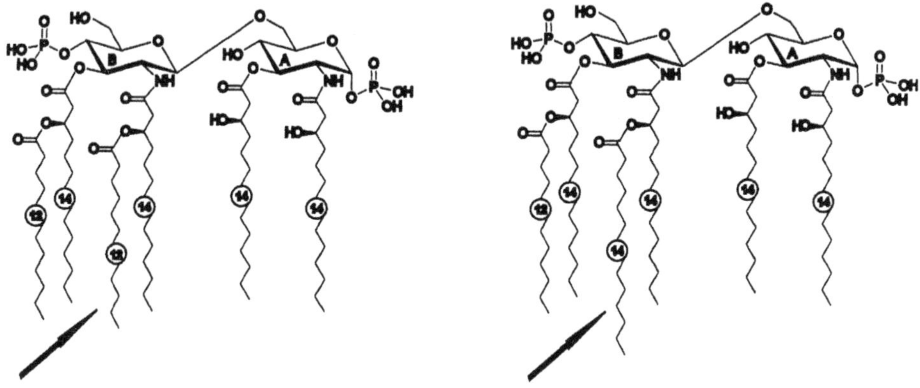

Figure 1. Structures of lipid A from LPSs of *Y. enterocolitica* O:11,23 and O:11,24, *Y. ruckeri* O:1 and O:2 (all left), and from *Y. enterocolitica* O:3 and O:9 (right). The arrows indicates the difference in fatty acid substitution.

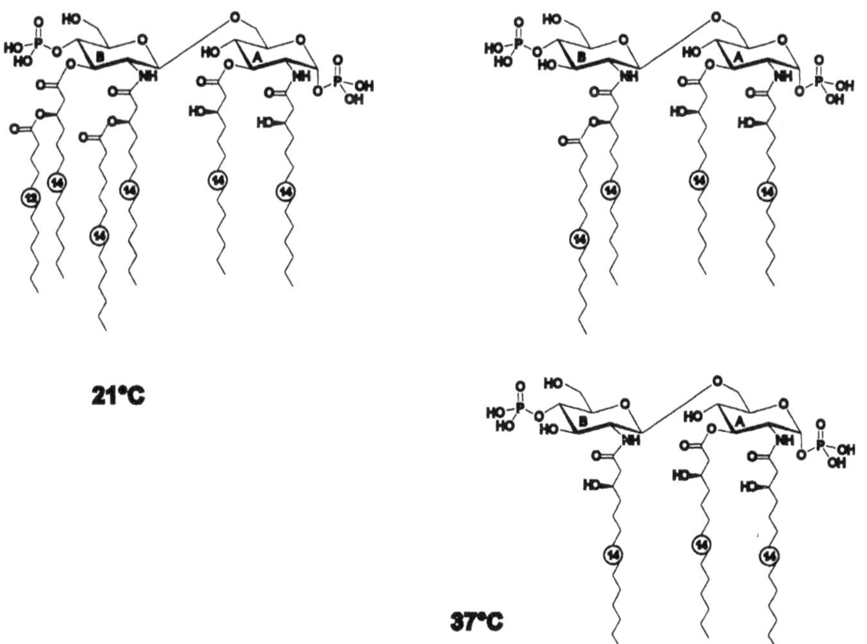

Figure 2. Structures of lipid A compounds of LPSs from *Y. enterocolitica* O:8 grown at 21°C (left) and at 37°C (right)

A first investigation of the biological properties of the different lipid A fractions showed that lipid A isolated from LPSs produced at 27°C induced in macrophages the synthesis of a significant higher amount of TNF-α than the other lipid A fraction (Kawahara *et al.*, 2002).

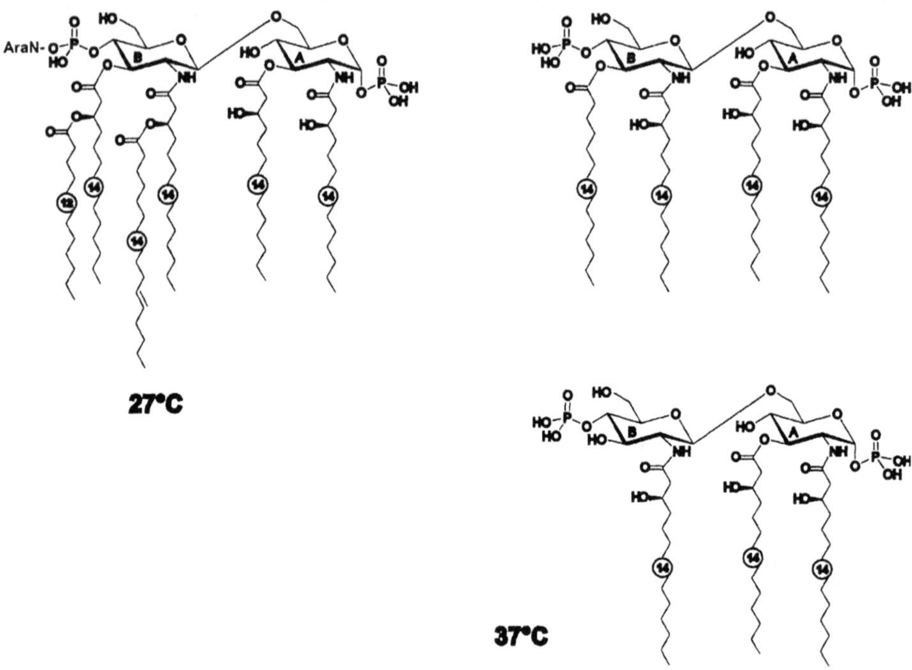

Figure 3. Structures of lipid A compounds of LPSs from *Y. pestis* strain Yreka grown at 27°C (left) and at 37°C (right). AraN, 4-amino-4-deoxy-L-arabinopyranose

2.2 The core structures

Core structures have been elucidated from LPSs of *Y. enterocolitica* O:3, O:8 and O:9 (Radziejewska-Lebrecht *et al.*, 1994; Radziejewska-Lebrecht *et al.*, 1998; Müller-Loennies *et al.*, 1999; Oertelt *et al.*, 2001), and from *Y. pestis* (Vinogradov *et al.*, 2002; Hitchen *et al.*, 2002). Common in all core regions from LPSs of *Y. enterocolitica* (Figure 4) is the hexasaccharide D-α-D-Hep-(1→7)-L-α-D-Hep-(1→7)-L-α-D-Hep-(1→3)-L-α-D-Hep-(1→5)-[α-Kdo-(2→4)]-α-Kdo (Kdo, 3-deoxy-D-*manno*-oct-2-ulopyranosonic acid; L,D-Hep, L-*glycero*-D-*manno*-heptopyranose; D,D-Hep, D-*glycero*-D-*manno*-heptopyranose) which is linked via Kdo I to lipid A. This hexasaccharide is in all three serotypes substituted at O-4 of Hep-I by a β-linked D-Glc*p* and at O-2 of Hep II by an α-linked D-Glc*p*. In a rough strain of serotype O:9, no further substituents were identified, and in another rough strain of serotype O:8 Hep II was additionally substituted at position O:3 by β-D-Glc*p*NAc in non-stoichiometric amounts. The core region of LPSs from serotype O:3 is the only one from *Yersinia* LPSs that carries (in non-stoichiometric amounts) a phosphate residue, at O-6 of the D,D-Hep residue. All other core regions are phosphate-free. Additionally, position O-3 of Hep II in the O:3 core region is substituted by a hexasaccharide of neutral and amino sugars that is sometimes called the outer core region. However, there is genetic evidence that this hexasaccharide represents a first repeating unit of an ancient O-specific polysaccharide which is not built up anymore. Instead, a homopolymer of L-6-deoxy-altropyranose is synthesized as O-antigen (Skurnik and Zhang, 1996).

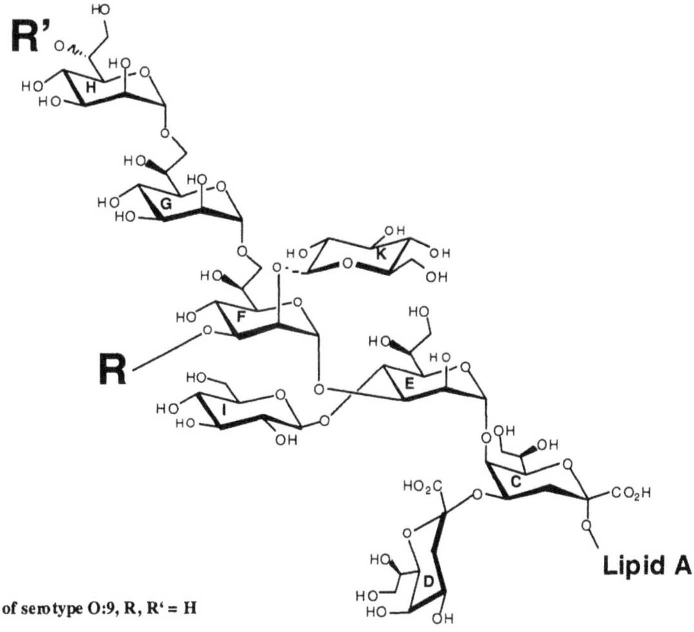

In a rough strain of serotype O:9, R, R' = H

In a rough strain of serotype O:8, R = GlcNAc and R' = H

In a rough strain of serotype O:3, R = ß-Glc-(1-6)-GalNAc-(1-6)-Gal-(1-4)-GalNAc-(1-3)-ß-FucNAc-(1-
3
|
1
ß-Glc

and R' = *P*

Figure 4. Core structures of LPSs from *Y. enterocolitica*

The core region of LPSs from *Y. pestis* strain KM218 was characterized completely (Vinogradov *et al.*, 2002) (Figure 5) and that of LPSs from strain BG tentatively (Hitchen *et al.*, 2002). In LPSs of bacteria grown at 37°C, the structure of the core region was similar to that of *Y. enterocolitica* O:8. Only the α-D-Glc*p* linked at O-2 of Hep II was missing. When the bacteria were grown at 25°C, the D,D-Hep residue was partly replaced by β-D-Galp and Kdo II by D-*glycero*-D-*talo*-oct-2-ulopyranosonic acid (Ko) (Gremyakova *et al.*, this book, p. 229-232). The biological meaning of these temperature-dependent structural changes is unclear.

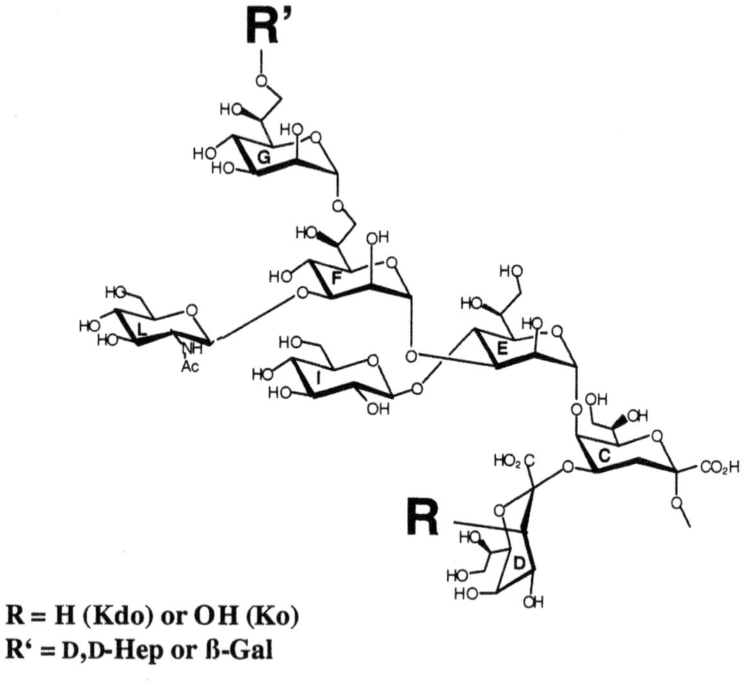

R = H (Kdo) or OH (Ko)
R' = D,D-Hep or ß-Gal

Figure 5. Core structures of LPSs from *Y. pestis* strain KM218. The β-Gal and Ko residues are partly substituting D,D-Hep and Kdo II, respectively, when bacteria are grown at 25°C

2.3 The O-specific polysaccharides

There is good evidence that the O-specific polysaccharides of LPSs from *Yersinia* play an important role in resistance to complement-mediated killing of the bacteria and in the colonization of host tissues. Also, resistance to cationic antimicrobial peptides which are part of the innate immune system is provided. A rather large number of structures of O-specific polysaccharides from LPSs of *Yersinia* are known (Knirel and Kochetkov, 1994; Jansson, 1999). Most of these structures are heteropolymers, i.e. consist of repeating units that are built up from various sugars. Such O-antigens were identified in *Y. enterocolitica*, *Y. pseudotuberculosis*, *Y. intermedia*, *Y. kristensenii*, *Y. frederiksenii*, *Y. aldovae*, *Y. ruckeri* and *Y. bercovieri*. Few O-specific polysaccharides represent homopolymers (composed of only one sugar), present e.g. in *Y. enterocolitica* O:3 (2-linked β-6-deoxy-L-altropyranose) and O:9 (2-linked 4-deoxy-4-formamido-α-D-rhamnopyranose, D-Rha*p*4NFo, perosamine). The O-specific polysaccharides of LPSs from *Yersinia* represent a mine of unusual sugars, like yersiniose A and B that differ only in the stereochemistry of their branch (Figure 6). Also, slight differences in the structures can be found that lead to new serotypes, confirming the high structural variability of the O-antigens (Figure 7). One example for this is the structural difference between the repeating units of the O-antigens from *Y. enterocolitica* O:11,23 and O:11,24 which solely is based on the *O*-acetylation at O-3 of the *N*-acetyl-galactosaminuronic acid residue (Gal*p*NAcA) which is present in serotype O:11,23 but absent in O:11,24.

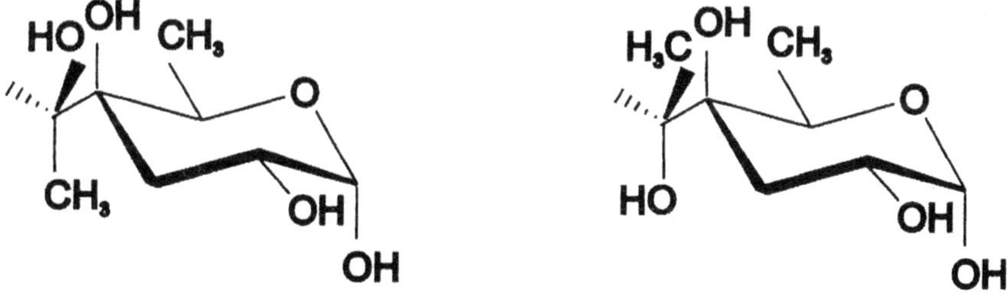

Figure 6. Structures of yersiniose A and B. The arrows indicate the difference between both sugars. Yersiniose A occurs e.g. in the O-specific polysaccharide of *Y. pseudotuberculosis* VI and yersiniose B in that of *Y. enterocolitica* O:4,32

Another example is he difference between the serotypes O:25,35 and O:12,25 of *Y. kristensenii* (Figure 8). The first carries a branching Glc*p* residue which is α-(1→4)-linked to Glc*p*NAc in the main chain. This Glcp residue is exchanged to a β-(1→4)-linked Glc*p*NAc residue in the second serotype.

Y. enterocolitica O:11,23

→3)-L-Qui*p*NAc-(1→4)-D-Gal*p*NAcA3Ac-(1→3)-L-Qui*p*NAc-(1→3)-β-D-Glc*p*NAc-(1

Y. enterocolitica O:11,24

→3)-L-Qui*p*NAc-(1→4)-D-Gal*p*NAcA-(1→3)-L-Qui*p*NAc-(1→3)-β-D-Glc*p*NAc-(1

Figure 7. Structural difference between the repeating units of the O-antigens from *Y. enterocolitica* O:11,23 and O:11,24. The only difference leading to another serotype is the O-acetylation at O-3 of the *N*-acetylgalactosaminuronic acid residue (GalpNAcA) in serotype O:11,23

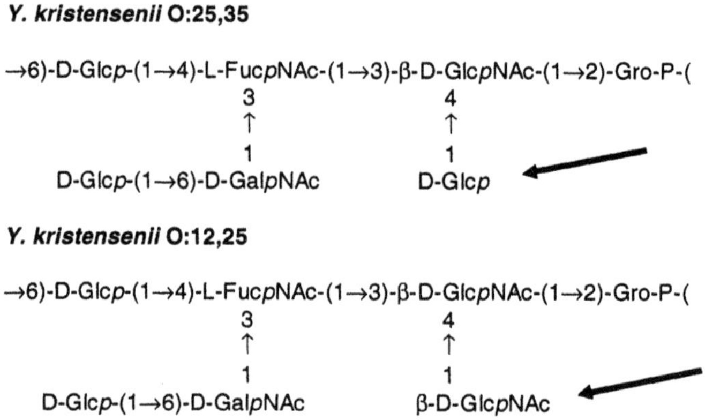

Y. kristensenii O:25,35

```
→6)-D-Glcp-(1→4)-L-FucpNAc-(1→3)-β-D-GlcpNAc-(1→2)-Gro-P-(
                    3                  4
                    ↑                  ↑
                    1                  1
   D-Glcp-(1→6)-D-GalpNAc            D-Glcp      ←
```

Y. kristensenii O:12,25

```
→6)-D-Glcp-(1→4)-L-FucpNAc-(1→3)-β-D-GlcpNAc-(1→2)-Gro-P-(
                    3                  4
                    ↑                  ↑
                    1                  1
   D-Glcp-(1→6)-D-GalpNAc            β-D-GlcpNAc      ←
```

Figure 8. Structural difference between the repeating units of the O-antigens from *Y. kristensenii* O:25,35 and O:12,25. The difference leading to another serotype is a β-D-GlcpNAc residue in the second serotype, exchanging the Glc of serotype O:25,35.

ACKNOWLEDGEMENTS

I thank all my past and present collaborators that have contributed to the actual knowledge on LPSs of *Yersinia*: Sabine Rund, Clemens Oertelt, Sven Müller-Loennies, Buko Lindner, Regina Engel, Sylvia Düpow (all of Research Center Borstel), Mikael Skurnik, Elise Pinta (University of Turku, Turku), Jose Antonio Bengoechea (Hospital Son Dureta, Palma, Mallorca), Yuriy Knirel (N. D. Zelinski Institute of Organic Chemistry, Moscow), Andrei Anisimov, Tat'yana Gremyakova (State Research Center of Applied Microbiology, Obolensk), Evgeni Vinogradov (NRCC, Institute for Biological Sciences, Ottawa), Kazuyoshi Kawahara (Kitasato Institute,

Tokyo), Timo Korhonen (University of Helsinki, Helsinki), and Joanna Radziejewska-Lebrecht (University of Katowice). Our research was founded in part by the *Deutscher Akademischer Austauschdienst.*

REFERENCES

Alexander, C. and Rietschel, E. Th., 2001, Bacterial lipopolysaccharides and innate immunity. *J. Endotoxin Res.* **7**: 167-202.

Aussel, L., Thérisod, H., Karibian, D., Perry, M. B., Bruneteau, M., and Caroff, M., 2000, Novel variation of lipid A structures in strains of different *Yersinia* species. *FEBS Lett.* **465**: 87-92.

Gremyakova, T.A., Vinogradov, E.V., Lindner, B, Kocharova, N.A., Senchenkova, S.N., Shashkov, A.S., Knirel, Y.A., Holst, O., Shaikhutdinova, R.Z. and Anisimov, A.P., 2003, The core structure of the lipopolysaccharide of *Yersinia pestis* strain KM218. In *The Genus Yersinia: entering the functional genomic era*, M. Skurnik, K. Granfors and J. A. Bengoechea, eds.: Kluwer Academic/Plenum Publishers, pp. 229-232.

Hitchen, P. G., Prior, J. L., Oyston, P. C. F., Panico, M., Wren, B. W., Titball, R. W., Morris, H. R., and Dell, A., 2002, Structural characterization of lipo-oligosaccharides (LOS) from *Yersinia pestis*: regulation of LOS structure by the PhoPQ system. *Mol. Microbiol.* **44**: 1637-1650.

Holst, O., Chemical structure of the core region of lipopolysaccharides. In *Endotoxin in Health and Disease*, H. Brade, S. M. Opal, S. N. Vogel, and D. C. Morrison, eds. NewYork, NY: Marcel Dekker, Inc., 1999.

Holst, O., 2002, Chemical structure of the core region of lipopolysaccharides – an update. *Trends Glycosci. Glycotechnol.* **14**: 87-103.

Jansson, P.-E., The chemistry of O-polysaccharide chains in bacterial lipopolysaccharides. In *Endotoxin in Health and Disease*, H. Brade, S. M. Opal, S. N. Vogel, and D. C. Morrison, eds. NewYork, NY: Marcel Dekker, Inc., 1999.

Kawahara, K., Tsukano, H., Watanabe, H., Lindner, B., and Matsuura, M., 2002, Modification of the structure and activity of lipid A in *Yersinia pestis* lipopolysaccharide by growth temperature. *Infect. Immun.* **70**: 4092-4098.

Knirel, Y. A. and Kochetkov, N. K., 1994, The structure of lipopolysaccharides of Gram-negative bacteria. III. The structure of O-antigens: A review. *Biochemistry (Moscow)* **59**: 1325-1383.

Müller-Loennies, S., Rund, S., Ervelä, E., Skurnik, M., and Holst, O., 1999, The structure of the carbohydrate backbone of the core-lipid A region of the lipopolysaccharide from a clinical isolate of *Yersinia enterocolitica* O:9. *Eur. J. Biochem.* **261**: 19-24.

Oertelt, C., Lindner, B., Skurnik, M., and Holst, O., 2001, Isolation and structural characterization of an R-form lipopolysaccharide from *Yersinia enterocolitica* serotype O:8. *Eur. J. Biochem.* **268**: 554-564.

Radziejewska-Lebrecht, J., Shashkov, A. S., Stroobant, V., Wartenberg, K., Warth, C., and Mayer, H., 1994, The inner core region of *Yersinia enterocolitica* YE75R (O:3) lipopolysaccharide. *Eur. J. Biochem.* **221**: 343-351.

Radziejewska-Lebrecht, J., Skurnik, M., Shashkov, A. S., Brade, L., Bartodziejska, B., and Mayer, H., 1998, Immunochemical studies on R mutants of *Yersinia enterocolitica* O:3. *Acta Biochim. Pol.* **45**: 1011-1019.

Skurnik, M. and Zhang, L., 1996, Molecular genetics and biochemistry of *Yersinia* lipopolysaccharide. *APMIS* **104**: 849-872.

Vinogradov, E. V., Lindner, B., Kocharova, N. A., Senchenkova, S. N., Shashkov, A. S., Knirel, Y. A., Holst, O., Gremyakova, T. A., Shaikhutdinova, R. Z., and Anisimov, A. P., 2002, The core structure of the lipopolysaccharide from the causative agent of plague, *Yersinia pestis*. *Carbohydr. Res.* **337**: 775-777.

Zähringer, U., Lindner, B., and Rietschel, E. Th., 1994, Molecular structure of lipid A, the endotoxic center of bacterial lipopolysaccharides. *Adv. Carbohydr. Chem. Biochem.* **50**: 211-276.

Chapter 44

The Core Structure of the Lipopolysaccharide of *Yersinia pestis* Strain KM218

Influence of Growth Temperature

Tat'yana A. GREMYAKOVA[1], Evgeny V. VINOGRADOV[2], Buko LINDNER[3], Nina A. KOCHAROVA[2], Sof'ya N. SENCHENKOVA[2], Aleksander S. SHASHKOV[2], Yuriy A. KNIREL[2], Otto HOLST[3], Rima Z. SHAIKHUTDINOVA[1], and Andrei P. ANISIMOV[1]
[1]State Research Center for Applied Microbiology, Obolensk, Moscow Region, Russia; [2]N. D. Zelinsky Institute of Organic Chemistry, Russian Academy of Sciences, Moscow, Russia; [3]Research Center Borstel, Center for Medicine und Biosciences, Borstel, Germany

1. INTRODUCTION

Pathogenicity of the causative agent of plague, the bacterium *Yersinia pestis*, is determined by a number of factors including a rough-type lipopolysaccharide (LPS) (Skurnik *et al.*, 2000), which mediates serum and cationic-antimicrobial-peptides resistance as well as infective toxic shock (Perry and Fetherston, 1997). Elucidation of the chemical structure of the LPS may usher in a new era in understanding pathogen-host interactions on the molecular level. Here we report on the full structure of the core region of *Y. pestis* LPS and the influence of growth temperature on the core structure.

2. ELUCIDATION OF THE LPS CORE STRUCTURE

Y. pestis strain KM218, a plasmidless derivative of the Russian vaccine strain EV line NIIEG, was grown at 25 and 37 °C (flea and mammalian host temperature, respectively) in liquid aerated media containing fish-flour hydrolysate and yeast autolysate. The lipopolysaccharides (LPS-25 and LPS-37) were isolated by phenol-water extraction and purified by treatment with DNAse, RNAse and Proteinase K followed by ultracentrifugation.

The Genus Yersinia, Edited by Skurnik *et al.*
Kluwer Academic/Plenum Publishers, New York, 2003

Each LPS was degraded with dilute HOAc to cleave the linkage between the core and lipid A moieties. Gel chromatography on Sephadex G-50 afforded core oligosaccharides (fraction II), which were fractionated by anion-exchange chromatography on HiTrap Q (Figure 1) and, after reduction with $NaBH_4$, on CarboPac PA1 (Dionex).

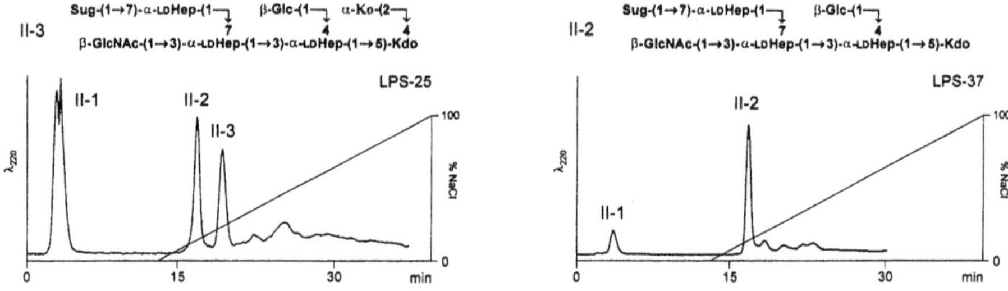

Figure 1. HiTrap Q chromatography of the core oligosaccharides from LPS-25 and LPS-37. II-1 is a contamination; major structures of II-2 and II-3 oligosaccharides are shown on the top. II-2 resulted from the cleavage of the terminal Kdo, whereas Ko in II-3 was not cleaved by acid. Sug is either β-Gal or α-DDHep. For abbreviations of sugars see the legend to Figure 2.

The isolated oligosaccharides were studied by sugar and methylation analyses, ion cyclotron resonance Fourier transform electrospray ionisation mass spectrometry and one- and two-dimensional ^{1}H and ^{13}C NMR spectroscopy. As a result, the structures of the core region of LPS-25 and LPS-37 were determined, and are shown in Figure 2.

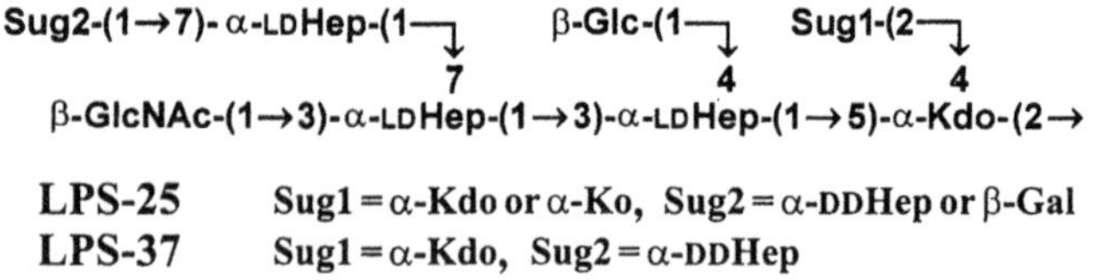

Sug2-(1→7)-α-LDHep-(1→7) β-Glc-(1→4) Sug1-(2→4)
β-GlcNAc-(1→3)-α-LDHep-(1→3)-α-LDHep-(1→5)-α-Kdo-(2→

LPS-25 **Sug1 = α-Kdo or α-Ko, Sug2 = α-DDHep or β-Gal**
LPS-37 **Sug1 = α-Kdo, Sug2 = α-DDHep**

Figure 2. Structures of the core region of LPS-25 and LPS-37. LDHep and DDHep stand for L-*glycero*- and D-*glycero*-D-*manno*-heptose, Kdo stands for 3-deoxy-D-*manno*-octulosonic acid and Ko for D-*glycero*-D-*talo*-octulosonic acid; a minority of the molecules lacks GlcNAc.

3. DISCUSSION

For the first time the full structure of the LPS core of *Y. pestis* has been established. It is distinguished by a structural heterogeneity and variability under different growth temperatures. At 25°C, four major core glycoforms are produced, which differ in the presence of either terminal Kdo or Ko attached to another Kdo residue and either terminal DDHep or Gal attached to an LDHep residue (LPS-25). At 37°C, there is only one glycoform containing terminal Kdo and DDHep (LPS-37). Hence, at the elevated temperature incorporation of Ko and Gal to the LPS core is suppressed. GlcNAc is present in non-stoichiometric amounts, independent of growth temperature, which further enhances the structural heterogeneity of the core.

Variation in the LPS structure is under the control of the two-component regulatory system PhoPQ (Hitchen *et al.*, 2002), which is involved in the global regulation of virulence gene expression in *Y. pestis*. Together with structural variations of lipid A (Kawahara *et al.*, 2002), the temperature-induced changes in the LPS core structure may protect the bacterium against defence mechanisms of both insect and mammalian hosts.

Remarkably, having one glucose residue less, the core of LPS-37 from *Y. pestis* KM218 represents a truncated form of the LPS core of *Y. enterocolitica* O:8 (Oertelt *et al.*, 2001). This similarity suggests that *Y. pestis* is closely related to not only *Y. pseudotuberculosis* (Skurnik *et al.*, 2000) but also to *Y. enterocolitica.*

ACKNOWLEDGEMENTS

This work was performed within the framework of the International Science and Technology Center Partner Project #1197p, supported by the Cooperative Threat Reduction Program of the US Department of Defense.

REFERENCES

Hitchen, P. G., Prior, J. L., Oyston, P. C., Panico, M., Wren, B. W., Titball, R. W., Morris, H. R., and Dell, A., 2002, Structural characterization of lipo-oligosaccharide (LOS) from *Yersinia pestis*: regulation of LOS structure by the PhoPQ system. *Mol. Microbiol.* **44**: 1637-1650.

Kawahara, K., Tsukano, H., Watanabe, H., Lindner, B., and Matsuura. M., 2002, Modification of the structure and activity of lipid A in *Yersinia pestis* lipopolysaccharide by growth temperature. *Infect. Immun.* **70**: 4092-4098.

Oertelt, C., Lindner, B., Skurnik, M., and Holst, O., 2001, Isolation and structural characterization of an R-form lipopolysaccharide from *Yersinia enterocolitica* serotype O:8. *Eur. J. Biochem.* **268**: 554-564.

Perry, R. D., and Fetherston, J. D., 1997, *Yersinia pestis* - etiologic agent of plague. *Clin. Microbiol. Rev.* **10**: 35-66.

Skurnik, M., Peippo, A., and Ervelä, E., 2000, Characterization of the O-antigen gene clusters of *Yersinia pseudotuberculosis* and the cryptic O-antigen gene cluster of *Yersinia pestis* shows that the plague bacillus is most closely related to and has evolved from *Y. pseudotuberculosis* serotype O:1b. *Mol. Microbiol.* **37**: 316-330.

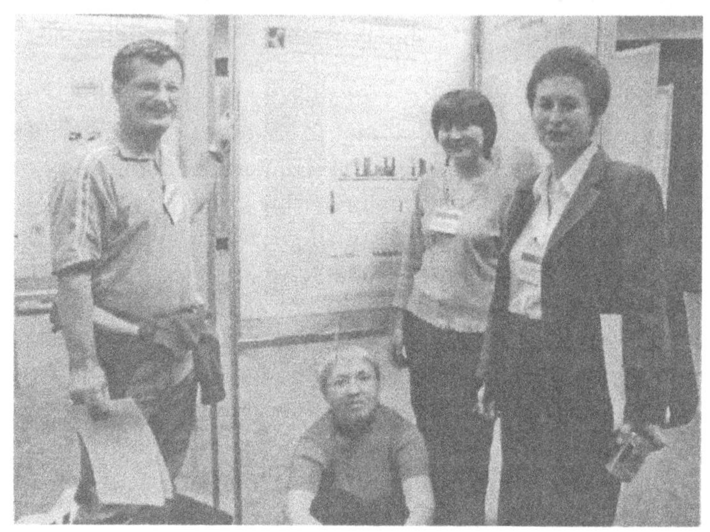

Picture 18. Andrei Anisimov with Tatiana Meka-Mechenko, Gulnaz Stybayeva and Larissa Nekrassova,

Picture 19. Rima Shaikhutdinova and Svetlana Dentovskaya at the City Reception

Chapter 45

Yersiniophages

Special reference to φYeO3-12

Maria I. PAJUNEN[1,2], Ian J. MOLINEUX[3] and Mikael SKURNIK[2,4]
[1]*Institute of Biotechnology, University of Helsinki, Finland;* [2]*Department of Medical Biochemistry and Molecular Biology, University of Turku, Finland;* [3]*Molecular Genetics and Microbiology, University of Texas, Austin, USA;* [4]*Department of Bacteriology and Immunology, Haartman Institute, University of Helsinki, Finland.*

1. INTRODUCTION

In the decade following the discovery bacteriophages (phages), research was driven by the hope that they might prove useful in combating bacterial diseases (phage therapy). Later, phages became favored objects for basic experiments of molecular biology. Nucleic acid replication, genetic recombination, regulation of gene expression, and assembly of complex structures are just a few examples of the fundamental biological processes elucidated. In addition, phages have an important role in modifying microbial communities by lysing their hosts, and transferring genetic material.

The study of bacterial pathogenicity revealed a frequent association between mobile DNA elements (insertion sequence (IS) elements, transposons, and phages) and many virulence functions. Bacteriophage-encoded virulence genes have been identified, these may code for proteins that enable the bacteria to colonize new hosts or evade immune responses as well as a range of toxins. Pathogenicity islands often share several characteristics with bacteriophages, *e.g.*, they encode virulence factors and integrases, their GC-content differs from the rest of the host genome, and they are flanked by repeat sequences (Hacker *et al.*, 1997). These similarities have led to the hypothesis that many pathogenicity islands might have a phage origin and have been acquired by new hosts by horizontal transfer. By

The Genus Yersinia, Edited by Skurnik *et al.*
·uwer Academic/Plenum Publishers, New York, 2003

studying phages we learn novel things about their hosts, the interaction between phage and host might also provide insight into the ways in which new and highly virulent pathogens evolve.

2. YERSINIOPHAGES

Based on their life cycles, phages have traditionally been divided into two groups: obligate lytic or virulent phages (like ϕYeO3-12 and ϕR1-37) and temperate phages (like PY20, Hertwig *et al.,* 1999) that are able either to grow lytically or to associate with the bacterial cell as prophages. Based on morphology, yersiniophages have been found to belong to all three families of double-stranded (ds) DNA tailed phages: *Myoviridae* (contractile tail), *Siphoviridae* (long, noncontractile tail), and *Podoviridae* (short tail). In addition, a filamentous prophage in *Y. pestis* biovar orientalis was recently characterized (Gonzalez *et al.,* 2002).

Molecular data on yersiniophages are limited (Novosel'tsev *et al.,* 1990). Traditionally, lytic phages have been used for typing and in epidemiological studies (Nicolle *et al.,* 1967; Bergan and Norris, 1978; Baker and Farmer III, 1982). More recently, a number of temperate bacteriophages from *Yersinia* were isolated and characterized to unravel their role in gene exchange by transduction between pathogenic and nonpathogenic strains (Hertwig *et al.,* 1999; Popp *et al.,* 2000). From these studies, it was concluded that phage mediated transfer of the virulence plasmid between strains is rather unlikely. However, transduction of other chromosomal loci that contribute to virulence may still be quite common. A phage-tail like bacteriocin, enterocoliticin, that shows inhibitory activity against enteropathogenic strains of *Y. enterocolitica* was recently characterized (Strauch *et al.,* 2001). These bacteriocins have been regarded as defective bacteriophages, likely originally temperate (Bradley, 1967; Daw and Falkiner, 1996); if these or their non-defective ancestors are prevalent horizontal gene transmission may be a significant factor in virulence acquisition.

Many phages use different parts of the lipopolysaccharide (LPS), the major surface component of Gram-negative bacteria, as receptors. In the Skurnik laboratory a number of lytic yersiniophages have been isolated from the raw incoming sewage of the Turku City sewage treatment plant, and these phages have extensively been used as tools in LPS genetics (Skurnik and Zhang, 1996). By employing different host strains for enrichment, phages with different specificities have been obtained. ϕYeO3-12 was obtained after enrichment using a smooth *Y. enterocolitica* serotype O:3 strain 6471/76-c (YeO3-c) (Al-Hendy *et al.,* 1991). ϕYeO3-12 was subsequently used to isolate a rough *Y. enterocolitica* O:3 mutant (YeO3-R1)

(Al-Hendy *et al.,* 1992); ϕR1-37 was then in turn obtained after enriching the sewage phages using YeO3-R1 (Skurnik *et al.,* 1995).

2.1 ϕR1-37

Yersiniophage ϕR1-37 can only infect YeO3-c when bacteria are grown at 37°C, indicating that the phage receptor is sterically blocked by abundant O-antigen in bacteria grown at 22°C. Characterization of various ϕR1-37 resistant mutants of YeO3-c showed that ϕR1-37 uses the outer core (OC) as its receptor (Skurnik *et al.,* 1995). Electron microscopy of negatively stained ϕR1-37 virions revealed particles with approximate dimensions of 100 nm for the head and 200 nm for the contractile tail. Normal capsids sometimes showed pentagonal outlines, indicating their icosahedral nature. Based on its morphology, ϕR1-37 belongs to the family *Myoviridae* (Matthews, 1981) resembling ϕKZ of *Pseudomonas aeruginosa* (Mesyanzhinov *et al.,* 2002). Moreover, pulse-field gel electrophoresis analysis showed that the ϕR1-37 genome consist of linear dsDNA with the estimated size of 270-280 kilobasepairs (kb) (Skurnik *et al.,* unpublished).

2.2 ϕYeO3-12

O-antigen of YeO3-c, a homopolymer of 6-deoxy-L-altrose, was shown to be the ϕYeO3-12 receptor, as spontaneous phage resistant mutants are missing the O-antigen. In addition, other *Yersinia* strains with a similar O-antigen to YeO3-c are ϕYeO3-12 sensitive (Pajunen *et al.,* 2000). ϕYeO3-12 also grows on *Escherichia coli* expressing the cloned O-antigen of YeO3-c; however, natural isolates of *E. coli* are phage resistant (Al-Hendy *et al.,* 1991; Al-Hendy *et al.,* 1992).

The physical and biological properties of ϕYeO3-12 suggested that it is closely related to coliphages T3 and T7, two of the original type phages defined by Demerec and Fano (1945). Dimensions of the icosahedral virion are ~57 nm in diameter for the head and 15 × 8 nm for the tail. Thus ϕYeO3-12 belongs to the family *Podoviridae.* One-step growth curves of ϕYeO3-12 propagated on YeO3-c at 22°C show eclipse and latent periods of 15 and 25 min, respectively, followed by a short rise of 10 min; the burst size is 100 to 140 p.f.u. per infected cell (Pajunen *et al.,* 2000). The size of the linear dsDNA ϕYeO3-12 genome is 39.6 kb (Pajunen *et al.,* 2001), which is in good agreement with that of T3 (Pajunen *et al.,* 2002) and T7 (Dunn and Studier, 1983).

2.2.1 Sequence analysis

Genome-wide comparisons with the closely related bacteriophages became possible after completing the nucleotide sequences of ϕYeO3-12 and T3. Because of extensive similarity between ϕYeO3-12 and T3 or T7, ϕYeO3-12 genes were named according to the T3/T7 nomenclature; the original 19 essential genes on the T7 genetic map have integer numbers, and genes added subsequently have decimal numbers (Figure 1). All genes are transcribed rightward; the leftmost genes being expressed at the earliest time. The early genes are transcribed by the host RNA polymerase (RNAP) from early σ^{70}-like promoters (A1, A2, and A3), with transcripts terminating at a common site, T_E. Middle and late genes are transcribed by the rifampicin-resistant phage promoter-specific RNAP, with transcripts terminating either at Tϕ or near the genome end. The specificity of the phage-encoded RNAP is one characteristic feature of all T7-like phages; although the RNAPs are similar, they do not efficiently transcribe heterologous DNA.

Comparative analysis shows that both the order and the arrangement of close-packed genes are conserved in ϕYeO3-12, T3, and T7 (Figure 1). Overall, ϕYeO3-12 is more similar to T3 than it is to T7; however, the genomes are not identical with respect to the total number of genes and regulatory genetic elements present. Many regions with high levels of similarity are flanked by regions of no similarity, suggesting that different genomes of T7 group have independently acquired genetic material by recombination. Relative to T3, T7 has acquired genes *0.4*, *0.5*, *1.4*, *2.8*, *4.7*, *7*, and *7.7* but lacks T3 gene *1.05* in equivalent genomic regions. Again relative to T3, ϕYeO3-12 has similarly gained genes *0.45*, *1.45*, and *13.5*. Several gene products in these divergent regions are thought to be related to group I introns or homing endonucleases that are thought to be mobile element often inserting into self-splicing elements, making the composite element mobile (Gimble, 2000). Between about 34 kb and 41.5 kb DNA can be efficiently into a T7 capsid. However, only about 32 kb is accounted for by essential and conditionally essential genes plus all known regulatory elements; at least 2 kb of DNA may need to be added to a 'basic' T7-like genome for efficient packaging. Some of the extra DNA may thus be derived from parasitic DNA elements that have found their ecological niche in a larger parasite. However, the question remains of the original sources of these genes that may not provide an advantage to the phage.

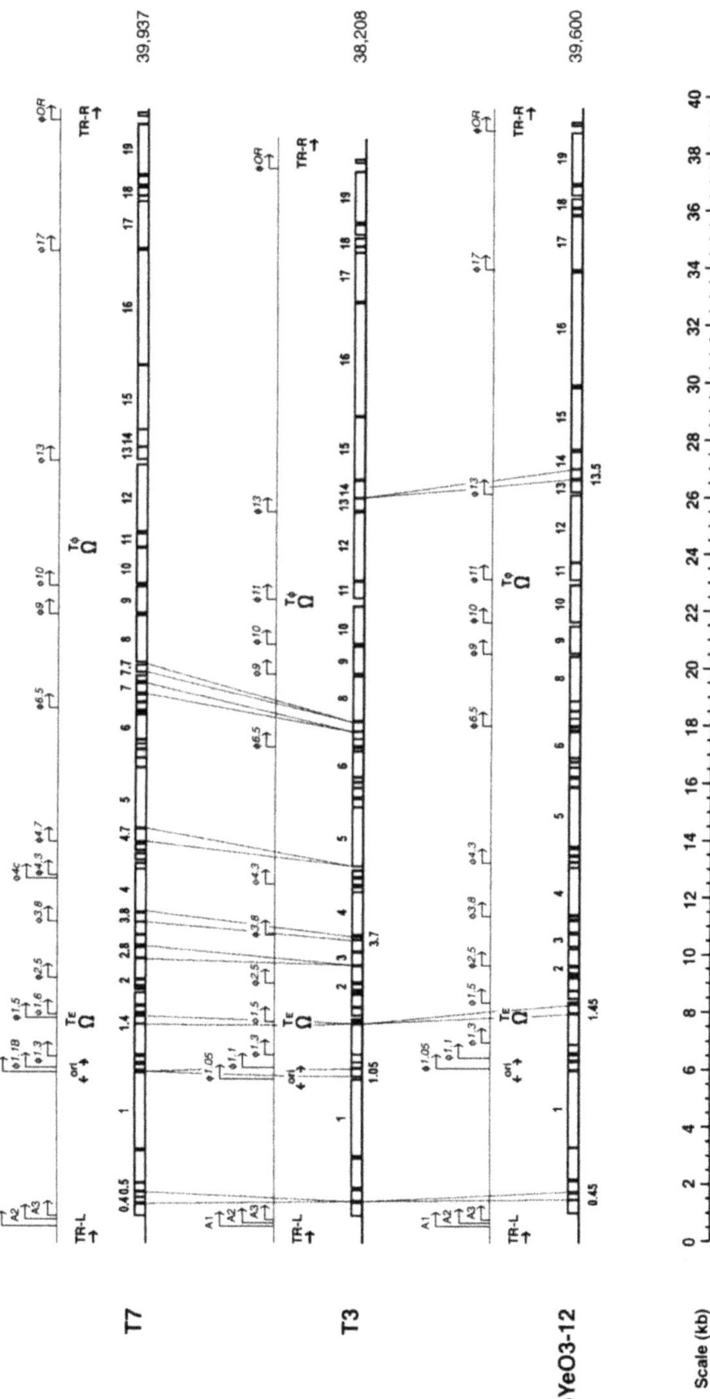

Figure 1. Comparison of φYeO3-12 (AJ251805), T3 (AJ318471), and T7 (V01146) genomes.

2.2.2 Genetic recombination

Bioinformatics revealed that the capsid and tail proteins of T3 seem to have close similarities to the proteins from different phage origins proposing a possible recombination event. T3 genes *6.3* through *14* are over 90% identical to ϕYeO3-12, whereas identity with T7 is ~70%; while T3 genes *15* through *17* are only ~65% identical to ϕYeO3-12 but over 90% identical to T7 (Pajunen *et al.*, 2002).

To assess the likely recombinatorial origin of T3, ϕYeO3-12 and T7 were allowed to recombine *in vivo* using an *E. coli* strain that expresses the cloned O-antigen of YeO3-c; coliphage progeny from this cross were selected that had many biological properties of T3 (Pajunen *et al.*, 2002). Two characterized recombinants appeared to be largely derived from T7 and to contain a single insertion of ϕYeO3-12 DNA. These data suggest that wild-type T3 is the product of recombination between a T7-like coliphage and a ϕYeO3-12 –like yersiniophage. However, this experimental data can only partially answer the question of the precise origin of T3, the actual wild-type bacteriophage is quite likely to have a more complex history. Nevertheless, this is the first experimentally controlled recombination between two obligate lytic phages whose normal hosts come from different bacterial genera. It also provides strong experimental support for the idea that genomes of the T7 group of phages are also mosaics taken from a natural gene pool, an idea proposed for unrelated phages (Hendrix *et al.*, 1999).

2.2.3 Host specificity

Tail fibers, encoded by gene *17*, are primarily responsible for phage attachment to the cell surface LPS. Six tail fibers, each consisting trimers of gp17 are attached to the phage tail just below its junction with the head-to-tail connector. ϕYeO3-12 gp17 is a protein of 645 amino acid residues, more than 100 residues larger than its T3 and T7 homologues. Although the N-terminal part of ϕYeO3-12 gp17 is homologous to T3 gp17 and T7 gp17, the C-terminal two-thirds are very different. This is consistent with the fact that in many dsDNA phages it has been demonstrated that the C-terminal parts of the tail fiber proteins evolve faster than other phage genes as a result of intense host selection (Haggård-Ljungquist *et al.*, 1992).

ϕYeO3-12 gp17 was shown to be sufficient to turn a coliphage into a yersiniophage when the tail fiber genes of ϕYeO3-12 and T3 were exchanged (Pajunen, 2001). Although gp17 was therefore shown to be the major determinant in host range specificity, other phage proteins may also contribute as the recombinants characterized were very unstable and mutations accumulated in phages that persisted. The location(s) of these mutations has not yet been determined.

ACKNOWLEDGEMENTS

Work at the University of Turku was supported by the Academy of Finland, the National Technology Agency of Finland, and the Turku Graduate School for Biomedical Sciences (M.I.P.), work at the University of Texas was supported by grant N00014-97-1-0295 from the Office of Naval Research, and GM32095 from the National Institutes of Health.

REFERENCES

Al-Hendy, A., Toivanen, P., and Skurnik, M., 1991, Expression cloning of the *Yersinia enterocolitica* O:3 *rfb* gene cluster in *Escherichia coli* K12. *Microbial Pathog.* **10**: 47-59.

Al-Hendy, A., Toivanen, P., and Skurnik, M., 1992, Lipopolysaccharide O side chain of *Yersinia enterocolitica* O:3 is an essential virulence factor in an orally infected murine model. *Infect. Immun.* **60**: 870-875.

Baker, P. M., and Farmer III, J. J., 1982, New bacteriophage typing system for *Yersinia enterocolitica, Yersinia kristensenii, Yersinia frederiksenii*, and *Yersinia intermedia*: correlation with serotyping, biotyping and antibiotic susceptibility. *J. Clin. Microbiol.* **15**: 491-502.

Bergan, T., and Norris, J. R., 1978, Bacteriophage typing of *Yersinia enterocolitica. Methods Microbiol.* **12**: 25-36.

Bradley, D. E., 1967, Ultrastructure of bacteriophage and bacteriocins. *Bacteriol. Rev.* **31**: 230-314.

Daw, M. A., and Falkiner, F. R., 1996, Bacteriocins: nature, function and structure. *Micron* **27**: 467-479.

Demerec, M., and Fano, U., 1945, Bacteriophage-resistant mutants in *E. coli. Genetics* **30**: 119-136.

Dunn, J. J., and Studier, F. W., 1983, Complete nucleotide sequence of bacteriophage T7 DNA and the locations of T7 genetic elements. *J. Mol. Biol.* **166**: 477-535.

Gimble, F. S., 2000, Invasion of a multitude of genetic niches by mobile endonuclease genes. *FEMS Microbiol. Letters* **185**: 99-107.

Gonzalez, M. D., Lichtensteiger, C. A., Caughlan, R., and Vimr, E. R., 2002, Conserved filamentous prophage in *Echerichia coli* O18:K1:H7 and *Yersinia pestis* biovar orientalis. *J. Bact.* **184**: 6050-6055.

Hacker, J., Blum-Oehler, G., Muhldorfer, I., and Tschape, H., 1997, Pathogenicity islands of virulent bacteria: structure, function and impact on microbial evolution. *Mol. Microbiol.* **23**: 1089-1097.

Haggård-Ljungquist, E., Halling, C., and Calendar, R., 1992, DNA sequences of the tail fiber genes of bacteriophage P2: evidence for horizontal transfer of tail fiber genes among unrelated bacteriophages. *J. Bacteriol.* **174**: 1462-1477.

Hendrix, R. W., Smith, M. C., Burns, R. N., Ford, M. E., and Hatfull, G. F., 1999, Evolutionary relationships among diverse bacteriophages and prophages: all the world's a phage. *Proc. Natl. Acad. Sci. USA* **96**: 2192-2197.

Hertwig, S., Popp, A., Freytag, B., Lurz, R., and Appel, B., 1999, Generalized transduction of small *Yersinia enterocolitica* plasmids. *Appl. Environ. Microbiol.* **65**: 3862-3866.

Matthews, R. E. F., 1981, The classification and nomenclature of viruses. *Intervirology* **16**: 53-60.

Mesyanzhinov, V. V., Robben, J., Grymonprez, B., Kostyuchenko, V. A., Bourkaltseva, M. V., Sykilinda, N. N., Krylov, V. N., and Volckaert, G., 2002, The genome of bacteriophage øKZ of *Pseudomonas aeruginosa. J. Mo. Biol.* **317**: 1-19.

Nicolle, P., Mollaret, H., Hamon, Y., and Vieu, J. F., 1967, Étude lysogenique, bacterio-cinogenique et lysotypique de l'espèce *Yersinia enterocolitica. Annales L'Inst. Pasteur* **112**: 86-92.

Novosel'tsev, N. N., Marchenkov, V. I., Sorokin, V. M., Kravtsov, A. N., and Degtiarev, B. M., 1990, Relation between the *Yersinia* phage and bacteriophages isolated from the environment. *Mol. Gen. Mikrobiol. Virusol.* **8**: 18-21.

Pajunen, M., Kiljunen, S., and Skurnik, M., 2000, Bacteriophage øYeO3-12, specific for *Yersinia enterocolitica* serotype O:3, is related to coliphages T3 and T7. *J. Bacteriol.* **182**: 5114-5120.

Pajunen, M. I., Kiljunen, S. J., Söderholm, M.-E. L., and Skurnik, M., 2001, Complete genomic sequence of the lytic bacteriophage øYeO3-12 of *Yersinia enterocolitica* serotype O:3. *J. Bacteriol.* **183**: 1928-1937.

Pajunen, M. I., 2001, PhD Thesis *Molecular analysis of Yersinia enterocolitica serotype O:3 –specific bacteriophage øYeO3-12.* University of Turku.

Pajunen, M. I., Elizondo, M. R., Skurnik, M., Kieleczawa, J., and Molineux, I. J., 2002, Complete nucleotide sequence and likely recombinatorial origin of bacteriophage T3. *J. Mol. Biol.* **319**: 1115-1132.

Popp, A., Hertwig, S., Lurz, R., and Appel, B., 2000, Comparative study of temperate bacteriophages isolated from *Yersinia. System. Appl. Microbiol.* **23**: 469-478.

Skurnik, M., Venho, R., Toivenen, P., and Al-Hendy, A., 1995, A novel locus of *Yersinia enterocolitica* serotype O:3 involved in lipopolysaccharide outer core biosynthesis. *Mol. Microbiol.* **17**: 575-594.

Skurnik, M., and Zhang, L., 1996, Molecular genetics and biochemistry of *Yersinia* lipopolysaccharide. *APMIS* **104**: 849-872.

Strauch, E., Kaspar, H., Schaudinn, C., Dersch, P., Madela, K., Gewinner, C., Hertwig, S., Wecke, J., and Appel, B., 2001, Characterization of enterocoliticin, a phage tail-like bacteriocin, and its effect on pathogenic *Yersinia enterocolitica* strains. *Appl. Environ. Microbiol.* **67**: 5634-5642.

Chapter 46

Properties of the Temperate *Yersinia enterocolitica* Bacteriophage PY54

Stefan HERTWIG, Iris KLEIN and Bernd APPEL
Robert Koch-Institut, Berlin, Germany

1. INTRODUCTION

In contrast to phages infecting other *Enterobacteriaceae*, there is only little information available about *Yersinia* phages. A number of phage typing sets containing temperate or virulent phages have been worked out (Kawaoka *et al.*, 1987). *Y. enterocolitica* strains have been reported to be frequently lysogenic (Tsubokura *et al.*, 1982), whereas there are only few reports about the isolation of temperate phages from *Y. pseudotuberculosis* and *Y. pestis*, Up to now, the characterization of *Yersinia* phages mainly focussed on host ranges, morphologies and DNA homologies (Popp *et al.*, 2000). The genomes of two virulent *Yersinia* phages (ϕA1122 and ϕYe03-12) have been analysed in detail (Garcia *et al.*, 2000; Pajunen *et al.*, 2001). Here, we report on characteristics of the temperate *Y. enterocolitica* phage PY54.

2. RESULTS AND DISCUSSION

PY54 isolated from a nonpathogenic *Y. enterocolitica* O:5 strain has a lambda-like morphology and infects O:5 strains as well as pathogenic O:5,27 strains. The phage genome comprises double-stranded DNA and contains 3′-protruding ends of 10 nucleotides. A rather unusual feature of PY54 is that its prophage replicates as a linear plasmid with covalently closed ends (telomeres). Besides *E. coli* phage N15, PY54 is the only member of this group so far identified. The phage genome and plasmid prophage have the

The Genus Yersinia, Edited by Skurnik *et al.*
...wer Academic/Plenum Publishers, New York, 2003

same size and both molecules are approximately 50% circularly permuted. The plasmid telomeres are generated by a phage encoded enzyme termed protelomerase (Tel) which cleaves a 42 bp palindromic sequence (*tel* site) located upstream from the *tel* gene (Hertwig *et al.*, 2003).

We have determined the complete DNA sequence (46339 bp) of the PY54 genome. The average G-C content is 44.6%, slightly lower than the 48.5±1.5% reported for *Y. enterocolitica*. Bioinformatic analysis revealed 67 open reading frames (ORFs) located on both DNA strands (Figure 1). For 44 deduced gene products, database matches were made. Through these homologies to known proteins of other phages and bacteria and by functional studies, possible functions of 25 genes could be assigned. As the genome of phage N15, the PY54 genome is divided into a left arm and a right arm separated by the *tel* site (Figure 1).

It appears that the left arm mainly contains late genes important for phage assembly. We identified a number of genes whose predicted products are similar to head and tail proteins of other phages. The order of these genes is reminiscent to lambdoid phages. However, in contrast to N15 head and tail proteins that are up to 97% identical to their lambda counterparts, only three out of 31 predicted products of the PY54 left arm are closely related to lambda proteins. Moreover, there are 15 ORFs in the left arm for which no homologous sequences could be found. In addition to genes encoding structural proteins, the PY54 left arm contains two genes obviously involved in plasmid partitioning. Their predicted products are closely related to the partitioning proteins SpyA and SpyB encoded by the *Yersinia* virulence plasmid pYV. These proteins interact with the centromere-like site *spy*C located on pYV downstream from the *spy* operon. Interestingely PY54 harbours eight identical copies of *spyC* scattered over the whole genome.

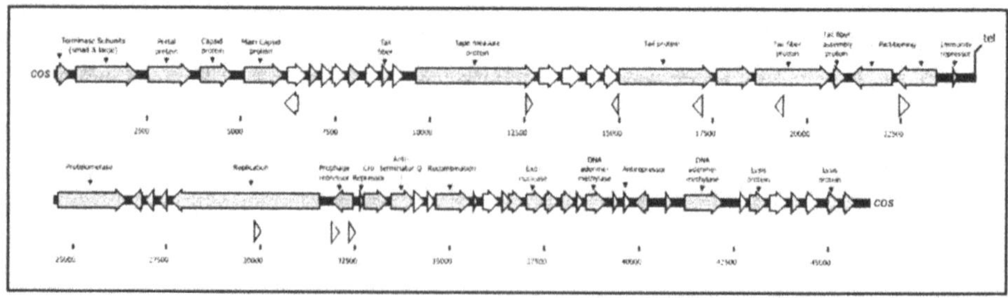

Figure 1: Genetic map of PY54

The PY54 right arm contains 36 ORFs. For most of the predicted proteins, database matches were made. Fifteen products reveal relationships to N15 proteins, most of which are implicated in the regulation of the lytic and lysogenic cycle or in plasmid prophage replication (Ravin *et al.,* 2000). As noted above, PY54 produces a protelomerase that is very similar to the corresponding enzyme of phage N15. Both proteins are essential for the conversion of the linear phage DNAs into the linear plasmid prophages. However, the target sequences (palindromes) for the protelomerases of PY54 and N15 are not identical. Upstream from the PY54 *tel* gene, there is a large ORF encoding a replication protein. This protein is suggested to be multifunctional, comprising primase, helicase, and origin binding activities and it is sufficient to drive the replication of miniplasmid derivatives of the prophage (Hertwig *et al.,* 2003). The PY54 right arm also harbours a number of regulatory genes. We have identified ORFs for products related to a prophage repressor, Cro-repressor, antirepressor, and antiterminator Q. Furthermore, this region of the PY54 genome contains genes probably encoding DNA modifying enzymes, an exonuclease, and lysis proteins.

REFERENCES

Garcia, E., Elliott, J.M., Macht, M., Stilwagen, S., Yockey, B., Andersen, G., and M.C. Chu, 2000, Analysis of *Yersinia pestis* bacteriophage phiA1122. *ASM General Meeting Abstract* 05/21/00-05/25/00.

Hertwig, S., Klein, I., Lurz, R., Lanka, E., and B. Appel, 2003, PY54, a linear plasmid prophage of *Yersinia enterocolitica* with covalently closed ends. *Mol. Microbiol.,* in press.

Kawaoka, Y., Mitani, T., Otsuki, K., and M. Tsubokura, 1987, Isolation and use of eight phages for typing *Yersinia enterocolitica* O3. *J. Med. Microbiol.* **23**: 349-352.

Pajunen, M.I., Kiljunen, S., Söderholm, E. L., and M. Skurnik, 2001, Complete genomic sequence of the lytic bacteriophage ϕYeO3-12 of *Yersinia enterocolitica* serotyp O:3. *J. Bacteriol.* **183**: 1928-1937.

Popp, A., Hertwig, S., Lurz, R., and B. Appel, 2000, Comparative study of temperate bacteriophages isolated from *Yersinia. Syst. Appl. Microbiol.* **23**: 469-478.

Ravin, V., Ravin, N., Casjens, S., Ford, M.E., Hatfull, G.F., and R.W. Hendrix, 2000, Genomic sequence and analysis of the atypical temperate bacteriophage N15. *J. Mol. Biol.* **299:** 53-73.

Tsubokura, M., Otsuki, K., and Y. Kawaoka, 1982, Lysogenicity and phage typing of *Yersinia enterocolitica* isolated in Japan. *Jpn. J. Vet. Sci.* **44**: 433-438.

Chapter 47

Transposon Mutagenesis of the Phage ϕYeO3-12

Saija KILJUNEN[1], Heikki VILEN[2], Harri SAVILAHTI[2] and Mikael SKURNIK[1,3]
[1]Department of Medical Biochemistry and Molecular Biology, University of Turku, Finland: [2]Program in Cellular Biotechnology, Institute of Biotechnology, Viikki Biocenter, University of Helsinki, Finland; [3]Department of Bacteriology and Immunology, Haartman Institute, University of Helsinki, Finland.

1. INTRODUCTION

ϕYeO3-12 is a lytic bacteriophage that infects *Yersinia enterocolitica* O:3 (YeO3). The phage is also able to infect and proliferate in *Escherichia coli* strains expressing the YeO3 O-antigen, which is the phage receptor (Al-Hendy *et al.*, 1991). The genome of ϕYeO3-12 is 39 600 bp long linear, double-stranded DNA molecule, coding for 58 putative genes (Pajunen *et al.*, 2001).

ϕYeO3-12 belongs into T7 – group of phages, T3 being its closest relative. These phages share the overall organization of the genome, with early, middle and late genes (Molineux, 1999). The products of early genes convert the metabolism of the host cell to the production of phage proteins, middle gene products are involved in the phage DNA metabolism, and late genes code for the structural proteins of the phage particle. Based on homology to T7 and T3, the functions of many ϕYeO3-12 genes have been deduced, but there are still open reading frames in the genome whose role is not known. Here, we used efficient *in vitro* transposon insertion mutagenesis to study the role of the phage genes.

The Genus Yersinia, Edited by Skurnik *et al.*
ıwer Academic/Plenum Publishers, New York, 2003

2. METHODS AND RESULTS

As the transposon mutagenesis system, we utilized the MuA transposase-catalysed *in vitro* transposition reaction, with *lacZ* as a marker gene in the transposon (Vilen *et al.*, 2003). Phage DNA was electroporated into *E. coli* JM109/pAY100, and mutants were identified by the blue color of the plaques. Altogether 17 transposon mutants were obtained, two of which had double-insertions (Figure 1).

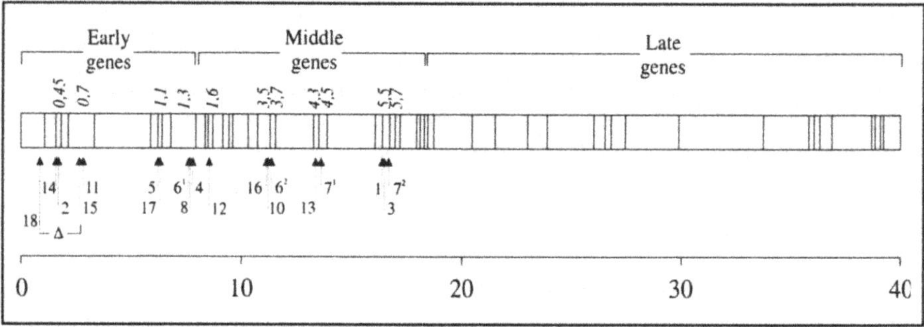

Figure 1. The ϕYeO3-12 genome. The sites of transposon insertions are marked with arrows with the corresponding mutant number. Only the genes containing insertions are indicated. The range of the deletion in mutant no. 18 is also shown.

Based on the plaque morphology and differences in the efficiency of plating (EOP), it was evident that some mutants had growth deficiency in *Y. enterocolitica* but not in *E. coli*. To study this further, mutants were analysed for their fitness (Rokyta *et al.,* 2002), i.e. phage titers were determined directly and 45 minutes after infection of bacteria (Figure 2). Mutants #4 and #8 (insertion in gene *1.3,* coding for DNA ligase), #10 (insertion in gene *3.5,* coding for lysozyme) and #11 (insertion in gene *0.7,* coding for protein kinase) had clearly lower fitness in YeO3-c than the wild-type phage. In *E. coli*, this difference was not detectable. Interestingly, mutant #18 (genes *0.3, 0.45, 0.6A* and *B* and half of the gene *0.7* deleted) had remarkably higher fitness in YeO3-c than in *E. coli*. On the other hand, EOP of this mutant in YeO3-c was only 1/100 of that in JM109/pAY100 (data not shown). Mutants #2, #14 (insertions in gene *0.45*), #5, #17 (insertions in gene *1.1*), and #15 (insertion in gene *0.7*) had clearly lower EOP in YeO3-c than in *E. coli*, even though there were no differences in their fitness. The EOPs of these mutants in *Shigella sonnei* and *Salmonella typhimurium* expressing YeO3 O-antigen were the same as in *E. coli*. Since Δpk, a ϕYeO3-12 mutant missing gp0.7, is fully virulent in YeO3-c (data not shown), the phenotypes of mutants #11 and #15 may be due to some second site mutation or polar effect.

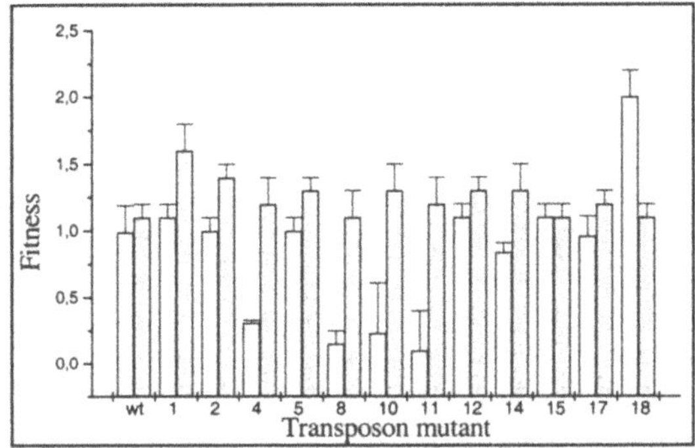

Figure 2. Fitness of transposon mutants. Fitness represents the number of doublings in phage numbers in 15 minutes, which is the approximate generation time of T7. White and gray bars show fitnesses of the mutants determined in YeO3-c and JM109/pAY100, respectively.

3. CONCLUSIONS

Transposon insertions were obtained in early and middle regions of the phage genome only, which is consistent with the essential nature of the late genes. Phage genes coding for DNA ligase and lysozyme were found to be needed for efficient propagation of φYeO3-12 in YeO3-c but not in *E. coli*. Also, in *Yersinia* but not in other bacteria tested, phage gene *0.45* seemed to be important in the early stages of infection. Still, further studies are needed to understand the biological basis for these phenotypes.

REFERENCES

Al-Hendy, A., Toivanen, P. and Skurnik, M., 1991, Expression cloning of the *Yersinia enterocolitia* O:3 *rfb* gene cluster in *Escherichia coli* K12. *Microb. Pathog*. **10**: 47-59.

Molineux, I.J., 1999, The T7 family of bacteriophages. In *Encyclopedia of molecular biology* (T.E. Creighton, ed.), John Wiley and Co., New York, N.Y., pp. 2495-2507.

Pajunen, M. I., Kiljunen, S. J., Söderholm, M. E.-L. and Skurnik, M., 2001, Complete genomic sequence of the lytic bacteriophage φYeO3-12 of *Yersinia enterocolitica* serotype O:3. *J. Bacteriol.* **183**: 1928-1937.

Rokyta, D., Badgett, M. R., Molineux, I. J. and Bull, J. J., 2002, Experimental genomic evolution: Extensive compensation for loss of DNA ligase activity in a virus. *Mol. Biol. Evol.,* **19**: 230-238.

Vilen, H., Aalto, J.-M., Kassinen, A., Paulin, L. and Savilahti, H., 2003, A direct transposon insertion tool for the modifiction and functional analysis of viral genomes. *J. Virol.* 77: 123-134.

Chapter 48

Analysis of Enterocoliticin, a Phage Tail-Like Bacteriocin

Eckhard STRAUCH, Heike KASPAR, Christoph SCHAUDINN, Christina DAMASKO, Antje KONIETZNY, Petra DERSCH, Mikael SKURNIK[1] and Bernd APPEL

Robert Koch-Institut and Freie Universität, Berlin, Germany; [1]Department of Bacteriology and Immunology, Haartman Institute, University of Helsinki and Department of Medical Biochemistry and Molecular Biology, University of Turku, Finland

1. INTRODUCTION

Bacteriocins have traditionally been defined as proteinaceous compounds produced by bacteria that inhibit or kill closely related bacteria. A special group of bacteriocins are high molecular weight particles, which can be sedimented by ultracentrifugation and are resolved by electron microscopy as phage tail-like particles. Bacteriocins of this type have been found in cultures of several Gram-negative bacteria, e.g. *Enterobacteriaceae, Vibrionaceae* and *Pseudomonadaceae* (Bradley, 1967).

The production of phage tail-like particles by strains of *Y. kristensenii, Y. frederiksenii* and *Y. intermedia* and one strain of *Y. enterocolitica* was reported earlier (Calvo *et al.*, 1986; Hamon *et al.*, 1967). We found a foodborne strain of *Y. enterocolitica*, which produced a bacteriocinogenic substance with an inhibitory activity against the pathogenic serotype strains O:3, O:5,27 and O:9 of *Y. enterocolitica.* Calorimetric measurements of the heat output of cultures of sensitive bacteria showed a complete loss of cellular metabolic activity immediately upon addition of enterocoliticin. Furthermore, a dose dependent efflux of K^+ ions into the medium was determined, indicating that enterocolicitin has a channel forming activity (Strauch *et al.*, 2001).

The Genus Yersinia, Edited by Skurnik *et al.*
'luwer Academic/Plenum Publishers, New York, 2003

2. MATERIAL AND METHODS

Y. enterocolitica 29930 (O:7,8 biotype 1A) is a foodborne isolate (Hoffmann *et al.*, 1998), *Y. enterocolitica* 13169 (O:3, biotype 4) was used as susceptible indicator strain for most experiments. *Yersinia* strains used for the determination of the inhibitory activity of enterocoliticin were from the strain collection of the Robert Koch Institute and include several strains initially obtained from the Institute Pasteur, Paris, France.

The antimicrobial activity of enterocoliticin was determined by spotting serial dilutions of enterocoliticin preparations (20 μl) on LB plates, which had been overlaid with LB soft agar containing approximately 1×10^6 CFU ml^{-1} of susceptible bacteria. Plates were incubated at 37°C overnight. The reciprocal of the highest dilution that formed a visible inhibition zone was defined as the relative activity (AU = activity units) of enterocoliticin. Enterocoliticin titers are expressed as activity units per ml (AU ml^{-1}).

3. RESULTS AND DISCUSSION

Enterocoliticin is a bacteriocin produced from *Y. enterocolitica* 29930 (serotype O:7,8, biotype 1A). The bacteriocin resembles a phage tail which contracts upon contact with sensitive bacterial cells. The particles are active against pathogenic *Y. enterocolitica* strains of the serotypes O:3; O:5,27 and O:9 and against some nonpathogenic *Yersinia* strains. *Y. enterocolitica* O:8 strains and *Y. pseudotuberculosis* strains are not susceptible.

The MIC value (minimal inhibitory concentration) of enterocoliticin determined was $8.2 \pm 0.4 \times 10^3$ AU ml^{-1} for the *Y. enterocolitica* strain 13169 (O:3) at 25 and 37 °C. The bactericidal activity of enterocoliticin was demonstrated by adding enterocoliticin to logarithmic cultures of the susceptible O:3 indicator strain. The highest concentration used (2.2×10^5 AU ml^{-1}) corresponding to approximately the 25-fold value of the MIC-concentration reduced the CFU close to four orders of magnitude within 60 minutes, while an enterocoliticin concentration of 2.5-fold of the MIC reduced the CFU titre by three orders of magnitude (more than 99 % killing).

To evaluate the efficacy of enterocoliticin for future application in an animal model, the antimicrobial activity was tested in a Hep-2 cell culture system, which had been infected with the sensitive indicator strain *Y. enterocolitica* 13169 (O:3). Depending on the incubation conditions of the HEp-2 cells the bacteria either invaded the cells or adhered to the surface of the eukaryotic cells. All bacteria remaining free in the cell culture medium were removed before enterocoliticin was added. In case of invasion no significant reduction of bacterial CFU was observed after addition of enterocoliticin indicating that the uptake into the eukaryotic cell rescued the bacteria from the activity of enterocoliticin. However, when bacteria adherent to eukaryotic cells were exposed to enterocoliticin, a time and dose dependent decrease of the CFU was measurable.

The receptor on the bacterial surface was determined by testing sensitivity of LPS and *ail* mutant strains of *Y. enterocolitica* (serotype O:3) to enterocoliticin. The results of the inhibition demonstrated that the outer core oligosaccharide is the bacteriocin receptor.

SDS-PAGE of enterocoliticin preparations were performed and two proteins with the size of ca. 50,000 kD and 15,000 kD were subjected to N-terminal sequencing. The amino acid sequences indicate a significant similarity to two putative proteins encoded by *Y. pestis* genes which are related to phage genes. A cosmid clone from a library of strain 29930 was identified by hybridization using a probe derived from the *Y. pestis* sequence. We have started sequencing the cosmid insert and the preliminary data indicates that it carries additional phage related genes.

REFERENCES

Bradley, D.E., 1967, Ultrastructure of bacteriophage and bacteriocins. *Bacteriol. Rev.* **31**:230-314.

Calvo, C., J. Brault, A. Ramos-Cormenzana, and H.H. Mollaret, 1986, Production of bacteriocin-like substances by *Yersinia frederiksenii*, *Y. kristensenii*, and *Y. intermedia* strains. *Folia Microbiol.* (Praha.) **31**:177-186.

Hamon, Y., P. Nicolle, J.F. Vieu, and H. Mollaret, 1966, Recherche de la Bacteriocinogenie Parmi les Souches de *Yersinia enterocolitica. Annales de l'Institut Pasteur* **111**:368-372.

Hoffmann, B., E. Strauch, C. Gewinner, H. Nattermann, and B. Appel, 1998, Characterization of plasmid regions of foodborne *Yersinia enterocolitica* biogroup 1A strains hybridizing to the *Yersinia enterocolitica* virulence plasmid. *Syst. Appl. Microbiol.* **21**:201-211.

Strauch E., H. Kaspar, C. Schaudinn, P. Dersch, K. Madela, C. Gewinner, S. Hertwig, J. Wecke, and B. Appel, 2001, Characterization of enterocoliticin, a phage tail-like bacteriocin, and its effect on pathogenic *Yersinia enterocolitica* strains. *Appl. Environ. Microbiol.* **2001**:5634-5642.

Picture 20. Michael Marceau gives a lecture.

Chapter 49

Function and Regulation of the *Salmonella*-Like *pmrF* Antimicrobial Peptide Resistance Operon in *Yersinia pseudotuberculosis*

Michael MARCEAU, Florent SEBBANE, Francois COLLYN and Michel SIMONET

Inserm EI9919, Université Lille 2 JE2225, Institut Fédératif de Recherche 17, Institut de Biologie de Lille, Institut Pasteur de Lille, F-59000 Lille, France

1. INTRODUCTION

Cationic antimicrobial peptides, which are mostly composed of 30 to 40 amino acid residues, are important effectors of nonspecific defences against microbes in plants and animals. Although they all are able to alter the properties of negatively charged microbial membranes, their mode of action slightly differs from one class of compounds to another. Therefore, bacterial pathogens have developed a variety of countermeasures in order to resist a broad range of these molecules (Hancock, 2000). The most studied mechanism contributes to the environment-induced resistance to alpha-helical peptides and polymyxin via the substitution of lipid A's 4' phosphate group with 4-amino-4-deoxy-L-Arabinose, reducing the lipopolysaccharide negative charge.

2. THE *pmrF* OPERON AND ITS REGULATION: THE *Salmonella* PARADIGM

In *Salmonella typhimurium*, lipid A substitution with 4-amino-4-deoxy-L-Arabinose requires the *pmrHFIJKLM* operon (also referred to as the *pmrF* operon) (Gunn *et al.,* 1998), which contains seven open reading frames

The Genus Yersinia, Edited by Skurnik *et al.*
'uwer Academic/Plenum Publishers, New York, 2003

(ORFs). In this species, transcription of the *pmrF* operon is induced by diverse environmental cues, such as mildly acidic environment (pH 6) and high iron concentrations. This regulation has been shown to involve the PmrA-PmrB two-component system, where PmrB is the integral membrane sensor-kinase and PmrA is the cognate regulatory protein that binds to the promoter region upstream of *pmrH.* A drop of extracellular Mg^{2+} ion concentration from the millimolar to micromolar range can also promote PmrA-mediated upregulation of the *pmrF* operon. In addition to PmrA-PmrB, this process requires PhoP-PhoQ, an other two-component system, known to be a pleiotropic regulator of bacterial virulence (Groisman, 2001). Recently, it has been demonstrated that the PhoP-PhoQ mediated transcriptional activation of the *pmrF* operon occurs via the PhoP-promoted biosynthesis of PmrD, a 85 amino-acid PmrA activator. The results obtained by our group with the *Salmonella* LT2 strain are consistent with the above model : resistance to polymyxin is induced by Ca^{2+} and Mg^{2+} ion limitation, but dramatically decreased at low Fe^{3+} concentrations (Figure 1A) and in *pmrA* or *phoP* mutants (Figure 2).

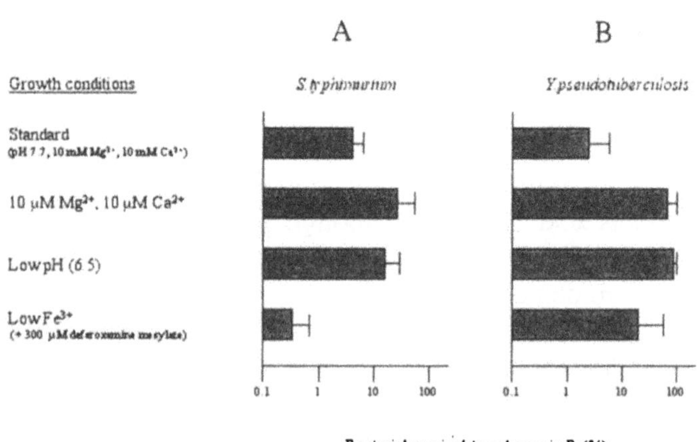

Figure 1. Influence of pH, Mg^{2+}/Ca^{2+} and Fe^{3+} ion concentrations on *S. typhimurium* (A) and *Y. pseudotuberculosis* (B) susceptibility to polymyxin B. Wild type strains of *S. typhimurium* (LT2) and *Y. pseudotuberculosis* (IP32777) were incubated at 37°C under different *in vitro* conditions with 5 µg/ml polymyxin B.

3. THE *Y. pseudotuberculosis pmrF* OPERON: SAME FUNCTION BUT DIFFERENT REGULATION COMPARED TO *S. typhimurium*

We have identified a chromosome-harbored *Salmonella pmrF*-like operon contributing to the resistance of *Y. pseudotuberculosis* to polymyxin and alpha-helical antimicrobial peptides. As expected, resistance to these compounds was enhanced by low Mg^{2+} and low pH environments, but, in contrast to *Salmonella,* it was not reduced when bacterial cells were grown in iron-chelated media (Figure 1B). In line with this observation, we found that PmrA-PmrB was not necessary for expression of the *Y. pseudotuberculosis pmrF* operon. In contrast, the PhoP-PhoQ two component system still appeared to be essential (Figure 2). These results suggested that, in comparison to *pmrF* in *Salmonella,* the homologous *Yersinia* operon played an identical role but was differently regulated.

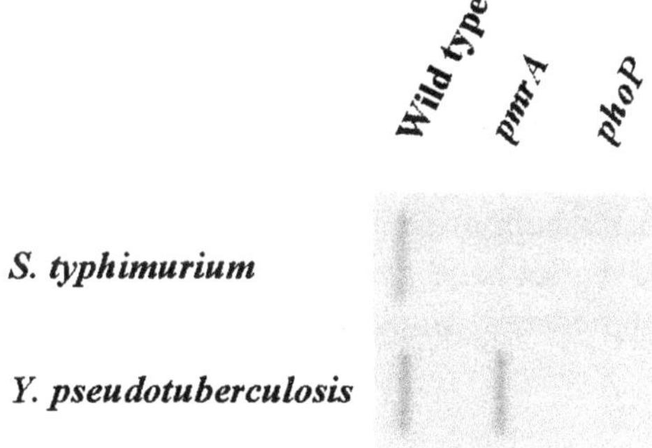

Figure 2. Compared transcription levels of the *pmrF* operon in *Y. peudotuberculosis*, *S. typhimurium* and their respective *pmrA*⁻ and *phoP*⁻ derivatives Total RNAs extracted from bacteria incubated for 30 minutes in antimicrobial peptide-free media were spotted onto nitrocellulose membranes and hybridized with a labeled 600bp *pmrF* probe.

4. UNLIKE IN *Salmonella*, THE *Y. pseudotuberculosis pmrF* OPERON IS NOT REQUIRED FOR VIRULENCE IN MICE

S. typhimurium survival within macrophages encoutered in the mouse intestine absolutely requires a functional *pmrF* operon (Gunn *et al.,* 2000). Pathogenic *Yersinia* resist phagocytosis and, therefore, remain extracellular.

We have measured the virulence of *pmrF* ⁻and *phoP* ⁻ mutants in an oral model of infection in the mouse. Whereas the pathogenicity of the *phoP* ⁻ mutant was considerably attenuated (≈100 fold), that of the *pmrF* ⁻ mutant was similar to the parental strain suggesting that, unlike *in Salmonella*, the *pmrF* operon minorly contributes to *Y. pseudotuberculosis* virulence in mice.

5. CONCLUSIONS

Our data indicate the high tolerance of *Y.pseudotuberculosis* to polymyxin and alpha-helical antimicrobial peptides requires a seven ORF *pmrF* operon, also found in other Gram negative bacteria like *S. typhimurium*. However, in *Yersinia*, although it has the same function, this operon appears to be differently regulated and to serve distinct purposes, with regards to the *Salmonella* model. This work suggests that a given bacterial mechanism can be differently used, according to the lifestyle of the microorganism.

ACKNOWLEDGEMENTS

This work was partly supported by the Conseil Régional Nord-Pas de Calais. F. Collyn and F Sebbane received a scholar fellowship from the Ministère de l'Enseignement Supérieur, de la Recherche et de la Technologie.

REFERENCES

Groisman, E.A., 2001, The pleiotropic two-component regulatory system. PhoP-PhoQ. *J Bacteriol.* **183**:1835-42.

Gunn, J.S., Lim, K.B., Krueger, J., Kim, K., Guo, L., Hackett, M., and Miller, S.I., 1998, PmrA-PmrB-regulated genes necessary for 4-aminoarabinose lipid A modification and polymyxin resistance. *Mol. Microbiol.* **27**:1171-82

Gunn, J.S., Ryan, S.S., Van Velkinburgh, J.C., Ernst, R.K., and Miller, S.I., 2000, Genetic and functional analysis of a PmrA-PmrB-regulated locus necessary for lipopolysaccharide modification, antimicrobial peptide resistance, and oral virulence of *Salmonella enterica* serovar *typhimurium. Infect. Immun. 68:6139-46.*

Hancock, R.E., and Scott, M.G., 2000, The role of antimicrobial peptides in animal defenses. *Proc Natl Acad Sci U S A.* **97**:8856-61.

Chapter 50

Porin from *Yersinia pseudotuberculosis*: Cloning and Analysis of Primary Structure

Marina P. ISSAEVA, Konstantin V. GUZEV, Olga D. NOVIKOVA, Tamara F. SOLOVJEVA, Sergei DEGTYAREV[1] and Valeri A. RASSKAZOV

Pacific Institute of Bioorganic Chemistry, FEBRAS, Vladivostok; [1]Scientific Research Institute of Molecular Biology and Biophysics, SBRAMS, Novosibirsk, Russia

1. INTRODUCTION

The porins are the major outer memrane proteins of Gram-negative bacteria. They form nonspecific water-filled channels to allow diffusion of small, polar molecules (< 600 Da) through the outer membrane. In general, porins are organized in trimers and, in conjunction with peptidoglycan and LPS, maintain the integrity of the cells. The well characterized pore-forming proteins are the two major non-specific constitutive porins of *Escherichia coli*, OmpF and OmpC, that are regulated by environmental stimuli such as osmotic pressure, pH, and temperature (Cowan, 1993). Porins play an important role in antibiotic resistance by decreasing the permeability of the outer membrane. Porins might also be involved in other functions, such as the invasion of epithelial cells by *Salmonella typhimurium* (Dorman *et al.*, 1989) and *Shigella flexneri* (Bernardini *et al.*, 1993). In addition to their functional properties, purified porins are immunogenic in either their trimeric or monomeric forms and they can form a complex with C1q, activating the complement classical pathway (Alberty *et al.*, 1993).

Yersinia pseudotuberculosis is an important agent of gastroenteritis throughout the world. Previous studies on the *Y. pseudotuberculosis* porins, called yersinins, indicated similarity to *E. coli* OmpF and OmpC on the basis of their modes isolation, molecular mass and N-terminal sequences.

The Genus Yersinia, Edited by Skurnik *et al.*
uwer Academic/Plenum Publishers, New York, 2003

2. CLONING AND SEQUENCING OF THE *ompF*-LIKE GENE OF *Y. pseudotuberculosis*

To isolate *ompF*-like gene the following primers were constructed. The first primer (VIMMKRN) was designed based on the hypothetical secondary structure of some 4.5S *micF* RNAs (*E. coli, Serratia marcescens* and *Y. pestis*) and their hybridization with *ompF* mRNAs. The second primer (GLVYQFstop) was directed to the C-terminus of enterobacterial porins.

```
YPS    AEIYNKDGNK LDLYGKVDAR HSFSDNNKQ- ----DGDKSY VRFGFKGETQ ITDQLTGYGQ 55
YPP    AEIYNKDGNK LDLYGKVDAR HSFSDNNKQ- ----DGDKSY VRFGFKGETQ ITDQLTGYGQ
SMA    AEIYNKDGNK LDLYGKVDGL HYFSKDKGN- ----DGDQTY VRFGFKGETQ ITDQLTGYGQ
ECF    AEIYNKDGNK VDLYGKAVGL HYFSKGNGEN SYGGNGDMTY ARLGFKGETQ INSDLTGYGQ
         β16        β1             L1            β2             β3

YPS    WEYNIQANNA EDSGAQDGNA TRLGFAGLKF AEFGSFDYGR NYGVIYDVNA WTDMLPVFGG 115
YPP    WEYNIQANNA EDTGAQDGNA TRLGFAGLKF AEFGSFDYGR NYGVIYDVNA WTDMLPVFGG
SMA    WEYNVQSNHA ESQGTE-GTK TRLGFAGLKF ADYGSFDYGR NYGVLYDVEG WTDMLPEFGG
ECF    WEYNFQGNNS EGADAQTGNK TRLAFAGLKY ADVGSFDYGR NYGVVYDALG YTDMLPEFGG
                 L2           β4           β5                    L3

YPS    DSISNSDNFM TGRSTGLATY RNNNFFGMVD GLNFALQYQG KNDRSEVKEA NGDGFGIG-- 173
YPP    DSISNSDNFM AGRSTGLATY RNNNFFGLVD GLNFALQYQG KNDRSEVKEA NGDGFGIG--
SMA    DTYTYSDNFM TGRTNGVATY RNNNFFGLVD GLNFALQYQG KNQNDGRDVK KQNGDGWGIS
ECF    DT-AYSDDFF VGRVGGVATY RNSNFFGLVD GLNFAVQYLG KNERDTARRS NGDGVGGSIS
                   β6                β7              L4        β8

YPS    STYDLGNGIN FGAGFSSSNR TLDQKYGSTA EGDKAQAWNV GAKYDANNVY LAVMYAETQN 233
YPP    STYDIGNGIN FGAGFSSSNR TLDQKYGSTA EGDKAQAWNV GAKYDANNVY LAVMYAETQN
SMA    STYDIGEGVS FGAAYASSNR TDDQKLRSNE RGDKADAWTV GAKYDANNVY LAAMYAETRN
ECF    YEYE-GFGIV --GAYGAADR TNLQEAQPLG NGKKAEQWAT GLKYDANNIY LAANYGETRN
                  β9           L5                β10          β11

YPS    LTPYGSFSDT ---------- -IANKTRDIE ITAQYQFDFG LRPSLGYVES KGKDLNN--V 280
YPP    LTPYGFYDFT ---------- -IANKTRDIE ITAQYQFDFG LRPSLGYVQS KGKDLND--V
SMA    MTPFGGGNFT NTCAATENCG GFASKTQNFE VTAQYQFDFG LRPEVSYLQS KGKNLNVPGV
ECF    ATPITN-KFT NTS------- GFANKTQDVL LVAQYQFDFG LRPSIAYTKS KAKDVE--GI
                  L6                  β12          β13              L7

YPS    DANHDLVKYV SVGTYYYFNK NMSTYVDYKI NLLDKDLFTE VNRIMTDDVV AVGLVYQF   338
YPP    DANHDLLKYV SVGTYYYFNK NMSTYVDYKI NLLDEDEFTI ANGLNTDNVV AVGLVYQF
SMA    GSDQDLVKYV SVGTTYYFNK NMSTYVDYKI NLLDDNDFTK ATGIATDDIV GVGLVYQF
ECF    G-DVDLVNYF EVGATYYFNK NMSTYVDYII NQIDSDNK-- -LGVGSDDTV AVGIVYQF
                  β14          β15              L8           β16
```

Figure 1. Sequence alignment of the OmpF-like porin of *Y. pseudotuberculosis* (YPS) with homologous enterobacterial porins. B-strands (β1–β16) and loops (L1-L8) are indicated as those of OmpF from *E. coli.* YPP, *Y. pestis* (Acc.no. CAC90240); *S. marcescens* (Acc.no.Q54471) and ECO, *E. coli* (Acc.no. P02931).

The 1.1-kb fragment was amplified from total *Y. pseudotuberculosis* DNA and cloned into pBluescript (SK-) to sequence it. The deduced amino acid sequence contained a 22-residue signal peptide and a 338-residue mature protein, with a calculated molecular mass of 37.7 kDa. Homology searches were performed using the BLAST program at NCBI server. From the multiple sequence alignment (Figure 1), OmpF-like porin belongs to a family of non-specific outer membrane porins and is more similar to *S. marcensens* OmpF (72%) rather than to *E. coli* OmpF (55%). The highest degree of sequence identity (92%) was found with putative OmpC protein of *Y. pestis*. This finding is expected because *Y. pestis* is considered as a recently emerged clone of *Y. pseudotuberculosis*. Notably, both these proteins have the same substitution, Glu replaced by Val, in a highly conserved pentapeptide, PEFGG, located in loop 3.

However, the cluster of three arginines (R42, R82, and R132) and acid residue D113 which are responsible for the pore size are very conserved. The highest similarity was mainly observed within transmembrane regions (β1–β16 strands) while sequence insertions and deletions were found in the loops (L1, L6, L7 and L8). A great difference between the *Yersinia* porins was found in loop 8 (only 58% of identity), suggesting a possible role of this loop in formation of specific cell surface-exposed epitopes.

3. A TOPOLOGY MODEL FOR *Y. pseudotuberculosis* OmpF-LIKE PORIN

X-ray analysis of *E. coli* OmpF porin at a resolution of 0.24 reveals the three-dimensional structure of 16 antiparallel β-strands forming a transmembrane barrel. The external segments of the barrel consist of seven surface exposed loops and one, loop L3, directed inside the barrel (Cowan, 1992). A two-dimensional model of OmpF-like porin was obtained employing the GOR secondary structure prediction from Southampton Bioinformatics Data Server (http://molbiol.soton.ac.uk/cgi-bin/GOR.pl) and using the X-ray structure of the homologous *E. coli* OmpF. From this comparison, a working folding model for *Y. pseudotuberculosis* OmpF-like porin was proposed (Figure 2).

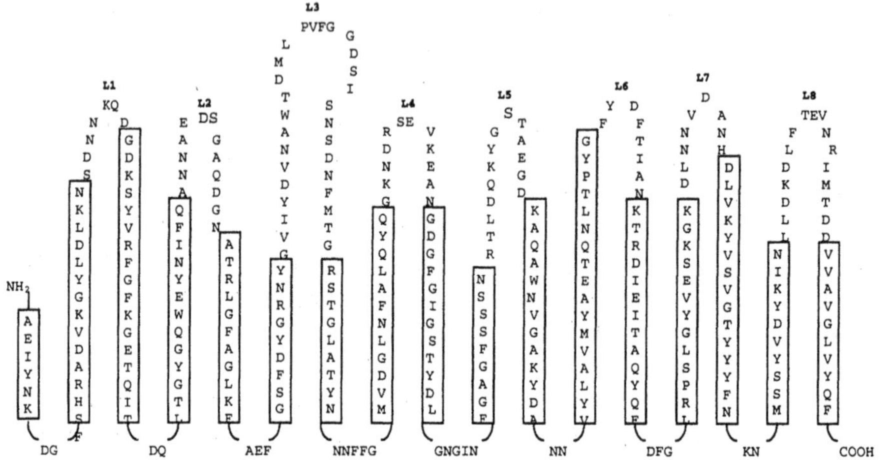

Figure 2. The predicted folding structure of Yersinin 1. The 338 amino acids of Yersinin 1 are shown in the one-letter amino acids code from the N-terminus (left) to the C-terminus (right). The 16 antiparallel β–sheets are in boxes. The outer loops are labeled according to the porin topology.

CONCLUSIONS

In summary, the deduced amino acid sequence of OmpF-like porin from *Y. pseudotuberculosis* is not very closely related to OmpF of *E. coli* but is very similar to OmpF of *S. marcescens*. Analysis of the upstream region of the putative *ompC* gene of *Y. pestis* reveals an *ompF*-like promotor motif. Thus the data presented here suggests that the *Y. pseudotuberculosis* porin belongs to OmpF-like porins.

REFERENCES

Alberty, S., G. Marques, S. Campruby, S. Merino, J. M. Tomas, F. Vivanco, and V. J. Benedy. 1993. C1q binding and activation of the complement classical pathway by *Klebsiella pneumoniae* outer membrane proteins. Infect. Immun. 61:852–860

Bernardini, M. L., M. G. Sanna, A. Fontaine, and P. Sansonetti. 1993. OmpC is involved in invasion of epithelial cells by *Shigella flexneri*. Infect. Immun. 61:3625-3635

Cowan, S. W., T. Schirmer, G. Rummel, M. Steiert, R. Ghosh, R. A. Pauptit, J. N. Jansonius, and J. P. Rosenbusch. 1992. Crystal structures explain functional properties of two *E. coli* porins. Nature 358:727-733

Cowan, S. W. 1993. Bacterial porins: lessons from three high-resolution structures. Curr. Opin. Struct. Biol. **3:**501–507.

Dorman, C. J., S. Chatfield, C. F. Higgins, C. Hayward, and G. Dougan. 1989. Characterization of porin and *ompR* mutants of a virulent strain of *Salmonella typhimurium*: *ompR* mutants are attenuated in vivo. Infect. Immun. 57:2136-2140.

Chapter 51

Pore-Forming Proteins of Genus *Yersinia*

Olga P. VOSTRIKOVA, Olga D. NOVIKOVA, Natalya Y. KIM, Galina N. LIKHATSKAYA and Tamara F. SOLOVJEVA
Pacific Institute of Bioorganic Chemistry, Far East Branch of the Russian Academy of Sciences, 690022, Vladivostok-22, Russia

1. INTRODUCTION

Amounts and types of porins expressed in Gram-negative bacteria vary in response to environmental changes. Little is known about effect of temperature on porin expression. Therefore, a comparative study of porins isolated from bacteria grown under different temperatures should clarify this problem. In this work, porins from three non-pathogenic *Yersinia* species, *Y. intermedia, Y. kristensenii, Y. frederiksenii,* cultivated at 6-8°C (the "cold" variants) and 37°C (the "warm" ones) were isolated and characterized.

2. CHARACTERIZATION

Porins were isolated from the cells using the procedure described by Rosenbush (Rosenbush, 1974). However, the *Yersinia* porins appeared to be less strongly associated with peptidoglycan than the *Escherichia coli* porins as they could be completely solubilized at low temperature (22-37°C). Final purification of the porins was achieved by gel-filtration. As a result, the monomeric and trimeric molecular forms were obtained. Thus, unlike the *E. coli* porins (Rosenbush, 1974) the *Yersinia* trimers partly dissociated during isolation and seemed to be less stable than the former ones. All isolated samples displayed a heat-modifiable property resembling that of other bacterial porins. Amino-acid compositions of all samples of the porins studied here were fairly similar. The porins were rich in acidic amino acids

The Genus Yersinia, Edited by Skurnik *et al.*
ˀluwer Academic/Plenum Publishers, New York, 2003

and had no cysteine residues. The molecular weights of the porin monomers estimated from their amino-acid compositions were in agreement with those obtained from the SDS-PAGE and MALDI-TOF mass-spectrometry data and were in the range 37.2 – 38.6 kDa.

3. STRUCTURE

In the far UV region (190-240 nm), the CD spectra of all monomeric and trimeric forms showed the peaks at 208 and 216 nm, characteristic of proteins having a significant amount of beta-sheeted structure. Secondary structure estimation obtained by the Provencher method (Provencher and Glockner, 1981) gave values of 3-13% alpha-helix, 59-75% beta-sheet, and 12-36% random and/or turn structures. All monomeric forms contained slightly less beta-sheet structure in comparison with the trimeric ones. This loss correlated with increase in the portion of random and/or turn structures. Some increase in content of alpha-helical structure in the heat-denaturated monomers from both bacterial variants was observed. In the near UV region (240-320 nm), the CD spectra of the trimers and monomers from both bacterial variants differed strongly. The spectra of the trimers showed several positive bands in region of 290-250 nm that reflected well-defined tertiary structure of these proteins. For comparison, the near-UV region CD spectra of the monomers showed low amplitude and loss of fine structure indicating unfolding of the tertiary structure upon dissociation. The tryptophan fluorescence spectra of the trimers from the "warm" and "cold" variants showed the emission maxima at 331-333±1 and at 337-338±1 nm, respectively, indicating a non-polar environment of tryptophan in these porins. The spectra of the *Yersinia* trimers were shifted by 1-8 nm towards longer wavelength compared to those of the classical *E. coli* porins (Markovic-Housley and Garavito, 1986) suggesting a higher degree of exposure of the tryptophans on the protein surface in the *Yersinia* porins. Upon heating the trimers, the tryptophan emission peaks shifted to 339-340±1 and 340-345±1nm for the "cold" and the "warm" variants respectively suggesting higher degree of tryptophan exposure to the solvent for the latter.

4. PORE-FORMING ACTIVITY

All porin trimers studied were reconstituted successfully into planar lipid bilayers to give stable channels (pore diameters of 0.96-1.56 nm). The pore conductance measurements revealed that the *Yersinia* porins were general diffusion porins that, however, had some peculiarities. The porin samples from the "cold" variants yielded a mixed population of two types of channels. The conductance values and sizes of the porin channels from the *Y. intermedia* and *Y. frederiksenii* "cold" variants were bigger than those from the "warm" ones. However, the reverse was true for the porin channels from the *Y. kristensenii* temperature variants.

REFERENCES

Rosenbusch, J.P., 1974, Characterization of the major protein from *Escherichia coli.* Regular arrangement on the peptidoglycan and unusual dodecylsulphate binding. J. Biol. Chem. 249: 8019-8029.

Provencher, S.W. and Glockner, J., 1981, Estimation of globular protein secondary stracture from circular dichroism. Biochemistry. 20: 33-37.

Markovic-Housley, Z. and Garavito, R.M., 1986, Effect of temperature and low pH on structure and stability of matrix porin in micellar detergent solutions. Biochim. Biophys. Acta. 869:158-170.

PART IV

GENE REGULATION

Picture 21. Robert Perry gives a lecture.

Picture 22. Questions after a lecture during a session chaired by Virginia L. Miller (standing at right) and José Antonio Bengoechea (sitting at right). Sitting in the front row are Susan and Joseph Straley and Jürgen Heesemann.

Chapter 52

Regulation of O-Antigen Biosynthesis in *Yersinia enterocolitica*

José Antonio BENGOECHEA
Unidad de Investigación, Hospital Son Dureta, Palma Mallorca, Spain

1. INTRODUCTION

Lipopolysaccharide (LPS) is a glycolipid present in the outer membrane (OM) of Gram-negative bacteria. The LPS molecule consists of three regions: (i) the lipid A, a polar glycolipid in which a disaccharide backbone is replaced with six or seven fatty acid; (ii) the core, an oligosaccharide often rich in negatively charged groups and, (iii) the O-antigen, a polysaccharide that protrudes into the surroundings. It is widely known that the lipid A part is responsible for most of the biological effects of LPS; however, there is increasing body of evidence showing that O-antigen modulates the biological activity of lipid A and also plays a critical role in the initial interaction of bacterium with the host.

The biosynthesis of heteropolymeric O-antigens whose O-units are composed of different sugar residues starts in the cytoplasm by the activation of sugar 1-phosphates in a reaction with one of the nucleoside triphosphates, followed by different biosynthetic pathways to give rise to cytoplasmically located individual nucleoside diphosphate (NDP)-activated sugar precursors. The assembly of the O-units takes place on the cytoplasmic face of the inner membrane and involves the transfer of sugar residues from the NDP-sugar precursors onto a lipid carrier molecule, undecaprenylphosphate (Und-P), by specific glycosyltransferases. The completed O-units still assembled on the Und-P are translocated by the O-unit flippase, Wzx, across the inner membrane to the periplasmic face. There the O-units are polymerized by the O-antigen polymerase, Wzy, into an O-antigen that is ligated to the lipid A-core structure by the O-antigen ligase encoded by the *waaL* gene of the core

The Genus Yersinia, Edited by Skurnik *et al.*
'uwer Academic/Plenum Publishers, New York, 2003

gene cluster (for a review see Whitfield, 1995). Based on this enzymatic mode of O-antigen biosynthesis LPS isolated from bacteria is a heterologous population of molecules; some do not carry any O-antigen while others that do have variation in the O-antigen chain lengths.

1.1 The O-antigen of *Yersinia enterocolitica* serotype O:8

The O-antigen polysaccharide of *Yersinia enterocolitica* serotype O:8 (hereafter named YeO8) is formed by branched pentasaccharide O-units that contain *N*-acetyl-galactosamine (GalNAc), L-fucose (Fuc), D- galactose (Gal), D-mannose (Man) and 6-deoxy-D-gulose (6d-Gul). The adjacent repeat units are joined together by a (1→4) glycosidic bond between GalNAc and Man residues (Tomshich *et al.,* 1987). The genes required for the O-antigen biosynthesis have been cloned and sequenced (GenBank Accession Number U46859). The YeO8 O-antigen gene cluster (hereafter *wb*-cluster) spanning 19 kb of the chromosome contains 18 genes that, similar to other *Yersinia*, is located between the *hemH* and *gsk* genes (Skurnik, 1999; Skurnik *et al.,* 2000). The genes code for enzymes involved in the biosynthesis of the O-antigen sugar precursors and their respective glycosyltransferases, the O-unit flippase, Wzx, the O-antigen polymerase, Wzy, and the O-antigen chain length determinant, Wzz. This protein is required to control the O-antigen length which, in YeO8, is of 7 to 10 repeats (Zhang *et al.,* 1997; Zhang *et al.,* 1996; Bengoechea *et al.,* 2002a). Promoter cloning experiments revealed that the wb-cluster contains two promoters, one promoter upstream of *ddhA*, the first gene in the *wb*-cluster, and another one between *manB* and *gne* genes (designated P_{wb1} and P_{wb2} respectively) (Bengoechea *et al.,* 2002b).

2. ROLE OF O-ANTIGEN IN VIRULENCE

The role of O-antigen in YeO8 virulence has been determined in intragastrically infected mice. The LD_{50} of the O-antigen negative mutant was about 100-fold higher than the wild type strain (hereafter named wt) (Zhang *et al.,* 1997). It is worth mentioning that this mutant was isolated using a bacteriophage whose receptor is the polymerised YeO8 O-antigen and that could be *trans*-complemented, *in vitro* and *in vivo,* with a cosmid containing the *wb*-cluster. The importance of O-antigen as virulence factor has been further corroborated by Miller and co-workers using signature-tagged transposon mutagenesis (STM) (Darwin *et al.,* 1999). In that study, 23 % of the attenuated mutants have insertions in the *wb*-cluster and since the transposon insertions were located in genes encoding proteins required to build up the O-unit all mutants lacked the O-antigen. Recently, we have tested whether the O-antigen is necessary for YeO8 virulence after the intestinal barrier has been bypassed. Interestingly, in intraperitonealy infected mice the LD_{50} of the O-antigen mutant was 1000-fold higher than the wt whereas in intravenously infected mice the LD_{50} was 100-fold higher than the wt. Altogether these results point out that the O-antigen is not only required in the initial stages of infection but also for YeO8 survival in deeper organs (Bengoechea *et al.*, unpublished).

3. TEMPERATURE-DEPENDENT O-ANTIGEN EXPRESSION IN YeO8

An interesting feature of *Y. enterocolitica* O-antigen is that its expression is temperature regulated (Al-Hendy *et al.*, 1991; Bengoechea *et al.*, 2002b; Skurnik, 1999). The optimum expression occurs when bacteria are grown at room temperature (RT, 22-25°C) whereas when they are grown at 37°C, the host temperature, only trace amounts of O-antigen are produced. (Figure 1A). Recently, we have started to elucidate the regulatory mechanisms behind the temperature-dependent expression of O-antigen (Bengoechea *et al.*, 2002b).

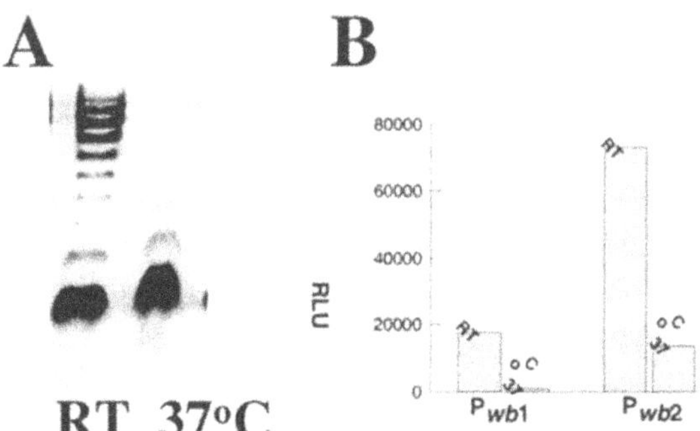

Figure 1. O-antigen expression in YeO8 is temperature dependent. **A**. DOC-PAGE analysis of LPS isolated from YeO8 bacteria grown at RT or 37oC. **B**. Promoter activities measured from YeO8 reporter strains carrying the *wb*P1::*lucFF* or *wb*P2::*lucFF* transcriptional fusions are given in relative light units (RLU, mean ± standard deviation).

By Northern blot analysis we could reveal that *wb*-cluster transcription was down-regulated at 37°C. This finding was confirmed using transcriptional fusions to a promoterless reporter gene, *lucFF*. Two reporter strains were constructed in which *lucFF* was introduced downstream of each of the two promoters identified in the *wb*-cluster. YeO8c-*wb*P1*::lucFF* bacteria grown at RT produced sixty-fold more light than bacteria grown at 37°C and YeO8c-*wb*P2*::lucFF* bacteria produced six-fold more light when grown at RT than at 37°C (Figure 1B) indicating that both promoters are under temperature-regulation and repressed in bacteria grown at 37°C.

3.1 Identification of loci involved in temperature regulation of O-antigen expression

To our surprise, *Escherichia coli* cosmid clones carrying the YeO8 *wb*-cluster also expressed the O-antigen in a temperature-dependent manner (Bengoechea *et al.,* 2002b). Thus we reasoned that these cosmids may contain loci involved in the regulation of O-antigen expression. Using a genetic approach, construction of cosmid deletions followed by *trans*-complementacion experiments, we mapped a 6 kb region downstream of the *wb*-cluster involved in temperature regulation. This region contains three complete ORFs of which two of them, *rosA* and *rosB* (for regulation of O-antigen synthesis), are arranged in an operon and confer O-antigen temperature dependent expression. However, it was somewhat unexpected to find out that the RosAB system required the expression of YeO8 Wzz, the O-antigen chain length determinant, to exert its regulatory effect. Expression of RosAB without Wzz did not down-regulate O-antigen expression at 37°C.

RosA is similar to drug resistance transport proteins and RosB is similar to a number of proteins involved in the glutathione-regulated potassium efflux. In good agreement with the computer predictions, functional analysis demonstrated that RosA is a proton motive force-driven efflux pump involved in resistance to amphipathic compound and RosB is a potassium antiporter that together with RosA is also involved in the resistance to antimicrobial peptides, key elements of the innate immune system (Bengoechea *et al.,* 2000).

However, all the previous findings were obtained in an *E. coli* background and it was possible that similar results would not be obtained in the *Yersinia* background. To address the role of RosAB and Wzz in O-antigen regulation in *Yersinia*, we constructed a *rosArosB* double mutant (YeO8-RosA$^-$B$^-$) and a *wzz* mutant (YeO8-ΔWzzGB) and analysed O-antigen expression (Bengoechea *et al.,* 2002b). The *rosAB* mutant expressed significantly more O-antigen than YeO8 when grown at 37°C whereas the differences were not so evident at RT (Figure 2A). The O-antigen expression in *rosA* or *rosB* non-polar mutants was similar to that of the double mutant indicating that both proteins are required for regulation of O-antigen expression. On the other hand, analysis of O-antigen expression by the YeO8 *wzz* mutant demonstrated that this mutant also expressed more O-antigen than the wt when grown at 37°C (Figure 2B). It is worth mentioning that the mutant also produced LPS without the modal distribution of O-antigen lengths which is a the typical phenotype of *wzz* mutants in other bacterial species.

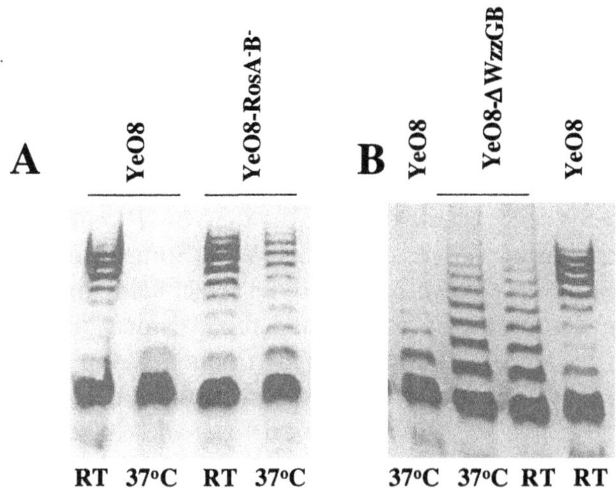

Figure 2. Role of RosAB (panel A) and Wzz (panel B) in the temperature dependent O-antigen expression in YeO8 determined by DOC-PAGE analysis of LPS.

In summary, the data showed that both RosAB and Wzz are involved in repression of O-antigen expression at 37°C both in *Yersinia* and *E. coli.*

3.2 RosAB and Wzz affect the transcriptional regulation of the *wb*-cluster

Taking into consideration the previous findings, it was quite feasible to postulate that RosAB affect the transcriptional regulation of the *wb*-cluster. Indeed, transcriptional analysis using the fusions *wb*P1::*lucFF* and *wb*P2::*lucFF* demonstrated the activity of the second promoter was 10-fold higher in the *rosAB* background than in the wt background whereas the activity of the first promoter was not affected (Bengoechea *et al.,* 2002b). It is worth noting that P_{wb1} drives the expression of most of the genes require to build up the O-antigen whereas P_{wb2} controls the expression of only two genes, *gne* and *wzz.* The *gne* gene encodes for an UDP-N-acetylglucosamine-4-epimerase (EC 5.1.3.7) that is involved in the synthesis of GalNAc, the first sugar of the O-unit (Bengoechea *et al.,* 2002a). Thus in the *rosAB* strain at 37°C the P_{wb2} activity is derepressed leading to over-expression of both Gne and Wzz which is linked to an increase in the O-antigen expression (see

Figure 1A). In good agreement, over-expression of *gne* or *wzz* in the wt strain caused an increase in O-antigen expression at 37°C. Since RosAB is an inner membrane efflux pump/K^+ antiporter it is unlikely that the proteins *per se* would have repressor or signal transduction functions. Instead, the results argue that the RosAB regulatory effect on the P_{wb2} promoter activity is linked to the signals generated by the RosAB system. These signals are the changes in the intracellular levels of H^+ and potassium (Bengoechea *et al.,* 2002b).

The effect of Wzz in the transcriptional regulation of the *wb*-cluster was tested by measuring light production of the *wb*P1::*lucFF* and *wb*P2::*lucFF* fusions in the *wzz* mutant background. For both promoter fusions a slightly higher light production was observed in the *wzz* mutant background than in the wt. Since Wzz is over-expressed in the *rosAB* mutant we studied the promoter activities in wt over-expressing Wzz. Somewhat unexpectedly both promoter fusions were down-regulated the effect being more pronounced for the P_{wb2} than for the P_{wb1} promoter and more evident in bacteria grown at 37°C than in bacteria grown at RT. This finding was in sharp contrast with the LPS phenotype of YeO8 over-expressing Wzz that produced more O-antigen when grown at 37°C (Bengoechea *et al.,* 2002b). Since more O-antigen is produced from a decreased number of transcripts the efficiency of the post-transcriptional events in the O-antigen biosynthesis must be better. At present, we find two non-exclusive explanations for this: (i) Since Wzz, Wzy, Wzx and perhaps also WaaL may form a complex involved in the O-antigen assembly/translocation it is possible that overexpression of Wzz could facilitate the Wzy polymerase function. Thus changes in the stoichiometry of the complex could modulate the efficiency of the O-antigen polymerization and translocation resulting in increased expression of O-antigen. (ii) Overexpression of Wzz could also affect the expression or function of systems that are involved in downregulation of the O-antigen expression at the post-transcriptional level. Supporting this explanation the expression of *rosAB* was affected by Wzz (Bengoechea *et al.,* 2002b). Thus it is tempting to speculate that, in addition to repress the P_{wb2} promoter activity, the presence of the RosAB system compromises the efficiency of the O-antigen biosynthetic machinery directly. In any case, at present we can not rule out the possibility that other systems are affected as well.

4. REGULATION OF O-ANTIGEN EXPRESSION BY OTHER SIGNALS THAN TEMPERATURE

Even though the temperature shift is an important signal that controls the expression of many *Yersinia* virulence factors, it is certain that *Yersinia* will encounter other environmental cues in the host which include, for example, low pH, iron availability or oxygen tension (Straley *et al.,* 1995). Furthermore, these other signals may alter the temperature-dependent regulation. Indeed, this has been demonstrated for *Yersinia* virulence factors *inv* and *ylpA*. It was then quite reasonable to postulate that other environmental signals could affect the thermoregulation of O-antigen expression. To assess this experimentally, we analysed the light production by the reporter strains YeO8c-*wb*P1*::lucFF* and YeO8c-*wb*P2*::lucFF* grown at different conditions. The results showed that acidic pH, low iron concentration or low oxygen tension at 37°C caused further down-regulation of the O-antigen expression and this was true for both promoter fusions although the effect was more pronounced for the P_{wb2} than for the P_{wb1} promoter (Bengoechea *et al.,* unpublished). Interestingly, some of these signals also regulate the RosAB system. For example, limiting amounts of iron up-regulated *rosAB* expression (Bengoechea *et al.,* unpublished). Keeping in mind that RosAB system is involved in O-antigen regulation, it is possible that *Yersinia* might modulate O-antigen expression in response to different signals by regulating RosAB expression.

5. CONCLUDING REMARKS

It is generally accepted that the LPS, and more specifically the O-antigen, plays an essential role in the virulence of Gram-negative pathogens. The O-antigen of *Y. enterocolitica* is no exception to this rule. We and others have demonstrated that O-antigen is essential for *Yersinia* full virulence and this has opened the possibility to use *Yersinia* O-antigen mutants as carrier strains for vaccination purposes. However, many questions are still open. For example, we still do not know the exact role of O-antigen in virulence and we have also started to realize how complex is the regulation of O-antigen expression. In this regard, it is tempting to speculate that the expression of the O-antigen is coordinated with that of other virulence factors. The fact that signals that regulate O-antigen expression also modulate the expression of other *Yersinia* virulence factors gives indirect support to this hypothesis. Future studies will hopefully answer these exciting questions.

ACKNOWLEDGEMENTS

I am specially grateful to Mikael Skurnik (University of Turku, Turku) who provided me with many insights into the field of *Yersinia* LPS. I also thank all the members of the *Yersinia* group in Turku and also Clemens Oertelt and Otto Holst from the Research Center Borstel. Part of this research has been funded by contract QLK-99-007800 of the European Commission and Fondo de Investigaciones Sanitarias (Spain).

REFERENCES

Al-Hendy, A., Toivanen, P., and Skurnik, M. (1991) The effect of growth temperature on the biosynthesis of *Yersinia enterocolitica* O:3 lipopolysaccharide: temperature regulates the transcription of the *rfb* but not of the *rfa* region. *Microb Pathog* 10: 81-86.

Bengoechea, J. A., Pinta, E., Salminen, T., Oertelt, C., Holst, O., Radziejewska-Lebrecht, J., Piotrowska-Seget, Z., Venho, R., and Skurnik, M. (2002a) Functional characterization of Gne (UDP-N-acetylglucosamine-4-epimerase), Wzz (chain length determinant), and Wzy (O-antigen polymerase) of *Yersinia enterocolitica* serotype O:8. *J Bacteriol* 184: 4277-87.

Bengoechea, J. A., and Skurnik, M. (2000) Temperature-regulated efflux pump / potassium antiporter system mediates resistance to cationic antimicrobial peptides in *Yersinia. Mol Microbiol* **37**: 67-80

Bengoechea, J. A., Zhang, L., Toivanen, P., and Skurnik, M. (2002b) Regulatory network of lipopolysaccharide O-antigen biosynthesis in *Yersinia enterocolitica* includes cell envelope-dependent signals. *Mol Microbiol* 44: 1045-1062.

Darwin, A.J. and Miller, V.L. (1999) Identification of *Yersinia enterocolitica* genes affecting survival in an animal host using signature-tagged transposon mutagenesis. *Mol Microbiol* **32**: 51-62.

Straley, S.C. and R.D. Perry. 1995. Environmental modulation of gene expression and pathogenesis in *Yersinia*. Trends Microbiol. **3**:310-317.

Skurnik, M. 1999. Molecular genetics of *Yersinia* lipopolysaccharide, p. 23-51. *In* J. Goldberg (ed.), Genetics of Bacterial Polysaccharides. CRC Press, Boca Raton, FL.

Skurnik, M., Peippo, A., and Ervelä, E. (2000) Characterization of the O-antigen gene clusters of *Yersinia pseudotuberculosis* and the cryptic O-antigen gene cluster of *Yersinia pestis* shows that the plague bacillus is most closely related to and has evolved from *Y. pseudotuberculosis* serotype O:1b. *Mol Microbiol* 37: 316-330.

Tomshich, S.V., R.P. Gorshkova and Y.S. Ovodov. 1987. Structural investigation of the lipopolysaccharide of *Y. enterocolitica* serovar O:8. Khimia Prirodnykh Soedinenii:657-664.

Whitfield, C. (1995) Biosynthesis of lipopolysaccharide O antigens. *Trends Microbiol* **3**: 178-185.

Zhang, L., Radziejewska-Lebrecht, J., Krajewska-Pietrasik, D., Toivanen, P., and Skurnik, M. (1997) Molecular and chemical characterization of the lipopolysaccharide O-antigen and its role in the virulence of *Yersinia enterocolitica* serotype O:8. *Mol Microbiol* 23: 63-76.

Zhang, L., Toivanen, P., and Skurnik, M. (1996) The gene cluster directing O-antigen biosynthesis in *Yersinia enterocolitica* serotype O:8: Identification of the genes for mannose and galactose biosynthesis and the gene for the O-antigen polymerase. *Microbiology* 142: 277-288.

Chapter 53

Regulation of the *Yersinia pestis* Yfe and Ybt Iron Transport Systems

Robert D. PERRY, Jennifer ABNEY, Ildefonso MIER, Jr., Yong LEE, Scott W. BEARDEN and Jacqueline D. FETHERSTON
Department of Microbiology, Immunology, and Molecular Genetics, University of Kentucky, Lexington, KY, USA

1. INTRODUCTION

Yersinia pestis encodes eleven proven or putative inorganic iron (Fe) or hemin/hemoprotein transport systems. However, the yersiniabactin (Ybt) siderophore-dependent and Yfe systems are the two primary mechanisms of iron acquisition. YfeA-D is an ABC transporter with both Fe and manganese (Mn) substrates (Perry *et al.*, 2001). A high-pathogenicity island (HPI) present in highly pathogenic species of *Yersinia* and other enteric bacteria encodes genes for 3 Ybt transport proteins, 6 Ybt biosynthetic enzymes, one transcriptional regulator (YbtA), and one protein of unknown function (YbtX). In *Y. pestis*, the HPI resides within a 102-kb chromosomal region subject to high-frequency deletion, the *pgm* locus (Buchrieser *et al.*, 1998; Gehring *et al.*, 1998; Rakin *et al.*, 1999).

Expression of both the Yfe and Ybt systems is repressed by Fe-surplus growth conditions through the interaction of Fur-Fe complexes with Fur binding sequences (FBS) (Perry *et al.*, 2001). Both systems have additional regulatory signals. Transcription of the *yfeA-D* promoter is also repressed by surplus manganese; Fur is required for this Mn repression (Bearden *et al.*, 1998). YbtA, an AraC-family transcriptional regulator, activates transcription of most *ybt* genes but represses its own transcription. The Ybt siderophore is also important for activation by YbtA (Fetherston *et al.*, 1996). This report further characterizes the regulatory mechanisms employed for expression of the Yfe and Ybt systems.

The Genus Yersinia, Edited by Skurnik *et al.*
'uwer Academic/Plenum Publishers, New York, 2003

2. MATERIALS AND METHODS

Table 1. Bacterial strains and plasmids used in this study[a]

Y. pestis strains	Relevant characteristics	Reference(s) or source
KIM6+	Ybt^+ Yfe^+	Fetherston *et al.*, 1995
KIM6	Pgm^- (*Δpgm* – Ybt^-) Yfe^+	Fetherston *et al*, 1995
KIM6-2069.1	Ybt^- $YfeE^-$ (*ΔyfeE*)	Bearden and Perry, 1999
KIM6-2067	Ybt^+ $YbtX^-$ (*ΔybtX*) Yfe^+	Fetherston *et al.*, 1999
KIM6-2045.1	Ybt^- (*Δpsn*) Yfe^+	Fetherston *et al*, 1995
KIM6-2064	Ybt^- (*ΔybtP*) Yfe^+	Fetherston *et al.*, 1999
KIM6-2073+	Ybt^- (*tonB::kan*) Yfe^+	Laboratory strain
KIM6-2055	Ybt^- (*ybtA::kan*) Yfe^+	Fetherston *et al.*, 1996
KIM6-2054.1	Ybt^- (*irp1::kan*) Yfe^+	Bearden *et al.*, 1997
KIM6-2046.1	Ybt^- (*irp2::kan*) Yfe^+	Fetherston *et al*, 1995
KIM6-2056.1	Ybt^- (*ΔybtE*) Yfe^+	Bearden *et al.*, 1997
KIM6-2071	Ybt^- (*ΔybtU*) Yfe^+	Geoffroy *et al.*, 2000
KIM6-2085+	Ybt^- (*ΔybtD*) Yfe^+	Bobrov *et al.*, 2002
KIM6-2070.1	Ybt^- (*ybtS:kan*) Yfe^+	Geoffroy *et al.*, 2000
KIM6-2072	Ybt^- (*ΔybtT*) Yfe^+	Geoffroy *et al.*, 2000
KIM6-2086	Ybt^- (*irp1-2086*) Yfe^+	Bobrov *et al.*, 2002
KIM6-2030+	Ybt^+ Yfe^+ Fur^- (*fur::kan*)	Staggs *et al.*, 1994
Plasmids		
pEU730	single copy number plasmid; promoterless *lacZ*; Spc^r	Froehlich *et al.*, 1994
pEUYfeA	*yfeA::lacZ* fusion; 270 bp PCR product in pEU730; Spc^r	This study
pEUYfeE	*yfeE::lacZ* fusion; 298 bp PCR product in pEU730; Spc^r	This study
pEUIrp2	*irp2::lacZ* fusion; 190 bp PCR product in pEU730; Spc^r	This study
pEUYbtP	*ybtP::lacZ* fusion; Spc^r	Fetherston *et al.*, 1999
pQE30	expression vector	Qiagen
pQEYbtA	1043 bp PCR product of *ybtA* in pQE30; IPTG-inducible *ybtA*	This study

[a] *Y. pestis* strains with a "+" possess an intact 102 kb *pgm* locus containing the Ybt system

2.1 Bacterial strains, plasmids, and culture conditions

All *Y. pestis* strains (Table 1) were grown for 6-8 generations in deferrated PMH2 in acid-cleaned glassware as previously described. For β-galactosidase assays mid-exponential phase cells were harvested. Ybt bioassays were performed as previously described (Bobrov *et al.*, 2002). Reporter plasmids were constructed by ligating PCR products of the *irp2*, *yfeA*, and *yfeE* promoter regions to a promoterless *lacZ* in the single-copy plasmid, pEU730. Cloned promoter regions were sequenced to ensure that PCR-errors were not introduced. pQEYbtA was constructed by ligating a 1043 bp PCR product encoding the entire *ybtA* ORF into pQE30 (Table 1).

Complementation analysis determined that a functional YbtA protein is expressed.

2.2 Assays and recombinant DNA techniques

Standard recombinant DNA methods were used and have been previously described (Bobrov *et al.*, 2002). Cells carrying reporter plasmids (Table 1) were grown to mid-exponential phase at 37°C in PMH2 with or without 10 μM $FeCl_3$ as previously described (Staggs *et al.*, 1994). β-galactosidase activity from cell lysates was measured with a Genesys5 spectrophotometer following cleavage of ONPG. Activities are expressed in Miller units (Bobrov *et al.*, 2002). Ybt siderophore was purified from culture supernatants by ethyl acetate extraction followed by HPLC as previously described (Perry *et al.*, 1999).

3. RESULTS

3.1 Regulation of the Yfe transport system

Table 2. β-Galactosidase activity from *yfeA::lacZ* and *yfeE::lacZ* reporters in *Y. pestis* derivatives grown in defined PMH2 medium

Strain	β-Galactosidase activity (Miller units) of cells grown with:				
	NA[a]	10 μM Fe	1 μM Fe	1 μM Mn	1 μM Zn
KIM6(pEUYfeA)	12,025±1,703[b]	1,746±192	2,441±165	4,777±267	10,282±2,748
KIM6-2069.1(pEUYfeA)	12,905±2,089	1,668±268	2,343±164	4,331± 33	12,162±2,199
KIM6(pEUYfeE)	769±41	592±53	764±58	655±59	ND[a]

[a] NA – no additions; ND – not determined
[b] Averages from a minimum of 4 assays ± standard deviations

Previously, we used a *yfeA::phoA* translational fusion to demonstrate repression of the promoter by surplus Fe and Mn via Fur. Studies with two other Fur-regulated promoters showed that Mn repression of such promoters was not a general phenomenon (Bearden *et al.*, 1998). However, to directly compare *lacZ* transcriptional reporters, we constructed a *yfeA::lacZ* reporter. Our results confirm that the *yfeA* promoter is repressed by Fe and Mn and that zinc (Zn) does not repress expression. The role of YfeE in iron acquisition remains undefined. Expression from the *yfeA::lacZ* reporter was not affected in a *ΔyfeE* strain (Table 2).

Regulation of *yfeE,* which has its own promoter with a potential FBS, had not been examined. Analysis of KIM6(pEUYfeE) indicates that the *yfeE* promoter has low transcriptional activity that is not repressed by surplus Fe, Mn, or Zn (Table 2).

3.2 Regulation of the Ybt siderophore-dependent system

We have previously characterized transcription from the *psn* (Ybt outer membrane [OM] receptor), *ybtA, ybtP,* and *ybtD* promoters (Bobrov *et al.*, 2002; Fetherston *et al.*, 1996, 1999). Although we and others have shown iron, YbtA, and/or siderophore regulation of several proteins encoded in the *irp2* operon of *Y. pestis* (Bearden *et al.*, 1997; Carniel *et al.*, 1987; Geoffroy *et al.*, 2000), transcription from the *Y. pestis irp2* promoter has not been directly examined. Consequently, we constructed an *irp2::lacZ* fusion and tested expression of this reporter in different *Y. pestis* strains grown with and without Fe. As expected, our results (Table 3) demonstrate that the *irp2* promoter is repressed by excess iron (~7-fold) and this repression requires a functional Fur protein. Expression under iron-starvation conditions is stimulated 16-fold by YbtA and 18-fold by Ybt (Table 3).

Table 3. β-Galactosidase activity (in Miller Units) from *irp2::lacZ* and *ybtP::lacZ* reporters in *Y. pestis* derivatives grown in defined PMH2 medium

Strain	No Additions	plus 10 μM $FeCl_3$
KIM6(pEUIrp2)+	25,663 ± 8,584[a]	3,240 ± 973
KIM6-2030(pEUIrp2)+	22,596 ± 10,266	40,678 ± 7,750
KIM6-2055(pEUIrp2)	1,561 ± 498	296 ± 157
KIM6-2046.1(pEUIrp2)	1,449 ± 186	368 ± 108
KIM6(pQEYbtA + pEUYbtP)	5,576 ± 1284	588 ± 188
KIM6(pQEYbtA + pEUYbtP) + IPTG	14,046 ± 1275	4,090 ± 494
KIM6(pQE30 + pEUYbtP)	423 ± 140	151 ± 42
KIM6(pQE30 + pEUYbtP) + IPTG	332 ± 64	107 ± 35

[b] Averages from 8 assays ± standard deviations

Maximal expression of the *ybt* promoters requires both YbtA and Ybt. To determine if YbtA can activate transcription in the absence of the siderophore, *ybtA* was cloned behind an IPTG-inducible promoter in the expression vector, pQE30, and electroporated into KIM6(pEUYbtP). This strain lacks the entire Ybt system but carries a *ybtP::lacZ* reporter plasmid. pQE30 alone does not induce expression of *ybtP::lacZ* while IPTG induction of *ybtA* during iron starvation stimulated β-galactosidase activity 42-fold compared to cells lacking YbtA. Expression of YbtA during iron-surplus growth still stimulated transcription but not to the extent observed with iron-starved cells. The modest stimulation in the absence of IPTG is likely due to low-level expression of YbtA even without the inducer (Table 3).

Addition of purified Ybt to a Ybt-biosynthetic mutant activated transcription of a *ybt* promoter at one hour of exposure (Perry *et al.*, 1999).

To assess the exposure time required to activate transcription, exogenous Ybt (at levels which stimulate growth of a Ybt biosynthetic mutant) was added to iron-starved KIM6-2046.1 cells carrying the *ybtP::lacZ* reporter. By 10 min, exogenous Ybt stimulated transcription 4-fold and reached a maximum of 7.5-fold by 30 min of exposure (Figure 1A). We next determined the minimum Ybt concentration required for transcriptional activation at 20 min of exposure. Ybt added at a concentration of 0.01 (in arbitrary units) stimulated transcription 2-fold (Figure 1B); this is 500-fold lower than the amount of Ybt required for growth stimulation in our bioassay.

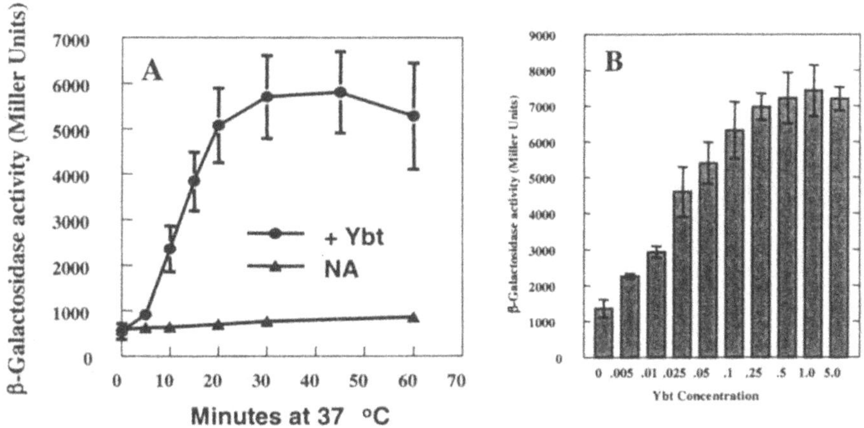

Figure 1. Stimulation of transcription from *ybtP::lacZ* by exogenous Ybt. A - exposure time course; B - Ybt concentration curve (in arbitrary units); 5.0 is the Ybt concentration required to stimulate growth of a Ybt biosynthetic mutant.

To directly assess the effects of 13 different mutations affecting the Ybt system on regulation of a *ybt* promoter, we electroporated pEUYbtP, the *ybtP::lacZ* reporter plasmid, into each mutant and assayed β-galactosidase activity under iron-deficient growth conditions. Our results (Figure 2) indicate that *ybt* mutations can be divided into 5 regulatory classes. Ybt^+ KIM6+ and the *ΔybtX* mutant (class I), which has no observed *in vitro* phenotype, have normal expression levels while $YbtA^-$ mutants (class II) have ~70-fold lower expression than KIM6+. Mutations affecting transport of Ybt into the bacterial cell (*Δpsn, ΔybtP, tonB::kan*; classIII) exhibit moderately higher transcription (~1.7-fold). Finally, mutations in genes encoding biosynthetic enzymes fall into two classes. Most Ybt biosynthetic mutants (*ΔybtD, ΔybtE, ΔybtU, irp1::kan, irp2::kan*; class IV) show reduced β-galactosidase activity (~18-fold) presumably because the siderophore is not produced. In an apparent paradox, three biosynthetic mutants (*ΔybtT, ybtS::kan, irp1-2086*; class V) show moderately increased expression (~1.4-fold) compared to KIM6+ (Figure 2).

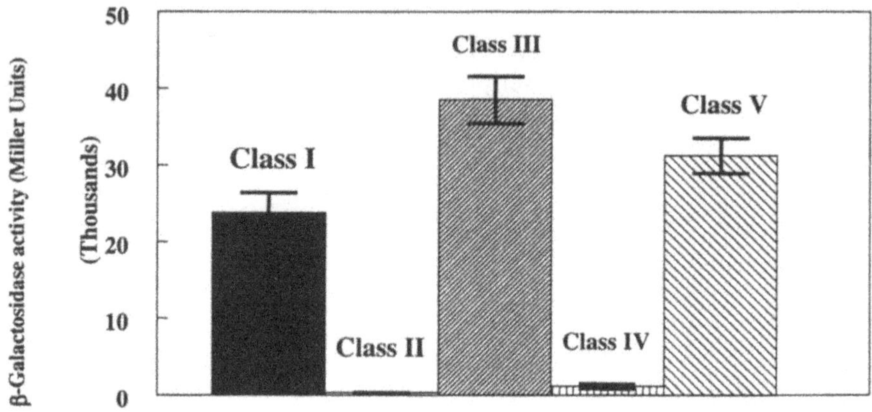

Figure 2. β-Galactosidase activity of *Y. pestis* derivatives carrying a *ybtP::lacZ* reporter. Only a representative from each regulatory class is shown: KIM6+ (Class I), KIM2055 (*ybtA::kan*) (Class II), KIM6-2073+ (*tonB::kan*) (Class III), KIM6-2056.1 (*ΔybtE*) (Class IV), and KIM6-2072 (*ΔybtT*) (Class V).

4. DISCUSSION

Fur regulation of the Yfe transport system is unique in that both Fe and Mn repress expression of the *yfeA-D* operon (Bearden *et al.*, 1998). We have confirmed Mn repression using a new *yfeA::lacZ* transcriptional reporter. Although the Yfe system may also accumulate Zn, expression of *yfeA::lacZ* is not affected by Zn availability (Table 2). We previously speculated that YfeE might regulate expression of the *yfeA-D* operon (Bearden and Perry, 1999). However, a *ΔyfeE* mutation did not affect expression of our *yfeA::lacZ* reporter. In addition, transcription of *yfeE* is low compared to *yfeA-D* and independent of Fe-, Mn-, or Zn-availability (Table 2). Thus YfeE does not regulate expression of the Yfe system and its function is undefined.

As expected, the *irp2* promoter is repressed by Fur-Fe and activated by YbtA plus Ybt, like the *psn* and *ybtP* promoters. In addition, overexpression of YbtA activates transcription of the *ybtP* promoter in the absence of siderophore or in the presence of Fe (Table 3; Fetherston *et al.*, 1996, 1999).

Analysis of the regulatory effects of various *ybt* mutations allows us to group these mutations into five classes. Class I exhibits normal regulation and contains Ybt+ KIM6+ and the *ΔybtX* mutant. *ybtX* lies within the *ybtPQXS* operon (Gehring *et al.*, 1998) and has modest similarity to *E. coli* EntS (Furrer *et al.*, 2002) which is required for the export of enterobactin. However, we have not detected an in vitro defect in this mutant (Fetherston *et al.*, 1999). Strains lacking YbtA (KIM6 [*Δpgm*] and KIM6-2055 [*ybtA::kan*]) show almost no transcriptional activity (class II). Class III mutants synthesize and secrete the siderophore but cannot use it for iron acquisition and have *ybtP::lacZ* activity modestly higher than wild-type levels (Figure 2). Presumably these mutants are more iron starved because the Ybt produced binds the trace levels of Fe in the medium preventing other Fe transport systems (e.g., Yfe) from acquiring it. Class IV mutants do not produce Ybt detectable by bioassay; transcriptional activity in these mutants (*ΔybtD, ΔybtE, ΔybtU, irp1::kan, irp2::kan*) is ~18-fold lower than in wild-type cells and ~4-fold higher than in class II mutants. Class V mutants also produce no detected Ybt but show enhanced transcription from the *ybtP* promoter - a 1.4-fold increase compared to wild-type. Class III and class V mutants show modest but reproducible differences in regulation (~1.4-fold).

In other systems, the binding of the siderophore to its OM receptor is involved in transcriptional activation. In *E. coli* (Braun, 1997) and *Pseudomonas putida* (Venturi *et al.*, 1995), TonB and OM receptors are essential for transcriptional activation of the ferric citrate and pseudobactin systems, respectively. The *P. aeruginosa* pyochelin system also requires an OM receptor for signal transduction (Heinrichs and Poole, 1996).

However, activation of *Y. pestis ybt* genes does not require the OM receptor, TonB, or an inner membrane permease (Class III mutants). Maximal expression of the Ybt system does require YbtA and Ybt (Class II and IV). Our results indicate that minute quantities of Ybt can activate transcription; a concentration 500-fold lower than that required for growth stimulation in our bioassay promoted transcription of *ybtP::lacZ* (Figure 1). This is similar to results obtained with alcaligin, the siderophore of *Bordetella bronchiseptica* (Brickman *et al.*, 2001). Thus, in transport mutants enough Ybt to activate transcription may enter the cell despite the lack of a specific transport system. Our model of YbtA activation proposes that a YbtA-Ybt complex stimulates transcription of *ybt* genes.

Class V biosynthetic mutants mutants may produce trace amounts of Ybt or an aberrant structure sufficient to interact with YbtA and stimulate transcription. The YbtT$^-$ and HMWP1-2086 mutants likely lack thioesterase activity for either removing aberrant structures or releasing the completed siderophore from the enzyme complex, respectively (Bobrov *et al.*, 2002; Geoffroy *et al.*, 2000). Consequently, these mutants could produce trace levels of Ybt sufficient for transcriptional activation but not for growth stimulation. The YbtS⁻ mutant cannot synthesize salicylate but may initiate Ybt synthesis with an alternate aromatic compound (e.g., 2,3-dihydroxybenzoate) to produce low levels of a modified Ybt structure that serves to stimulate transcription. Experiments are ongoing to determine whether these three mutants produce Ybt or Ybt-like molecules with signaling activities.

ACKNOWLEDGEMENTS

This work was supported by Public Health Service grants AI33481.

REFERENCES

Bearden, S.W., Fetherston, J.D., and Perry, R.D., 1997, Genetic organization of the yersiniabactin biosynthetic region and construction of avirulent mutants in *Yersinia pestis*. *Infect. Immun.* **65**: 1659-1668.

Bearden, S.W., Staggs, T.M., and Perry, R.D., 1998, An ABC transporter system of *Yersinia pestis* allows utilization of chelated iron by *Escherichia coli* SAB11. *J. Bacteriol.* **180**: 1135-1147.

Bearden, S.W., and Perry, R.D., 1999, The Yfe system of *Yersinia pestis* transports iron and manganese and is required for full virulence of plague. *Mol. Microbiol.* **32**: 403-414.

Bobrov, A.G., Geoffroy, V.A., and Perry, R.D., 2002, Yersiniabactin production requires the thioesterase domain of HMWP2 and YbtD, a putative phosphopantetheinylate transferase. *Infect. Immun.* **70**: 4204-4214.

Braun, V., 1997, Surface signaling: novel transcription initiation mechanism starting from the cell surface. *Archives of Microbiology* **167**: 325-331.

Brickman, T.J., Kang, H.Y., and Armstrong, S.K., 2001, Transcriptional activation of *Bordetella* alcaligin siderophore genes requires the AlcR regulator with alcaligin as inducer. *J. Bacteriol.* **183**: 483-489.

Buchrieser, C., Prentice, M., and Carniel, E., 1998, The 102-kilobase unstable region of *Yersinia pestis* comprises a high-pathogenicity island linked to a pigmentation segment which undergoes internal rearrangement. *J. Bacteriol.* **180**: 2321-2329.

Carniel, E., D. Mazigh, D., and H. H. Mollaret, H.H., 1987, Expression of iron-regulated proteins in *Yersinia* species and their relation to virulence. *Infect. Immun.* **55**: 277-280.

Fetherston, J.D., Lillard, J.W., Jr., and Perry, R.D., 1995, Analysis of the pesticin receptor from *Yersinia pestis*: role in iron-deficient growth and possible regulation by its siderophore. *J. Bacteriol.* **177**: 1824-1833.

Fetherston, J.D., Bearden, S.W., and Perry, R.D., 1996, YbtA, an AraC-type regulator of the *Yersinia pestis* pesticin/yersiniabactin receptor. *Mol. Microbiol.* **22**: 315-325.

Fetherston, J.D., Bertolino, V.J., and Perry, R.D., 1999, YbtP and YbtQ: two ABC transporters required for iron uptake in *Yersinia pestis*. *Mol. Microbiol.***32**: 289-299.

Froehlich, B., Husmann, L., Caron, J., and Scott, J.R., 1994, Regulation of *rns*, a positive regulatory factor for pili of enterotoxigenic *Escherichia coli*. *J. Bacteriol.* **176**: 5385-5392.

Furrer, J.L., Sanders, D.N., Hook-Barnard, I.G., and McIntosh, M.A., 2002, Export of the siderophore enterobactin in *Escherichia coli*: involvement of a 43 kDa membrane exporter. *Mol. Microbiol.* **44**: 1225-1234.

Gehring, A.M., DeMoll, E., Fetherston, J.D., Mori, I., Mayhew, G.F., Blattner, F.R., Walsh, C.T., and Perry, R.D., 1998, Iron acquisition in plague: modular logic in enzymatic biogenesis of yersiniabactin by *Yersinia pestis*. *Chem. Biol.* **5**: 573-586.

Geoffroy, V.A., Fetherston, J.D., and Perry, R.D., 2000. *Yersinia pestis* YbtU and YbtT are involved in synthesis of the siderophore yersiniabactin but have different effects on regulation. *Infect. Immun.* **68**: 4452-4461.

Heinrichs, D.E., and Poole, K., 1996, PchR, a regulator of ferripyochelin receptor gene (*fptA*) expression in *Pseudomonas aeruginosa*, functions both as an activator and as a repressor. *J. Bacteriol.* **178**: 2586-2592.

Perry, R.D., Balbo, P.B., Jones, H.A., Fetherston, J.D., and DeMoll, E., 1999, Yersiniabactin from *Yersinia pestis*: biochemical characterization of the siderophore and its role in iron transport and regulation. *Microbiology* **145**: 1181-1190.

Perry, R.D., Bearden, S.W., and Fetherston, J.D., 2001, Iron and heme acquisition and storage systems of *Yersinia pestis*. *Recent Res. Devel. Microbiol.* **5**: 13-27.

Rakin, A., Noelting, C., Schubert, S., and Heesemann, J., 1999, Common and specific characteristics of the high-pathogenicity island of *Yersinia enterocolitica*. *Infect. Immun.* **67**: 5265-5274.

Staggs, T.M., Fetherston, J.D., and Perry, R.D., 1994, Pleiotropic effects of a *Yersinia pestis fur* mutation. *J. Bacteriol.* **176**: 7614-7624.

Venturi, V., Weisbeek, P., and Koster, M., 1995, Gene regulation of siderophore-mediated iron acquisition in *Pseudomonas*: not only the Fur repressor. *Mol. Microbiol.* **17**: 603-610.

Chapter 54

Function and Regulation of the Transcriptional Activator RovA of *Yersinia pseudotuberculosis*

Geraldine NAGEL, Ann Kathrin HEROVEN, Julia EITEL and Petra DERSCH

Department of Microbiology, Freie Universität Berlin, Germany

1. THE TRANSCRIPTIONAL ACTIVATOR RovA

During the course of infection *Yersinia pseudotuberculosis* invades and translocates the epithelial cell layer in the ileum through M-cells and subsequently spreads to other organs. For transcytosis *Y. pseudotuberculosis* uses the outer membrane protein invasin, which mediates tight association and internalization by host cells.

Invasin expression is under environmental control and is selectively induced at moderate temperature in bacteria grown to late stationary phase in rich media (Pepe *et al.*, 1994; Nagel *et al.*, 2001). We found that transcription of *inv* requires the RovA protein. The *rovA* gene was identified by a genetic complementation strategy, which restored temperature regulation of an unexpressed *inv-phoA* fusion in *E. coli* K-12. Maximal RovA-mediated transcription of invasin requires the binding of RovA to an AT-rich DNA segment extending the 207 bp upstream of the transcriptional start site of *rovA*. Gel retardation assays showed that RovA preferentially interacts with this promoter fragment and suggested two independent RovA binding sites (Nagel *et al.*, 2001).

The RovA protein is a member of the SlyA/Hor family of transcriptional activators and homologs of this protein were found in several *Enterobacteriaceae*, where they play a role in pathogenesis, i.e. invasion, intracellular survival and virulence gene expression. We showed that the homologous SlyA protein of enteropathogenic *E. coli,* can also induce *inv* expression, and vice versa RovA is capable to induce the SlyA dependent

The Genus Yersinia, Edited by Skurnik *et al.*
ıwer Academic/Plenum Publishers, New York, 2003

clyA gene, encoding a cryptic contact-depending pore forming hemolysin in *E. coli* K-12 (Oscarsson *et al.*, 1996). Thus, RovA and SlyA are heterogenically cross-functional.

2. ROLE OF RovA IN BACTERIAL VIRULENCE

In order to study the function of RovA we introduced a resistance cassette into the *rovA* gene and introduced the *rovA* mutant allele into the *Y. pseudotuberculosis* chromosome. The analysis of the resulting *rovA* deletion mutant demonstrated that RovA plays a significant role in cell invasion and virulence. We performed adhesion and invasion assays and found that the *Y. pseudotuberculosis* wild type strain and the *rovA* mutant strain adhered equally well to mammalian epithelial cells, however, the *rovA* mutant entered about 7-fold less efficiently than the wild type strain. In the mouse infection model, coinfection experiments demonstrated that the *rovA* mutant strain of *Y. pseudotuberculosis* is significantly attenuated in virulence. 24 hours post infection, less bacteria of the *rovA* mutant strain were recovered from the Peyer´s patches of the mice, and no or very low amounts of *Yersinia* were found in the deeper organs such as liver and spleen. A similar decrease of the dissemination has been observed with a *rovA* mutant strain of *Y. enterocolitica* (Revell and Miller, 2000). One possible explanation for the reduction in virulence might be the fact that RovA is involved in the resistance of *Yersinia* to oxidative and osmotic stress. In survival experiments we could demonstrate that the *rovA* mutant strain is hypersensitive to hydrogen peroxide and high concentrations of salt. Recently, it has also become evident that RovA is a global regulator, which is involved in the regulation of numerous genes of *Yersinia*. Two-dimensional gel electrophoresis demonstrated that the expression of at least 50 proteins, some of which might be involved in pathogenesis, was changed in a *rovA* mutant strain. The RovA and the SlyA protein both activate *inv* and *clyA* transcription, respectively. However, the mechanism how the RovA protein changes global gene expression is unclear, since the amount of some proteins was reduced, while the amount of other proteins was significantly increased in the *rovA* mutant strain.

3. REGULATION OF *rovA* EXPRESSION

Studies with a chromosomal *rovA-lacZ* fusion strain showed that *rovA* follows the same expression pattern as invasin, indicating that environmental regulation of *inv* is mainly mediated by the control of RovA synthesis.

Furthermore, we showed that a *rovA-lacZ* fusion is not expressed in a *rovA* mutant strain, demonstrating that a positive autoregulatory mechanism is also involved in *rovA* regulation. A σ^{70}-dependent promoter of *rovA* was determined by primer extention, and deletions of the *rovA* promoter sequences revealed an extended regulatory region required for maximal *rovA* expression. RovA preferentially interacts with AT-rich sequences of the promoter fragments and generates three different RovA-DNA complexes. Additional regulatory studies expressing *rovA* from an independent *lac* promoter showed that the expression of an *inv-phoA* fusion in *E. coli* was still under temperature and growth phase control. This demonstrated that the environmental control of RovA occurs on a post-transcriptional level. To gain further insight in the regulatory network of RovA we performed a mutant hunt and we are currently investigating potential regulatory proteins controlling *rovA* synthesis.

ACKNOWLEDGEMENTS

The work from our laboratory was supported by a grant from the Deutsche Forschungsgemeinschaft DE616/2-1.

REFERENCES

Nagel, G., A. Lahrz, and P. Dersch. 2001. Environmental control of invasin expression in *Yersinia pseudotuberculosis* is mediated by regulation of RovA, a transcriptional activator of the SlyA/Hor family. Mol Microbiol. 41:1249-69.

Oscarsson, J., Y. Mizunoe, B. E. Uhlin, and D. J. Haydon. 1996. Induction of haemolytic activity in *Escherichia coli* by the *slyA* gene product. Mol Microbiol. 20:191-9.

Pepe, J. C., J. L. Badger, and V. L. Miller. 1994. Growth phase and low pH affect the thermal regulation of the *Yersinia enterocolitica inv* gene. Mol Microbiol. 11:123-35.

Revell, P. A., and V. L. Miller. 2000. A chromosomally encoded regulator is required for expression of the *Yersinia enterocolitica inv* gene and for virulence. Mol Microbiol. 35:677-85.

Chapter 55

Temperature and Growth Phase Regulate the Transcription of the O-Antigen Gene Cluster of *Yersinia enterocolitica* O:3

Pia LAHTINEN, Agnieszka BRZEZINSKA and Mikael SKURNIK[1]
Department of Medical Biochemistry and Molecular Biology, University of Turku and [1]*Department of Bacteriology and Immunology, Haartman Institute, University of Helsinki, Finland*

1. INTRODUCTION

Temperature regulates very precisely the expression of *Y. enterocolitica* virulence factors *in vitro* including that of YadA, Ail, invasin, urease, lipopolysaccharide (LPS) etc., such that some factors are optimally expressed at 37°C and others, at 25°C (Skurnik and Toivanen, 1993). In addition to change in temperature some other environmental factors such as pH and ionic strength are modulating the gene expression in *Yersinia* (Pepe *et al.*, 1994; Revell and Miller, 2000). This modulation must also be very important *in vivo* and therefore it is important to know the molecular details of the regulatory mechanisms to be able to study their role in virulence.

LPS is a complex outer membrane molecule of Gram-negative bacteria. It consists of three moieties: lipid A, the core and O-side chain (O-antigen). The O-antigen is a virulence factor, which functions as a barrier against killing by alternative complement pathway and thus plays an important role in effective colonization of host tissues. The O-antigen of *Y. enterocolitica* serotype O:3 is a homopolymer composed of 6-deoxy-L-altrose repeating units, linked together by 1,2 linkages (Skurnik, 1999).

The expression of the homopolymeric O-antigen is regulated by temperature at transcriptional level (Al-Hendy *et al.*, 1991) such that it is optimally expressed at 25°C and less at 37°C, i.e. the number of repeating units per LPS molecule is higher when bacteria are grown at 25°C than at

'he Genus Yersinia, Edited by Skurnik *et al.*
ıwer Academic/Plenum Publishers, New York, 2003

37°C. The mechanism behind this temperature regulation is still unknown. Theoretically, there are two options for the regulatory mechanism; first, there might be an activator for the transcription of the O-antigen gene cluster promoters that is functional at lower temperatures and inactive at 37°C, or second, there might be a repressor for the promoters that is active at 37°C. Since it is thought that the temperature regulation of *Y. enterocolitica* virulence factors plays a very important role in the pathogenesis of *Y. enterocolitica* infection, a deeper understanding of the regulatory pathways is needed.

The O-antigen gene cluster of *Y. enterocolitica* serotype O:3 (YeO3) contains two operons (Zhang *et al.*, 1993) both of which are preceeded by tandem promoters to initiate transcription (Figure 1). We constructed two reporter strains where a firefly luciferase gene *luc*FF was introduced into both O-antigen gene cluster operons of *Y. enterocolitica* serotype O:3 strain 6471/76-c (YeO3-c, a virulence plasmid-cured strain) and monitored the transcriptional activity of both operons (Figure 1).

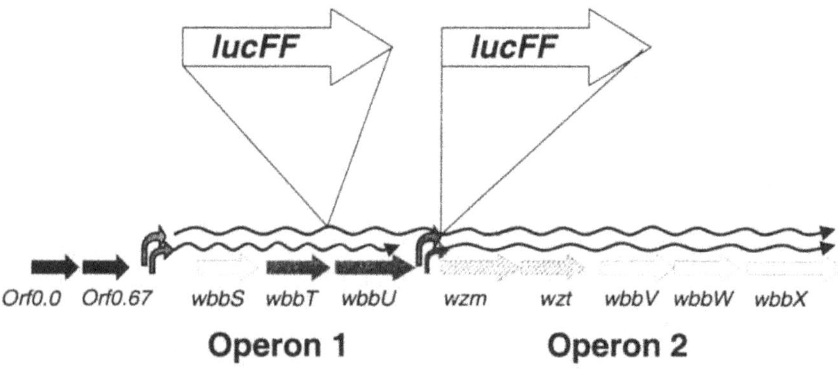

Figure 1. Construction of reporter strains. The *lucFF* gene was fused into the O-antigen gene cluster of *Y. enterocolitica* O:3 between the *wbbT* and *wbbU* genes (operon 1 reporter) and between the tandem promoters and the *wzm* gene (operon 2 reporter). In both reporters the O-antigen expression was not affected.

2. RESULTS

To characterize the regulation of the O-antigen gene cluster expression light production was monitored from both reporter strains incubated in different conditions including such where no growth took place (bacteria were incubated in PBS). In addition, incubation temperature shifts were made between 22°C (RT) and 37°C.

In the first experiments luciferase activity of the reporter strains was measured after 24 hrs of incubation at RT and 37°C. The results showed that the light production was affected by both the growth temperature and growth phase (Figure 2). Temperature influenced the O-antigen expression more strongly in stationary phase bacteria. At that stage a clearly visible decrease in the luciferase activity was observed. Noteworthy, operon 1 had about 10-fold higher activity than operon 2 (not shown).

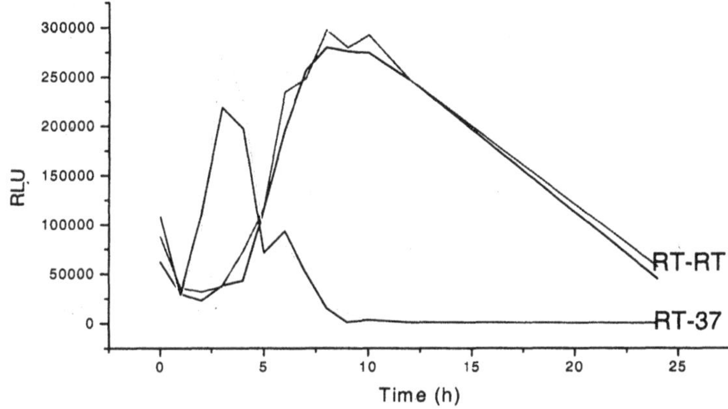

Figure 2. Operon 1 reporter activity during a 24 h growth experiment.

In exponentially growing bacteria temperature had almost no effect on the magnitude of the luciferase activity. Note that in the experiment of Figure 2 the RT grown bacteria reached the exponential growth phase a little later that bacteria growing at 37°C. This observation was verified by supplementing the culture with fresh medium to keep its OD_{600} at 0.5 and the bacteria constantly in the exponential phase. In such a culture the light production remained at constant level and only if the culture was left unadjusted the stationary phase temperature-dependent effects became apparent.

3. CONCLUSIONS

Temperature seems to regulate both O-antigen gene cluster operons of *Y. enterocolitica* O:3 mainly during the stationary phase of growth. In exponential phase bacteria both operons are actively transcribed independent of temperature since temperature shift during exponential phase did not have signifigant effect on the luciferase activity. Dilution of stationary phase bacteria grown at RT decreases the light production very rapidly and dramatically. Our results suggest that in stationary phase *Y. enterocolitica* O:3 bacteria the O-antigen expression is controlled by a regulatory circuit which may include an activator at RT and/or a repressor at 37°C. Further work is needed to elucidate regulatory mechanisms.

REFERENCES

Al-Hendy, A., Toivanen, P., and Skurnik, M. (1991) The effect of growth temperature on the biosynthesis of *Yersinia enterocolitica* O:3 lipopolysaccharide: temperature regulates the transcription of the rfb but not of the rfa region. Microb Pathog 10: 81-86.

Pepe, J. C., Badger, J. L., and Miller, V. L. (1994) Growth phase and low pH affect the thermal regulation of the *Yersinia enterocolitica inv* gene. Mol Microbiol 11: 123-135.

Revell, P. A., and Miller, V. L. (2000) A chromosomally encoded regulator is required for expression of the *Yersinia enterocolitica inv* gene and for virulence. Mol Microbiol 35: 677-685.

Skurnik, M. (1999) Molecular genetics of *Yersinia lipopolysaccharide*. In Genetics of Bacterial Polysaccharides, J. Goldberg, ed. Boca Raton, FL: CRC Press, pp. 23-51.

Skurnik, M., and Toivanen, P. (1993) *Yersinia enterocolitica* lipopolysaccharide: genetics and virulence. Trends Microbiol 1: 148-152.

Zhang, L., Al-Hendy, A., Toivanen, P., and Skurnik, M. (1993) Genetic organization and sequence of the rfb gene cluster of *Yersinia enterocolitica* serotype O:3: Similarities to the dTDP-L-rhamnose biosynthesis pathway of *Salmonella* and to the bacterial polysaccharide transport systems. Mol Microbiol 9: 309-321.

PART V

EPIDEMIOLOGY

Chapter 56

Molecular Epidemiology of *Yersinia enterocolitica* 4/O:3

Maria FREDRIKSSON-AHOMAA[1,2] and Hannu KORKEALA[2]
[1]*Institute of Hygiene and Technology of Food of Animal Origin, Ludwig-Maximilians-University, Munich, Germany;* [2]*Department of Food and Environmental Hygiene, University of Helsinki, Finland.*

1. INTRODUCTION

The epidemiology of *Yersinia enterocolitica* 4/O:3 (YE 4:O3) infections is complex and poorly understood. Most cases of yersiniosis occur sporadically without an apparent source (Bottone, 1999). YE is thought to be a significant foodborne pathogen, even though pathogenic isolates have seldom been recovered from foods. Although pigs have been shown to be a major reservoir for human pathogenic YE 4:O3 strains, the transmission route from pigs to humans remains unproven. Indirect evidence suggests that food, particularly pork, is an important link between the swine reservoir and human infections. In case-control studies, a correlation has been demonstrated between the consumption of raw or under-cooked pork and the prevalence of yersiniosis (Tauxe *et al.*, 1987; Ostroff *et al.*, 1994). In order to identify reservoirs of infections, transmission vehicles and associations between clinical cases, several DNA-based methods have been used to characterise YE 4:O3 strains (Iteman *et al.*, 1996). However, the high genetic similarity between strains and the predominating genotypes of YE 4:O3 have limited the benefit of these methods in epidemiological studies. Thus many factors relating to the epidemiology of YE, such as the sources and transmission routes of YE infections, have remained obscure.

ιe *Genus Yersinia*, Edited by Skurnik *et al.*
wer Academic/Plenum Publishers, New York, 2003

2. YE 4:O3 IN FINLAND AND GERMANY

2.1 Incidence in humans

YE 4:O3 is the most common cause of human yersiniosis globally (Bottone, 1999). This bioserotype is also the most frequently isolated type in humans in Germany and Finland. Yersiniosis has been a recognised disease in Germany since 2001, the number of reported cases varying from 4.9 to 26.9 per 100 000 inhabitants between different regions in 2001 (RKI, 2002) (Table 1).

Table 1. Incidence of yersiniosis in Germany

State	2000 Total no.	2000 Per 100 000	2001 Total no.	2001 Per 100 000
Baden-Württemberg			566	5.4
Bayern			594	4.9
Berlin	58	1.7	253	7.4
Brandenburg	72	2.8	313	12.0
Bremen	16	2.4	48	6.9
Hamburg	35	2.1	189	11.1
Hessen	83	1.4	364	6.1
Mecklenburg-Vorpommen	63	3.5	343	19.1
Niedersachsen			670	8.5
Nordrhein-Westfalen	344	1.9	1133	6.3
Rheinland-Pfalz			351	8.9
Saarland	23	2.1	74	6.7
Sachsen	258	5.8	722	16.0
Sachsen-Anhalt	125	4.7	591	21.9
Schleswig-Holstein			229	8.2
Thüringen	116	4.7	673	26.9
Total			7113	8.7

Table 2. Incidence of yersiniosis in Nordic countries

Country	1999 Total no	1999 Per 100 000	2000 Total no	2000 Per 100 000	2001 Total no	2001 Per 100 000
Finland	634	12.3	641	12.4	728	14.1
Sweden	478	5.4	632	7.1	579	6.5
Norway	125	2.8	140	3.1	123	2.7
Denmark	339	6.4	265	5.0	286	5.3

In Finland, yersiniosis has been reported since 1995 and the number of reported cases has ranged between 12.4 and 17.9 per 100 000 inhabitants during 1995-2001 (KTL, 2000, 2001) (Table 2). The incidence is probably much higher since only the most serious cases are registered.

2.2 Prevalence in animals

Human pathogenic YE 4:O3 strains have frequently been isolated from slaughtered pigs in Finland and Southern Germany (Fredriksson-Ahomaa *et al.*, 2000a, 2001a). The highest prevalence has been obtained in pig tonsils (Table 3). YE 4:O3 strains have also been recovered from pet animals (Fredriksson-Ahomaa *et al.*, 2001c).

Table 3. Isolation rate of YE 4:O3 in some animals in Finland and Southern Germany

		Finland			Southern Germany		
Sample	No. of samples	Year	No. of samples	Isolation rate	Year	No. of samples	Isolation rate
Pigs	Tonsils	1995	185	26%	2000	50	60%
		1999	210	51%			
	Feces	1997	133	4%	2000	50	10%
Cattle	Tonsils				2001	50	0
	Feces				2001	50	0
Birds	Oral cavity				2001	50	0
	Feces	1997	222	0	2001	50	0
Dogs	Feces	1998	95	1%			
Cats	Feces	1998	97	2%			

2.3 Occurrence in pig slaughterhouses

Occurrence of virulence plasmid (pYV) -positive YE has been studied in samples from pig carcasses, offal and the environment with both PCR and culture methods in Finland (Fredriksson-Ahomaa *et al.*, 2000b) (Table 4).

Table 4. Detection rate of pYV-positive YE in a Finnish slaughterhouse

		PCR method		Culture method	
Sample	No. of samples	No. of pos. samples	Detection rate	No. of pos. samples	Detection rate
Carcass	80	17	21%	5	6%
Ear	17	4	24%	2	12%
Liver	13	5	38%	4	31%
Kidney	13	11	85%	9	69%
Heart	8	5	63%	4	25%
Environment	89	12	13%	5	6%

The detection rate was high on pig offal (ears, heart, liver and kidneys) with both methods. Pathogenic YE was also detected in the slaughterhouse environment from different sites, including the brisket saw, the hook from which the pluck set (heart, lungs, oesophagus, trachea, diaphragm, liver, kidneys, and tongue with tonsils) hung, the knife used for evisceration, the floor in the eviscerating and the weighing area, the meat-cutting table, the aprons used by trimming workers, the computer keyboard used in the meat inspection area, the handle of the coffeemaker used by slaughterhouse workers, and the air in the bleeding area. YE 4:O3 was the only pathogenic bioserotype found in the slaughterhouse.

Table 5. Isolation rate of YE 4:O3 on pig offal in Southern Germany

Samples	No. of samples	No. of pos. samples	Isolation rate
Pig pluck sets	20	20	100%
Tonsils	20	17	85%
Tongues	20	15	75%
Lungs	20	14	70%
Hearts	20	14	70%
Diaphragms	20	10	50%
Livers	20	5	25%
Pig kidneys	20	3	15%

Occurrence of YE 4:O3 on pig offal has also been studied in Southern Germany (Fredriksson-Ahomaa *et al.*, 2001a). The pluck sets (tonsils, tongue, heart, lungs, esophagus, trachea, diaphragm and liver) hanging on racks and kidneys in containers were studied with culture method. The highest isolation rate of YE 4:O3 was obtained from tonsil samples and the lowest from kidney samples (Table 5). By contrast, in Finland, kidneys were highly contaminated with YE 4:O3. This may be due to the slaughtering process. In Finland, in many slaughterhouses, kidneys are removed along with the pluck set and then hung together on a hook.

2.4 Prevalence in foods

Table 6. Detection rate of pYV-positive YE in foods

Sample	No. of samples	PCR method: No. of pos. samples	PCR method: Detection rate	Culture method: No. of pos. samples	Culture method: Detection rate
Pig tongue	99	82	83%	79	80%
Pig liver	33	24	73%	10	30%
Pig heart	41	29	71%	18	44%
Pig kidney	36	24	67%	10	28%
Minced pork	255	63	25%	4	2%
Fish	200	0	0	0	0
Chicken	43	0	0	0	0
Lettuce	101	3	3%	0	0

Prevalence of pYV-positive YE in foods has been studied in Finland with both PCR and culture methods (Fredriksson-Ahomaa *et al.*, 1999b, 2001d). The highest detection rate was obtained from pig offal (tongues, livers, hearts and kidneys) with both methods (Table 6).

The detection rate was clearly higher in minced pork with PCR method compared to culture method. No pathogenic YE was recovered from fish, chicken and lettuce samples, however, three positive (3%) lettuce samples were found with PCR.

The prevalence of YE 4:O3 in minced meat containing pork has also been studied in Germany (Fredriksson-Ahomaa et al, 2001a). The isolation rate of 12% was clearly higher in Southern Germany compared to 2% in Finland (Fredriksson-Ahomaa *et al.*, 1999b).

3. EPIDEMIOLOGY OF YE 4:O3

3.1 Pulsed-field gel electrophoresis typing

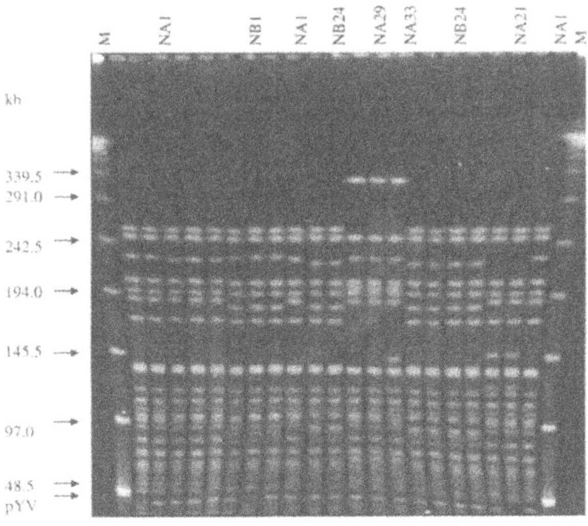

Figure 1. NotI profiles obtained from YE 4:O3 isolates recovered from tonsils of Finnish slaughter pig

Genetic diversity is limited in bioserotype 4:O3, but with PFGE using *Not*I, *Apa*I and *Xho*I enzymes, this group can efficiently be divided into several genotypes with a discriminatory index of 0.93 (Fredriksson-Ahomaa *et al.*, 1999a). These enzymes were selected after a pilot study where 35 enzymes were tested. *Not*I was shown to be a good screening enzyme (DI=0.74), producing a clear banding pattern without smearing and partial digestion (Figure 1). The presence of pYV is easily confirmed from the *Not*I patterns. PFGE has been used to identify sources and possible transmission routes of YE 4:O3 in Finland.

3.2 Diversity of different genotypes of YE 4:O3

The most common genotypes found in pig tonsils in Finland have been widely distributed amongst the slaughterhouses (Fredriksson-Ahomaa *et al.*, 2000a, 2001b). These genotypes have also frequently been found on pig carcasses and offal (ears, livers, kidneys and hearts), and in the slaughterhouse environment (Fredriksson-Ahomaa *et al.*, 2000b, 2001b). The same genotypes have been found at retail level on pig edible offal and in pork.

Genotypes commonly found in human YE 4:O3 infections in Finland were recovered from many sources of pig origin in slaughterhouses and retail shops. Altogether 80% of human strains were indistinguishable from strains of pig origin when 212 human strains were compared with 334 non-human strains. In all, 39 genotypes found in human infections were found from different sources of pig origin (Fredriksson-Ahomaa *et al.*, 2001b). The two most common genotypes found in humans and porcine samples were also found in dogs and cats, which had been fed raw offal and pork (Fredriksson-Ahomaa *et al.*, 2001c).

Diversity of different genotypes of YE 4:O3 strains recovered from pig tonsils in Southern Germany and Finland has also been studied. All genotypes found in German strains differed from Finnish strains (unpublished data) indicating that YE 4:O3 genotypes have a differential geographical distribution.

3.3 Possible transmission routes of YE 4:O3 from pigs to humans

The YE 4:O3 –positive pig tonsils and feces may contaminate the carcass, offal and the environment during the slaughtering process. YE 4:O3 can be transmitted from slaughterhouses to meat processing plants and then to retail level via contaminated pig carcasses and offal (Fredriksson-Ahomaa *et al.*, 2001b). Contaminated pig offal and pork may be important transmission vehicles of YE 4:O3 from retail shops to man. Cross-contamination from offal and pork will occur directly or indirectly via equipment, air and food handlers in slaughterhouses, retail shops and residential kitchens. As a psychrotrophic microbe, YE is able to multiply along the cold-chain from the slaughterhouse to home refrigerator. YE 4:O3 will also be transmitted to humans, especially to young children, indirectly from offal and pork via dogs and cats (Fredriksson-Ahomaa *et al.,* 2001c). A possible transmission route of YE 4:O3 from pigs to humans is demonstrated in Figure 2.

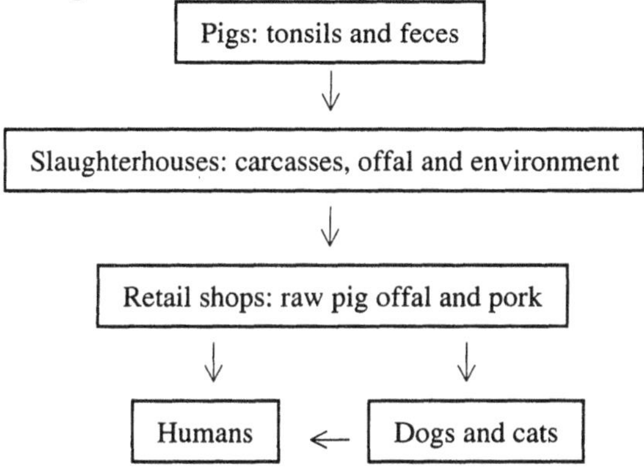

Figure 2. A possible transmission route of YE 4:O3 from pigs to humans

4. CONCLUSIONS

The main source of sporadic human YE 4:O3 infections in Finland is the pig. YE 4:O3 is a common bacterium in slaughtered pigs, especially in tonsils. During evisceration, the tonsils are removed in conjunction with the pluck set (tongue, oesophagus, trachea, heart, lungs, diaphragm, liver and kidneys). The spread of YE 4:O3 from highly contaminated tonsils to the offal can not be avoided, especially when they hang together on a hook. Contaminated pig tongues, hearts, livers and kidneys have shown to be an important transmission vehicle of this bacterium from pigs to man. The only way to prevent the spread of pathogenic YE from pig tonsils to offal, carcasses and the environment is to modify the European Union legislation (64/433/EEC), to mandate that the head, containing the tonsils and tongue, should be removed prior to evisceration and that inspection and handling of the head, tonsils and tongue should occur in a separate room.

REFERENCES

Bottone, E. J., 1999, *Yersinia enterocolitica*: overview and epidemiologic correlates. *Microb. Infect.* **1**: 323-333.

Fredriksson-Ahomaa, M., Autio, T., and Korkeala, H., 1999a, Efficient subtyping of *Yersinia enterocolitica* bioserotype 4/O:3 with pulsed-field gel electrophoresis. *Lett. Appl. Microbiol.* **29**: 308-312.

Fredriksson-Ahomaa, M., Björkroth, J., Hielm, S., and Korkeala, H., 2000a, Prevalence and characterisation of pathogenic *Yersinia enterocolitica* in pig tonsils from different slaughterhouses. *Food Microbiol.* **17**: 93-101.

Fredriksson-Ahomaa, M., Bucher, M., Hank, C., Stolle, A., and Korkeala, H., 2001a, High prevalence of *Yersinia enterocolitica* 4:O3 on pig offal in Southern Germany: a slaughtering technique problem. *System. Appl. Microbiol.* **24**: 457-463.

Fredriksson-Ahomaa, M., Hallanvuo, S., Korte, T., Siitonen, A., and Korkeala, H., 2001b, Correspondence of genotypes of sporadic *Yersinia enterocolitica* 4/O:3 strains from human and porcine origin. *Epidemiol. Infect.* **127**: 37-47.

Fredriksson-Ahomaa, M., Hielm, S., and Korkeala, H., 1999b, High prevalence of *yadA*-positive *Yersinia enterocolitica* in pig tongues and minced meat at retail level. *J. Food Prot.* **62**: 123-127.

Fredriksson-Ahomaa, M., Korte, T., and Korkeala, H., 2000b, Contamination of carcasses, offals and the environment with *yadA*-positive *Yersinia enterocolitica* in a pig slaughterhouse. *J. Food Prot.* **63**: 31-35.

Fredriksson-Ahomaa, M., Korte, T., and Korkeala, H., 2001c, Transmission of *Yersinia enterocolitica* 4/O:3 to pets via contaminated pork, *Lett. Appl. Microbiol.* **32**: 375-378.

Fredriksson-Ahomaa, M., Lyhs, U., Korte, T., and Korkeala, H., 2001d, Prevalence of pathogenic *Yersinia enterocolitica* in food samples at retail level in Finland. *Arch. Lebensmittelhyg.* **52**: 66-68.

Iteman, I., Guiyoule, A., and Carniel, E., 1996, Comparison of three molecular methods for typing and subtyping pathogenic *Yersinia enterocolitica* strains. *J. Med. Microbiol.* **45**: 48-56.

KTL, 2000, Infectious disease in Finland 1995-1999. KTL B4/2000, National Public Health Institute, Helsinki, Finland.

KTL, 2001, Infectious disease in Finland 2000. KTL B8/2001, National Public Health Institute, Helsinki, Finland.

Ostroff, S. M., Kapperud, G., Huteagner, L. C., Nesbakken, T., Bean, N. H., Lassen, J., and Tauxe, R. V., 1994, Sources of sporadic *Yersinia enterocolitica* infections in Norway: a prospective case-control study. *Epidemiol. Infect.* **112**: 133-141.

RKI, 2002, Epidemiologisches Bulletin No. 29. Robert Koch Institute, Berlin, Germany.

Tauxe, R. V., Wauters, G., Goossens, V., van Noyen, R., Vandepitte, J., Martin, S. M., de Moel, P., and Tiers, G., 1987, *Yersinia enterocolitica* infections and pork: the missing link. *Lancet.* **i**: 1129-1132.

Chapter 57

Occurrence of *Y. enterocolitica* in Slaughter Pigs and Consequences for Meat Inspection, Slaughtering and Dressing Procedures

Truls NESBAKKEN, Karl ECKNER, Hilde Kristin HØIDAL and Ole-Johan RØTTERUD

Norwegian Meat Research Centre, P. O. Box 396, Økern, 0513 Oslo, Norway; The Norwegian School of Veterinary Science, Oslo, Norway; Norwegian Institute for Food and Environmental Analysis, Oslo, Norway; Genpoint AS, Oslo, Norway

1. INTRODUCTION

It is impossible to reject pigs contaminated with *Yersinia enterocolitica* by traditional postmortem meat inspection. Meat inspection procedures concerning the carcass head also seem to represent a cross-contamination risk: According to the EU regulations, incision of the submaxillary lymph nodes is a compulsory procedure (European Commission, 1995). However, this may result in *Y. enterocolitica* being transported from the tonsillar region to other parts of the carcass by the knives and hands of the meat inspection personnel. In view of the fact that the incidence of tuberculosis in pigs and humans has been reduced to a very low level in many parts of the world, it may be possible to reconsider regulations that require incision of the submaxillary lymph nodes to detect tubercular nodes.

The present investigation had two goals:

1. to assess the occurrence of *Y. enterocolitica* in lymphoid tissues and the intestinal tract in pigs, together with the risk for contamination of carcasses during the compulsory meat inspection procedures and the procedures during slaughtering and dressing.
2. to compare traditional isolation and Polymerase Chain Reaction (PCR) (BUGS'n BEADS™ bacterial DNA isolation kit) methods as tools for risk management in slaughterhouses.

The Genus Yersinia, Edited by Skurnik *et al.*
Kluwer Academic/Plenum Publishers, New York, 2003

2. MATERIALS AND METHODS

2.1 Collection of samples

All samples were collected from one federally inspected slaughterhouse in south-eastern Norway where about 90 pigs were slaughtered per h. Enclosure of the rectum with a plastic bag is a common practice during evisceration.

Sub-maxillary lymph nodes were collected from 97 randomly selected, freshly eviscerated, healthy slaughter pigs. The samples originated from 12 different conventional slaughter pig herds, each represented by 2 – 15 (mean 8.1) individuals. Seven herds totalling 63 pigs were categorised as herds with a specialised slaughter pig production, while five herds with 34 pigs were farrow-to-finish herds.

Blood samples, tonsils, sub-maxillary lymph nodes, mesenteric lymph nodes, contents of stomach, ileum, caecum, colon, faeces and carcass samples from a total of 24 freshly eviscerated, healthy slaughter pigs were collected. Eight individuals represented each herd. The samples originated from three different conventional slaughter pig herds that were selected based on recent isolation of *Y. enterocolitica* O:3 from tonsils of slaughtered pigs from these herds.

Carcass surface samples were collected just subsequent to meat inspection before removal of the head and the final dressing of the carcass. Each of four sample sites (about 200 cm^2, 100 cm^2 from each of the carcass halves) from the ham, pelvic duct, kidney region and medial neck of the carcasses was rubbed thoroughly with sterile swabs dipped into sterile peptone water (Mölnlycke HealthCare AB, type 157300, Gothenburg, Sweden) and put into small, sterile, plastic bags.

2.2 Serological methods

Sera were analysed for antibodies against *Y. enterocolitica* O:3 by an indirect pig immunoglobulin (Ig) lipo-polysaccharide enzyme-linked immunosorbent assay (LPS-ELISA). A basic cut-off of optical density (OD %) > 10 was used to maximise the specificity of the ELISA.

2.3 Isolation and detection

Swabs were diluted in 10 ml peptone water (Oxoid L34 (Oxoid Ltd., Basingstoke, Hampshire, England)), homogenised in a peristaltic blender for 30 sec and prepared for enrichment. Samples were analysed for *Y.*

enterocolitica by an International Organization for Standardization (1994) method (ISO 10273). Samples of tonsils (10 g), lymph nodes (5 g), and gastrointestinal tract contents/faeces (10 g) were diluted 1:1 with peptone water (Oxoid L34), homogenised in the peristaltic blender for 30 sec, and 1 ml was transferred to 99 ml Irgasan – Ticarcillin – potassium chlorate (ITC) enrichment broth (Irgasan DP 300, Ciba-Geigy, Basle, Switzerland; Ticarcillin, Beecham Research Laboratories, Brentford, England). The ITC enrichment broth was incubated for two days at 25°C when a 10-µl volume was streaked for selective isolation on *Salmonella-Shigella*-desoxycholate-calcium chloride (SSDC, *Yersinia*-agar; E. Merck AG, Darmstadt, Germany, 1.11443) and Cefsulodin – Irgasan – Novobiocin (CIN) agar (Oxoid CM653, SR109) plates. SSDC plates were incubated for 24 h and CIN plates for 18 – 24 h at 30 °C. Colonies characteristic for *Yersinia* were biochemically confirmed, first selecting only lactose-negative, urease positive colonies, and later with Vitek (Bio-Merieux, Marcy l'Etoile, France) using the revised bio-grouping scheme for *Y. enterocolitica* (Wauters *et al.,* 1987) as a key and serologically for O:3 and O:9 reactivity (Sanofi Diagnostics-Pasteur, 63501, 63502, Marnes la Coquette, France). Detection of pathogenic *yersiniae* was performed using BUGS'n BEADS™ bacterial DNA isolation kit (Genpoint AS, Oslo, Norway). According to the instruction manual, 100 µl of enriched sample was added to a 1.5-ml Eppendorf tube containing 800 µl of binding and washing buffer (BW) and 200 µg magnetic Bacteria Binding Beads (Chemagen Biopolymer-Technologie, Baeswiler, Germany). Contents were mixed by pipetting and left at room temperature for 5 min. The tube was subsequently placed in a magnetic separator (ABgene®, Epsom, England) to facilitate the separation of beads (with bound bacteria) from the medium. After careful removal of the supernatant by pipetting, 50 µl of lysis buffer was added to the bacteria-bead complex followed by incubation at 80 °C for 5 min. Released DNA was then precipitated onto the beads by addition of 150 µl refrigerated 96 % ethanol and incubation another 5 min at room temperature. To wash the DNA-bead complex, the supernatant was removed using the magnet, followed by two additions of 1 ml 70 % ethanol (using the magnet in between). The DNA-bead complex was finally resuspended in 40 µl sdH_2O and incubated at 80 °C for 10 min to evaporate remaining ethanol.

PCR assay of *Yersinia* targeted the virulence plasmid of *Yersinia* species (Cornelis *et al.,* 1998; Roggenkamp *et al.,* 1995). The PCR reactions (50 µl) contained 3 µl of DNA-bead-complex, 1 x buffer (1.5 mM $MgCl_2$), 200 µM each dNTP, 10 pmol each primer, and 1 U of DyNAzyme™ Thermostable DNA polymerase (Finnzymes Oy, Espoo, Finland). The amplification profile was as follows; 37 cycles of 94°C for 15 s, 58°C for 30 s, 72°C for 1 min, with a prior denaturation at 94°C for 4 min, and a final extension at 72°C for 7 min. PCR-products were visualised by 2% agarose gel electrophoresis, stained by ethidium bromide, and analysed using a GeneGenius gel documentation system (Syngene, Cambridge, England).

2.4 Statistical analysis

Methods were compared using simple tabular analysis and simple and multiple logistic regression using the statistical package Intercool Stata for Windows 7.0.

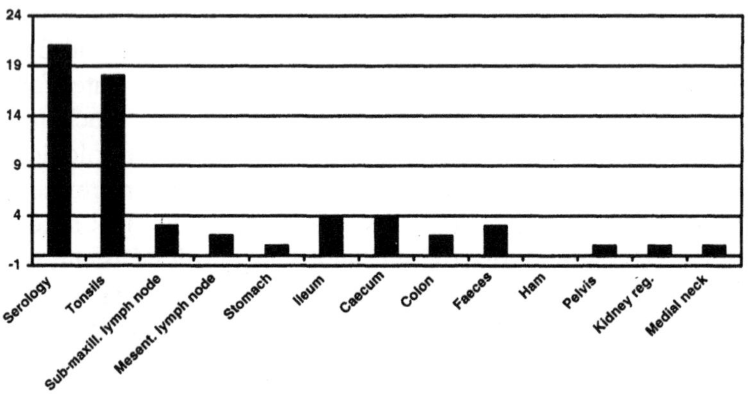

Figure 1. Y. enterocolitica in lymphoid tissues, intestinal contents and carcass surfaces of pigs (n=24)

3. RESULTS AND DISCUSSION

Specific antibody titres against *Y. enterocolitica* O:3 were found in 22 (91.7 %) of the 24 samples investigated, with 21 (87.5%) animals showing an OD% > 10 (Figure 1). Virulent yersiniae were detected in the tonsils of one animal with negative titres, but not in any other tissue sample from negative animals. *Y. enterocolitica* O:3 was not isolated from the tonsils, but from the carcass of one animal with positive titres. *Y. enterocolitica* O:3 was isolated from sub-maxillary lymph nodes in animals from three out of seven specialised slaughter herds, but not from any animal from five farrow-to-finish herds (Table 1).

Table 1. Occurrence of *Y. enterocolitica* in the sub-maxillary lymph nodes in 97 pigs from 12 randomly selected herds

Herd type	No. of animals/ number of herds	No. of positive animals (%)	No. of positive herds (%)
Specialised slaughter production	63/7	5 (7.9)	3 (42.9)
Farrow-to-finish herds	34/5	0	0
All herds	97/12	5 (5.2)	3 (25.0)

More detailed follow up of three selected specialised slaughter pig herds showed that virulent *yersiniae* were detected from tonsils in 18 (75%) out of the 24 animals. As shown in Figure 1, virulent *yersiniae* were also detected on various other sites of many animals. Cross-contamination was illustrated by the fact that *Y. enterocolitica* O:3 was also isolated from one carcass where *Y. enterocolitica* O:3 was not isolated from tonsils, lymph nodes or gastrointestinal tract.

The ISO method and BUGS'n BEADS™ were compared lumping results from lymphoid tissues, intestinal tract and carcass together. Adjusting for sampling sites, there was no statistical difference between the methods in multiple logistic regression. The odds ratio for BUGS'n BEADS™ was 1.10 (95% CI= 0.67-1.91).

Table 2. Comparison of the International Organization for Standardization (1994) (ISO 10273) isolation procedure and the modified BUGS'n BEADS™ method for detection of virulent *yersiniae* from various sources of slaughter pigs. No. of positive samples (%)

Procedure	Lymphoid tissues (n = 72)	Intestinal contents (n = 120)	Carcass surface sites (n = 96)	Total (n = 288)
ISO 10273	17 (23.6%)	13 (10.8%)	3 (3.1%)	33 (11.5%)
BUGS'n BEADS™	23 (31.9%)	13 (10.8%)	0	36 (12.5%)
Both methods	23 (31.9%)	14 (11.7%)	3 (3.1%)	40 (13.9%)

The results indicate that the compulsory procedure of incisions of the sub-maxillary lymph nodes (European Commission, 1995) represents a cross-contamination risk for virulent *Yersinia* (Figure 1 and Table 1).

The mesenteric lymph nodes might represent a cross-contamination risk since 8.3% of the samples were positive. These lymph nodes should also be investigated when the meat inspection personnel have specific reasons to do so (European Commission, 1995).

The association between positive titres and occurrence of virulent yersiniae in the tonsils (21 to 18) was striking. In our study the relative difference of finding virulent *yersiniae* in the tonsils and the faeces was 18 to 3 (6 to 1). The relatively low level of *Y. enterocolitica* O:3 in faeces in our study (12.5%) also agrees with Nielsen *et al.*, (1996).

Stomach, ileum, caecum, and colon contents also represent contamination risks for *Y. enterocolitica* O:3 if the slaughterhouse personnel cut into the viscera with their knives by accident; the frequency of virulent *Yersinia* varied from 4.2% to 16.7% within these sections.

REFERENCES

Cornelis G.R., Boland A., Boyd A.P., Geuijen C., Iriarte M., Neyt C., Sory M.P., Stainier I. The virulence plasmid of *Yersinia*, an antihost genome. Microbiol. Mol. Biol. Rev. 1998; 62:1315-52.

European Commission. Council directive 64/433/EEC on health condition for the production and marketing of fresh meat. Brussels, Belgium, 34pp, 1995.

International Organization for Standardization. Microbiology - General guidance for the detection of presumptive pathogenic *Yersinia enterocolitica* (ISO 10273). International Organization for Standardization, Genève, Switzerland, 16 pp., 1994.

Nielsen B., Heisel C., Wingstrand A. Time course of the serological response to *Yersinia enterocolitica* O:3 in experimentally infected pigs. Vet. Microbiol. 1996; 48:293-303.

Roggenkamp A., Schubert S., Jacobi C.A., Heesemann J. Dissection of the *Yersinia enterocolitica* virulence plasmid pYV08 into an operating unit and virulence gene modules. FEMS Microbiol. Lett. 1995; 134:69-73.

Wauters G., Kandolo K., Janssens M. Revised grouping scheme of *Yersinia enterocolitica*. Contrib. Microbiol. Immunol. 1987; 9:14-21.

Chapter 58

Molecular Epidemiology of the Five Recent Outbreaks of *Yersinia pseudotuberculosis* in Finland

Saija HALLANVUO[1], Pekka NUORTI[2], Ulla-Maija NAKARI[1] and Anja SIITONEN[1]
[1]Department of Microbiology, Laboratory of Enteric Pathogens and [2]Department of Infectious Diseases Epidemiology, National Public Health Institute (KTL), Helsinki, Finland.

1. INTRODUCTION

Yersinia pseudotuberculosis is an increasingly important causal agent of zoonotic enteric infections in humans especially in cold climates. It is widespread in nature causing infections also in a variety of domestic and wild animals including sheep, pig and deer (Sanford, 1995; Slee and Button, 1990). With occasional systemic dissemination of the bacteria, the infections are often more severe than *Y. enterocolitica* infections (Iteman *et al.*, 1995).

In Finland, *Y. pseudotuberculosis* has recently emerged as an outbreak-associated pathogen. The number of sporadic *Y. pseudotuberculosis* infections has normally varied from 30 to 40 cases annually. However, since 1997, five outbreaks with altogether about 280 infections have occurred.

Due to the homogeneity of the species, phenotypic markers are inadequate for epidemiological tracing in outbreak investigations. Therefore, we used pulsed-field gel electrophoresis (PFGE) to evaluate the applicability of genetic subtyping in the investigations of *Y. pseudotuberculosis* outbreaks.

2. METHODS

Y. pseudotuberculosis strains isolated in the Finnish clinical microbiology laboratories and identified according to the standard protocols

The Genus Yersinia, Edited by Skurnik *et al.*
'uwer Academic/Plenum Publishers, New York, 2003

(Aleksic and Bockemühl, 1999) were serotyped by slide agglutination (Denka-Seiken, Tokyo, Japan).

For PFGE, 80 strains were chosen to represent five different outbreaks. The strains were grown on nutrient agar plates at 30°C overnight. In plug preparation, a rapid method (Graves and Swaminathan, 2001) was adjusted to *Y. pseudotuberculosis* isolates. Plug-DNA digested with the enzymes *Spe*I and *Not*I were run in 1.2% gels for 24h at pulse time of 7-15s.

3. RESULTS

In the first four outbreaks in August, 1997 (A), September, 1998 (B), November, 1998 (C) and October, 1999 (D), the causative serotype was O:3, whereas in the last outbreak in 2001 (E), serotypes O:3 and O:1 were simultaneously involved.

The outbreaks A and B occurred in closely located regions. The isolates from these outbreaks had indistinguishable PFGE pattern with both enzymes used.

The outbreak C comprised of several geographically separated clusters of cases. The PFGE patterns of the isolates from these clusters were indistinguishable, but different from the pattern of the outbreaks A and B.

The outbreak D also comprised of geographically separated clusters. In PFGE, the isolates from these clusters had indistinguishable pattern from the outbreak C isolates.

In the outbreak E, most of the serotype O:3 infections appeared in three clusters. The isolates from these clusters had distinct PFGE patterns, one of them being indistinguishable from the pattern of the outbreaks C and D. The serotype O:1 infections in the outbreak E were timely and geographically more separate than the serotype O:3 infections of the same outbreak. The serotype O:1 isolates diverged into two main PFGE types.

4. DISCUSSION

Infections caused by *Y. pseudotuberculosis* have mainly been sporadic in Finland. Occasionally, however, small outbreaks have occurred (Tertti *et al.*, 1984). Since 1994, the Finnish clinical microbiology laboratories have been obligated to report their *Y. pseudotuberculosis* findings to the National Infectious Diseases Registry of the National Public Health Institute (KTL). Most of the laboratories have also routinely submitted their isolates to KTL for serotyping and, when necessary, the isolates have been genotyped by PFGE. The active laboratory based surveillance was a prerequisite for the detection of the increased number of infections caused by a certain serotype. In addition, the detailed genomic typing was necessary for efficient investigation of the outbreaks and for recognizing the geographically dispersed infection clusters.

In the outbreaks A and B, the PFGE type was unique and has not reappeared in more recent outbreaks. In the outbreak C, the serotyping and PFGE rapidly demonstrated the geographically dispersed infection clusters as parts of the same outbreak. Subsequent case-control study pointed out iceberg lettuce as a possible source of these infections (Nuorti *et al.,* 1999). The identical PFGE pattern of the isolates from the outbreaks C and D helped in the investigations of the outbreak D and the possibility of the same vehicle as in the outbreak C was rapidly noticed.

The serotype O:3 isolates in the outbreak E belonged to three distinct PFGE types that can be regarded as possibly related to each other and to the outbreak type C and D (Tenover *et al.,* 1995). Serotype O:1 PFGE types were very closely related to each other. The strains of serotype O:1 have been isolated more rarely from human stool specimens than strains of serotype O:3 in Finland. After the outbreak E, the number of O:1 isolates decreased again.

In conclusion, the laboratory based surveillance including serotyping and genotyping enabled recognizing the outbreaks at an early stage by demonstrating that the geographically separate clusters were part of the same outbreak.

REFERENCES

Aleksic, S., and Bockemühl, J., 1999, *Yersinia* and other *Enterobacteriaceae.* In: Manual of Clinical Microbiology, 7th ed (Murray, P.R., Baron, E. J., Pfaller, M. A., Tenover, F.C., Yolken, R.H., eds.), pp. 483-491. American Society for Microbiology, Washington, DC.

Graves, L.M., and Swaminathan, B., 2001, PulseNet standardized protocol for subtyping *Listeria monocytogenes* by macrorestriction and pulsed-field gel electrophoresis. Int. J. Food. Microbiol. **65**:55-62.

Iteman, I., Najdenski, H., and Carniel, E., 1995, High genomic polymorphism in *Yersinia pseudotuberculosis.* Contrib. Microbiol. Immunol. **13**:106-11.

Nuorti, P., Mikkola, J., Hallanvuo, S., Siitonen, A., Lyytikäinen, O., Ruutu, P., 1999, An outbreak of *Yersinia pseudotuberculosis* O:3 infections associated with consumption of iceberg lettuce in Finland. Abstract, presented at the 39th Interscience Conference on Antimicrobial Agents and Chemotherapy (ICAAC) San Francisco.

Sanford, S.E., 1995., Outbreaks of Yersiniosis Caused by *Yersinia pseudotuberculosis* in Farmed Cervids. J. Vet. Diagn. Invest. **7**:78-81.

Slee K.J., and Button C., 1990, Enteritis in sheep, goats and pigs due to *Yersinia pseudotuberculosis* Infection. Aust.Vet. J. **67**:320-2.

Tenover, F.C., *et al.*, 1995, Interpreting chromosomal DNA restriction patterns produced by pulsed-field gel electrophoresis: Criteria for bacterial strain typing. J. Clin. Microbiol. **33**:2233-2239.

Tertti, R., Granfors, K., Lehtonen, O.P., Mertsola, J., Mäkelä, A.L., Välimäki, I., Hänninen, P., and Toivanen, A., 1984, An outbreak of *Yersinia pseudotuberculosis* infection. J. Infect. Dis. **149**:245-250.

Chapter 59

Yersinia pestis from Natural Foci

Vladimir V. KUTYREV, Ol'ga A. PROTSENKO, George B. SMIRNOV[1], Elena BOOLGAKOVA, Lubov M. KUKLEVA, Irina V. ZUDINA, Nadezhda A. VIDYAEVA and Inna KOOZMICHENKO
Russian Research Anti-Plague Institute "Microbe", Saratov, Russia; [1]Gamaleya Research Institute of Epidemiology and Microbiology, Mocsow, Russia.

1. INTRODUCTION

Three historical pandemics of plague (Justinian plague, "black death" and the pandemics of the year 1894) were caused by three different biovars of *Yersinia pestis,* namely antiqua, medievalis and orientalis (Devignat, 1951). These biovars form the main, or classical, variants of the causative agent of plague and differ from each other by nitrification and denitrification activities and by the ability to ferment glycerol. The putative properties of the classical variants are virulence for guinea pigs and white mice, inability to ferment rhamnose and melibiose, uniformity of serological characteristics, sensitivity to specific bacteriophages, and presence of three plasmids, namely pFra/Tox (60 – 65 Md), pCad (45 Md), and pPst (6 Md). The strains of the classical variants have been isolated in natural foci of Africa, America, Middle Asia, Mongolia, Vietnam, and former Soviet Union where the marmots, ground squirrels, and sandpipers have been identified as the carriers.

At the beginning of the XX century new foci of plague were found on the territories of USSR and Mongolia where the small rodents, voles and field-voles were found to be constant carriers. The strains isolated from these foci harbored the main properties of *Y. pestis* but differ from the classical variants by reduced virulence for guinea pigs while possessing standard virulence for white mice (selective virulence). Based on the proposal of L. Timofeeva (1972) the strains from the new foci of plague were grouped into non-

The Genus Yersinia, Edited by Skurnik *et al.*
...wer Academic/Plenum Publishers, New York, 2003

classical subspecies and named according to the geographical location of the natural foci, namely *Y. pestis caucasica, Y. pestis altaica, Y. pestis hissarica,* and *Y. pestis ulegeica.*

Strong proof for the descendance of *Y. pestis* from *Y. pseudotuberculosis* serovar O:1b was obtained in different molecular studies (Achtman *et al.,* 1999; Skurnik *et al.,* 2000). Reductive evolution has resulted in the loss of activity of certain genes in the *Y. pseudotuberculosis* genome. These mutant genes in *Y. pestis* may be considered as silent genes or pseudogenes (Buchrieser *et al.,* 1999; Simonet *et al.,* 1999, Skurnik *et al.,* 2000).

The molecular mechanisms leading to pseudogenes include insertions of mobile genetic elements, deletions and frameshift mutations. It is believed that the pseudogenes facilitated the adaptation of the pathogen to survive in the animal host instead of the environment.

To learn more about the origin of the non-classical *Y. pestis* subspecies we compared phenotypic properties between non-classical *Y. pestis* subspecies, classical *Y. pestis* and *Y. pseudotuberculosis* strains. Three strains of each *Y. pestis* subspecies and four *Y. pseudotuberculosis* strains were studied.

2. RESULTS

The non-classical subspecies retained the species specific properties characteristic to the classical subspecies, namely the sensitivity to specific bacteriophages, lack of mobility and the presence of the three plasmid replicons. However, the large plasmids of the strains of the non-classical subspecies were bigger in size than the plasmids of the classical subspecies (pFra in different strains - 67-82 Md, pCad - 47-50 Md); the pPst plasmid had the same molecular size in the non-classical subspecies strains, however, this plasmid was not found in *Y. pestis caucasica* (Filippov *et al.,* 1990).

However, while the classical *Y. pestis* subspecies had lost the ability to utilize rhamnose and melibiose, the non-classical subspecies had retained these abilities that are also characteristic to *Y. pseudotuberculosis* as well as the oxidation of phenylalanine and tryptophane. On the other hand, the non-classical *Y. pestis* subspecies as well as the classical one had lost the galactokinase activity, and the ability to synthesize O-antigen present in *Y. pseudotuberculosis.*

Thus, the same silent (mutant) genes were present in the classical and non-classical *Y. pestis* subspecies strains (galaktokinase, O-antigen). A number of functional genes retained in the non-classical *Y. pestis* subspecies (rhamnose and melibiose fermentation and oxidation of amino acids) were silent genes in the classical *Y. pestis* subspecies. The non-classical *Y. pestis* subspecies and *Y. pseudotuberculosis* lack both the isocitratelyase activity that is present in classical *Y. pestis* subspecies.

Differences in the frequency and types of mutations affecting pigmentation and pesticin sensitivity were found between the studied *Y. pestis* subspecies. The chromosomal region responsible for pigmentation consists of two loci, *hms* and *ybt* (Fetherston and Perry, 1997). The loss of virulence due to the simultaneous loss of pigmentation ability and pesticin sensitivity occurs in the classical *Y. pestis* subspecies via deletion of the whole *pgm* region mediated by the IS100-dependent recombination. The non-classical *Y. pestis* subspecies often retained virulence despite the loss of pigmentation; they were pesticin sensitive (Kutyrev *et al.*, 1992). Their hms^- ybt^+ genotype suggests stability for the *ybt* region. The same is true for *Y. pseudotuberculosis* and may be explained by the loss of one of the flanking IS100-elements in this organism (Buchrieser, 1998).

3. CONCLUSIONS

Our results showed that the non-classical *Y. pestis* subspecies strains possessed properties of both the classical *Y. pestis* subspecies and those of *Y. pseudotuberculosis*, i.e. in this respect they could be placed between the causative agents of plague and pseudotuberculosis suggesting that they may form a step in the process of evolution of clasical *Y. pestis*.

ACKNOWLEDGEMENTS

The Russian Fond of Fundamental Research, grant 00-04-48796, supported the work.

REFERENCES

Achtman, M., Zurth, K., Morelli, G., Torrea, G., Guiyoule, A., and Carniel, E. (1999) *Yersinia pestis*, the cause of plague, is a recently emerged clone of *Yersinia pseudotuberculosis*. *Proc Natl Acad Sci USA* **96**: 14043-14048.

Buchrieser, C., Brosch, R., Bach, S., Guiyoule, A., and Carniel, E. (1998) The high-pathogenicity island of *Yersinia pseudotuberculosis* can be inserted into any of the three chromosomal asn tRNA genes. *Mol Microbiol* **30**: 965-978.

Buchrieser, C., Rusniok, C., Frangeul, L., Couve, E., Billault, A., Kunst, F., Carniel, E., and Glaser, P. (1999) The 102-kilobase *pgm* locus of *Yersinia pestis*: sequence analysis and comparison of selected regions among different *Yersinia pestis* and *Yersinia pseudotuberculosis* strains. *Infect Immun* **67**: 4851-4861.

Devignant, R. (1951) Variétés de l'éspèce *Pasteurella pestis*. Nouvelle hypothèse. *Bulletin WHO* **4**: 247-263.

Fetherston, J. D., Schuetze, P., and Perry, R. D. (1992) Loss of the pigmentation phenotype in *Yersinia pestis* is due to the spontaneous deletion of 102 kb of chromosomal DNA which is flanked by a repetitive element. *Mol Microbiol* **6**: 2693-2704.

Filippov, A. A., Solodovnikov, N. S., Kookleva, L. M., and Protsenko, O. A. (1990) Plasmid content in *Yersinia pestis* strains of different origin. *FEMS Microbiol Lett* **55**: 45-48.

Kutyrev, V. V., Filippov, A. A., Oparina, O. S., and Protsenko, O. A. (1992) Analysis of *Yersinia pestis* chromosomal determinants Pgm(+) and Pst(s) associated with virulence. *Microb Pathog* **12**: 177-186.

Perry, R. D., and Fetherston, J. D. (1997) *Yersinia pestis* - etiologic agent of plague. *Clin Microbiol Rev* **10**: 35-66.

Simonet, M., Riot, B., Fortineau, N., and Berche, P. (1996) Invasin production by *Yersinia pestis* is abolished by insertion of an IS200-like element within the inv gene. *Infect Immun* **64**: 375-379.

Skurnik, M., Peippo, A., and Ervelä, E. (2000) Characterization of the O-antigen gene clusters of *Yersinia pseudotuberculosis* and the cryptic O-antigen gene cluster of *Yersinia pestis* shows that the plague bacillus is most closely related to and has evolved from *Y. pseudotuberculosis* serotype O:1b. *Mol Microbiol* **37**: 316-330.

Timofeeva L.A. Problems of extremely dangerous infections. Saratov.-1972.-1:15-23.

Chapter 60

A Virulence Study of *Yersinia enterocolitica* O:3 Isolated from Sick Humans and Animals in Brazil Using PCR and Phenotypic Tests

Juliana P. FALCÃO[1], Deise P. FALCÃO[2], Ednéia F. CORRÊA[2] and Marcelo BROCCHI[1]
[1]*Faculdade de Medicina, Ribeirão Preto, USP, Brazil.* [2]*Faculdade de Ciências Farmaceuticas,Araraquara, UNESP, Brazil.*

1. INTRODUCTION

Yersinia enterocolitica is recognized all over the world as an enteric pathogen causing a wide spectrum of clinical and immunological manifestations. In Brazil strains belonging to biotype 4/serotype O:3 (4/O:3) are associated with human and animal diseases, although the isolation of this species from human and animal clinical cases is not usual in this country. The objective of this work was to determine drug resistance profiles and some phenotypic and molecular characteristics of *Y. enterocolitica* 4/O:3 strains isolated from sick humans and animals.

2. MATERIAL AND METHODS

A collection of 71 *Y. enterocolitica* strains originally isolated from human (37) and animal (34) clinical cases from different regions of Brazil was studied.

All the isolates were tested for drug resistance using the disc diffusion technique (Bauer *et al.*, 1966). The following antimicrobial drugs were used: cephalotin (Cfl), ampicillin (Amp), tetracycline (Tet), ofloxacin (Ofx), norfloxacin (Nor), cefotaxime (Ctx), gentamicin (Gen), cefoxitin (Cfo), kanamycin (Kn), streptomycin (St), penicillin (Pen), amikacin (Ami), chloramphenicol (Clo), nalidixic acid (Nal).

The isolates were tested for temperature-dependent autoagglutination, calcium-dependent growth at 37°C, Congo Red uptake, pyrazinamidase production, salicin fermentation and esculin hydrolysis (Farmer *et al.*, 1992).

Detection of the *inv, ail, yst* and *virF* genes was performed using PCR technique. DNA extraction was performed according to Harnett *et al.*, (1996) and PCR carried out according to Saiki *et al.*, (1988). Primers for the amplification of the *inv, ail, yst* and *virF* genes were used as described by Rasmussen *et al.*, (1994), Nakajima *et al.*, (1992), Ibrahim *et al.*, (1997) and Wren and Tabaqchali (1990), respectively.

3. RESULTS

All the 71 *Y. enterocolitica* 4/O:3 strains were resistant to, at least, three of the 14 antimicrobials used .

Altogether 72.97% of the 37 *Y.enterocolitica* strains isolated from human clinical cases and 41.18% of the 34 strains isolated from sick animals were positive for temperature-dependent autoagglutination, calcium-dependent growth at 37°C and Congo Red uptake .

Almost all strains (99.0%) isolated from human clinical cases and 100% of those isolated from animals were virulent according to the esculin hydrolysis, salicin fermentation and pyrazinamidase production tests. The *inv*, *ail* and *yst* virulence genes were detected in all the studied strains.

The *vir*F gene was detected in 94.60% of the 37 *Y. enterocolitica* isolated from human clinical cases and in 61.76% of the 34 strains isolated from animals. Although 78.87% of the studied *Y. enterocolitica* strains possessed the *vir*F gene, only 57.75% were positive for the phenotypic tests related to the expression of this gene.

4. CONCLUSIONS

Y. enterocolitica 4/O:3 strains have been a source of infection for humans and animals in Brazil. The potential virulence of the isolated strains was confirmed by the presence of virulence genes and by phenotypic virulence tests and in addition, the strains showed multiple drug resistance profiles. The disagreement between the PCR for the *vir*F detection and the phenotypic tests related to the expression of this gene could be due to a loss of activity of the *vir*F gene in some strains.

ACKNOWLEDGEMENTS

This work was supported by grants from FAPESP (00/09175-8)

REFERENCES

Bauer, A.W., Kirby, W.M., Sherris, J.C. and Turck, M., 1966, Antibiotic susceptibility testing by a standardized single disc method. Am. J. Clin. Pathol. 45: 493-496.

Farmer III, J.J., Carter, G.P., Miller, V.L., Falkow, S. and Wachsmuth, I.K., 1992, Pyrazinamidase, CR-MOX agar, salicin fermentation-esculin hydrolysis, and D-xylose fermentation for identifying pathogenic serotypes of *Yersinia enterocolitica*. J.Clin. Microbiol. 30: 2589-2594.

Harnett, N., Lin, Y.P. and Krishnan, C.K. , 1996, Detection of pathogenic *Yersinia enterocolitica* using the multiplex polymerase chain reaction. Epidemiol. Infect. 117: 59-67.

Ibrahim, A., Liesack, W., Griffiths, M.W. and Robins-Browne, R.M., 1997, Development of a highly specific assay for rapid identification of pathogenic strains of *Yersinia enterocolitica* based on PCR amplification of *Yersinia* heat-stable enterotoxin gene (*yst*). J. Clin. Microbiol. 35:1636-1638.

Nakajima, H., Inoue, M., Mori, T., Itoh, K.I., Arakawa, E. and Watanabe,H., 1992, Detection and identification of *Yersinia pseudotuberculosis* and pathogenic *Yersinia enterocolitica* by an improved polymerase chain reaction method. J. Clin. Microbiol. 30: 2484-2486.

Rasmussen, H.N., Rasmussen, O.F., Andersen, J.K. and Olsen, J.E., 1994, Specific detection of pathogenic *Yersinia enterocolitica* by two-step PCR using hot-start and DMSO. Mol. Cell. Probes 8: 99-108.

Saiki, R.K., Gelfand, D.H., Stoffel, S., Higuchi, R., Horn, G.T., Mullis, K.B. and Erlich,H. A., 1988, Primer-directed enzymatic amplification of DNA with a thermostable DNA polymerase. Science 239: 487-491.

Wren, B.W. and Tabaqchali S., 1990, Detection of pathogenic *Yersinia enterocolitica* by the polymerase chain reaction. Lancet 336: 693.

Chapter 61

Molecular Virulence Characteristics and Kinetics of Infection of *Yersinia pseudotuberculosis* Isolated from Sick and Healthy Animals

Carlos Henrique G. MARTINS[1] and Deise P. FALCÃO[2]
[1]Universidade de Franca, Franca, SP.; [2]Faculdade de Ciências Farmacêuticas - UNESP - Araraquara - SP, Brazil

1. INTRODUCTION

In Brazil *Yersinia pseudotuberculosis* is a zoonotic species that has never been reported from human cases. The bacteria were isolated only from animals in the states of Paraná and Rio Grande do Sul. The majority of the animals showed symptoms of yersiniosis that was associated with an alarming rate of morbidity during this period of time. A previously study (Martins *et al.*, 1998) showed that 80.0% of these strains, no matter whether they were isolated from sick or healthy animals, possessed plasmids with molecular weights varying from 23.9 to 110 MDa. The objective of this study was

- to use PCR to analyse the strains for the presence of the virulence plasmid gene *lcrF* and for the chromosomal *inv*, *irp1*, *irp2*, *ybtE* and *psn* genes and for the IS*100* element.
- to use experimental animal model to evaluate the pathogenicity of the strains "*in vivo*" by intragastrical and intravenous inoculation routes.

2. MATERIALS AND METHODS

A total of 105 *Y. pseudotuberculosis* strains isolated from sick and healthy buffaloes (34), cattle (70) and swine (1) were studied. They were

The Genus Yersinia, Edited by Skurnik *et al.*
'uwer Academic/Plenum Publishers, New York, 2003

isolated between 1982 and 1990 in the southern region of Brazil. 102 strains belonged to biotype 2 / serogroup O:3 and three to biotype 1 / serogroup O:1.

The strains were tested for the presence of the *inv*, *lcrF*, *irp1*, *irp2*, *ybtE* genes and IS*100* by the PCR technique (Fukushima *et al.*, 2001). To this end three sets of multiplex PCRs were used: *inv* + *lcrF*, IS*100* + *psn* and *ybtE* + *irp1* + *irp2*.

Five *Y. pseudotuberculosis* strains were chosen for animal experiments to study kinetics of infection "*in vivo*" (Table 1). The strains were inoculated intragastrically and intravenously into Swiss mice. The animals were observed daily for clinical alterations. Groups of five mice were killed at different time points after inoculation: after 6 hr and 3, 6, 10, 15, 21 and 30 days. The organs and tissues were checked for possible macroscopic alterations and progress of infection was followed by performing viable bacterial counts on homogenates of selected tissues (Falcão *et al.*, 1984)

Table 1. General characteristics of the strains used in animal experiments

Strains	Virulence markers	Origin	Clinical Material
61 Yp 2/O:3	pYV (47.0 MDa), *lcrF*, *irp1*, *irp2*	Cattle	Diarrheal feces
21 Yp 2/O:3	pYV (52.0 MDa), *lcrF*, *irp1*, *irp2*	Swine	Diarrheal feces
32 Yp 1/O:1	pYV (52.0 MDa) *lcrF*, *irp1*, *irp2*, *psn*	Buffalo	Mesenteric lymph nodes
94 Yp 2/O:3	No pYV, *irp1*, *irp2*	Buffalo	Normal feces
36 Yp 2/O:3	No pYV, *irp1*, *irp2*	Cattle	Diarrheal feces

3. RESULTS

The PCR amplification demonstrated that the *inv*, *irp1* and *irp2* genes were present in all the isolates. The *lcrF* gene was detected only in the strains harboring the pYV plasmid, regardless of the plasmid size.

All the three serogroup O:1 strains were positive for the *psn* gene.

None of the strains were positive for the *ybtE* gene or for IS*100*.

The animal experiments showed that bacteria carrying the virulence plasmid infected almost all organs and tissues of the inoculated animals independent of the route of infection. The results also showed that the three *Y. pseudotuberculosis* biotype 1 / serogroup O:1 strains were the most invasive. In addition, clinical and pathological alterations occurred only in animals inoculated with bacteria carrying the pYV plasmid.

4. CONCLUSION

These results confirmed that for the full expression of the disease the *Y. pseudotuberculosis* strains depend on virulence genes present in both the pYV plasmid and the chromosome.

ACKNOWLEDGEMENTS

This work was supported in part by a grant from the Fundação de Amparo à Pesquisa do Estado de São Paulo (FAPESP – 98/00984-9).

REFERENCES

Falcão, D.P., Shimizu, M.T., and Trabulsi, L.R., 1984, Kinetics of infection induced by *Yersinia. Curr. Microbiology* **11**: 303-308.

Fukushima, H., Matsuda, Y., Seki, R., Tsubokura, M., Takeda, N., Shubin, F.N., Paik, I.K., and Zheng, X.B., 2001, Geographical heterogeneity between far eastern and western countries in prevalence of the virulence plasmid, the superantigen *Yersinia pseudotuberculosis*-derived mitogen, and the high-pathogenicity island among *Yersinia pseudotuberculosis* strains. *J. Clin. Microbiology* **39**: 3541-3547.

Martins, C.H.G., Bauab, T.M., Falcão, D.P., 1998, Characteristics of *Yersinia pseudotuberculosis* from animals in Brazil. *J. Appl. Microbiology* **85**: 703-707.

Chapter 62

Detection of *Yersinia enterocolitica* in Slaughter Pigs

Viktoria ATANASSOVA, Jutta HUGENBERG and Christian RING
School of Veterinary Medicine Hannover, Centre of Food Science, Department of Food Hygiene and Microbiology

During recent years there has been an increase of human infections caused by the consumption of pig meat and meat products contaminated by *Yersinia enterocolitica.* With regard to this problem a successful bacteriological detection of this disease agent is necessary for the rapid and specific diagnosis of food contaminations/ infections and food intoxications. The aim of the present study was to monitor the occurrence of *Y. enterocolitica* in slaughter pigs in Northern Germany and the comparison of different culture methods for the detection of the bacterium.

1. MATERIALS AND METHODS

A total of 540 samples were collected under sterile conditions from slaughtered pigs previously inspected without objection. The samples were transported and stored at 2-4°C before processing in the laboratory.

The prevalence of *Yersinia* spp. was determined by four different culture methods as outlined in Table 1.

- In method **A** the samples were incubated for 24 h at 30°C in the *Yersinia*-Selective-Enrichment-Bouillon according to Ossmer (Merck KGaA, Darmstadt).
- In method **B** the bacteria were enriched for 48 h at 22-25°C in a bouillon according to ISO/TC34/SC9 (1994).
- In method **C** the samples were enriched for up to 21 d at 4°C according to Mair and Fox (1986).
- In method **D** the samples were first pre-enriched for up to 21 d at 4°C and subsequently enriched for further 48 h at 22°C (Harmon *et al.,* 1984).

Following all the above described enrichment procedures the bouillons were plated on the *Yersinia*-Selective-Agars according to Schiemann (CIN-agar) or Wauters (SSDC-agar) and incubated for 24 h at 28°C or 24-48 h at 30°C.

Table 1. Details of methods used for the determination of *Y. enterocolitica*

Step/Method	A	B[1]	C	D
Pre-enrichment				PBS + sorbitol + bile salts (1:10) for 7, 14 or 21 d at 4°C
Enrichment	Ossmer bouillon (1:10) for 24 h at 30°C	PSB-bouillon (1:10) ITC-bouillon (1:100) for 48 h at 22-25°C (with shaking), subsequent treatment with KOH (0.25%) for 20 s	PBS (1:10) for 7, 14 or 21 d at 4°C	Modified Rappaport medium (1:10) for 48 h at 22°C, subsequent treatment with KOH (0.25%) for 20 s
Selective plates	*Yersinia* selective agar CIN 24 h at 28 °C and/or SSDC, 24 –48 h at 30 °C followed by serological and biochemical differentiation			

[1]PSB, Pepton-Sorbit-Bile; ITC, Irgasan-Ticarcillin-potassium chlorate

2. RESULTS

Enriching the samples in the *Yersinia*-Selective-Enrichment-Bouillon after Ossmer (MERCK) resulted in no isolates of *Yersinia* spp. (n=150). After using the ISO/TC34/SC9 technique (1994) it was also not possible to detect *Yersinia* spp. (n=390).

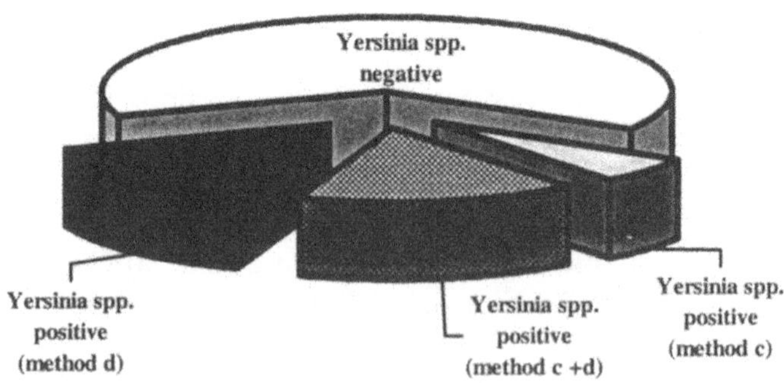

Figure 1. Proportion of *Yersinia* spp. positive samples detected with the method c, d and c +d

In the following experiments 70 tonsils were either analysed by the cold enrichment method after Mair and Fox (1986) (method **C**) or the two-step-cold enrichment after Harmon *et al.*, (1984) (method **D**). Using method **C** *Yersinia* spp. was detected in 21.4% (15/70) of the tonsils (86.7% (13/70) positive for *Y. enterocolitica*), whereas method **D** resulted in 30.0% (21/70) *Y. enterocolitica* positive samples (Figure 1). 15.7% (11/70) of the tonsils were positive for *Y. enterocolitica* using either method **C** or method **D** (Figure1).

After biochemical and serological differentiation, the following sero-/ biotypes were confirmed: *Y. enterocolitica* serotype 0:3/biotype 4 (20/25); *Y. enterocolitica* serotype 0:6/biotype 1A (2/25); *Y. enterocolitica* serotype 0:41,43/biotype 1A (1/25); *Y. intermedia* (2/25).

3. CONCLUSIONS

Using the methods **A** and **B** *Yersinia* spp. could not be detected in any samples. This could be due to the high enrichment temperatures during these procedures.

Comparing the results of the cold-enrichment methods according to Mair and Fox (1986) and to Harmon *et al.,* (1984) with the latter method more *Yersinia* spp. positive samples were detected. Including a high percentage of positive cases detected in both cold-enrichment methods (15.7%), our experiments showed that the combination of the methods gave the best results (35.7% *Yersinia* spp. positive samples).

Assuming that the warm-enrichment methods are frequently used in the laboratories it could be supposed that in an unknown number of human food intoxications and regular food investigations *Y. enterocolitica* is detected false negative. However, due to the disadvantage that the cold-enrichment methods resulting in many *Yersinia* spp. positive cases are very time-consuming it is necessary on the one hand to evaluate molecular-mikrobiological techniques, on the other hand to take special care in preventing the entry of the pathogen into the food chain.

REFERENCES

Harmon M.C., Swaminathan B., Forrest J.C., Isolation of *Yersinia enterocolitica* and related species from porcine samples obtained from an abattoir. J. Appl. Bacteriol., 56, 1984, 421-427

Mair N.S., Fox E., Yersiniosis: Laboratory Diagnosis, Clinical Features and Epidemiology, PHLS, 1986

Schiemann D.A., Synthesis of a selective agar medium for *Yersinia enterocolitica.* Can. J. Microbiol., 25, 1979, 1298-1304.

Wauters G., Correlation between ecology, biochemical behaviour and antigenic properties of *Yersinia enterocolitica.* Contrib. Microbiol. Immunol., 2, 1973, 68-70

Picture 23. Bakhtiyar Suleimenov, Gulnaz Stybayeva, Igor Domaradsky, Bakyt Atshabar, Tatiana Gremyakova and Tatyana Meka-Mechenko entering the City Reception.

Chapter 63

Mechanism of Formation of a Population Level of Virulence of *Yersinia pestis*

Bakyt B. ATSHABAR
M. Aikimbayev's Kazakh Scientific Center for Quarantine and Zoonotic Diseases, 14, Kapalskaya str., Almaty, Kazakhstan

A natural focus of plague is a complex and dynamic biological system that provides for the spatial and temporal existence of *Yersinia pestis*. The stability and size of plague foci gives an idea of the specificity of the ecological niche for plague microbe and the level of systemic interaction of the plague agent with its natural host. Selective mechanisms result in a dynamic equilibrium between bacterial and rodent populations.

The continued existence of the plague focus depends on a combination of a genetically determined resistance in the main rodent host, the possibility of chronic infection among these resistant hosts, the virulence level of *Y. pestis* strains in the focus and the availability of competent flea vectors to transmit the bacteria. There is experimental evidence that resistance of mammals depends on the level of specific and functional oxygen-dependent metabolism of phagocytes.

The populations of virulent *Y. pestis* are formed as a result of mutations, competition and selection. Experimental work has provided evidence that phagocytic cells increase the rate of mutations in *Y. pestis* (1). Therefore, we speculated that this is the primary mechanism for generating mutation-related variability among *Y. pestis*. Heterogeneity within the *Y. pestis* populations lays basis for the evolutionary variability. The resulting variants within the population will then be subjected to selection within the main rodent host.

Therefore, the term "primary host" in natural focus has a deep biological meaning, because the evolutionary variability of populations of the plague bacterium corresponds to a particular species of primary host (2). The leading factor in the interaction in the parasite-host system is the integrative combination of the pathogenic functions of the microbe, which are

The Genus Yersinia, Edited by Skurnik *et al.*
Kluwer Academic/Plenum Publishers, New York, 2003

manifested as virulence, and the morphological-functional status of the macroorganism, or host, which is manifested as resistance.

Virulence is not only required for the survival of pathogenic bacteria in host populations, it is also an important ecological factor of selection, which appears in the interactions between the causative agent of plague and its natural hosts. The genetically determined level of resistance to plague in the primary host population is a selective factor for the elimination of many mutants exhibiting a wide range of virulence in natural populations of *Y. pestis*.

Changes in the host-parasite system (for example, a change of primary host) will lead to changes in the *Y. pestis* population. Secondary and occasional hosts, which are more sensitive to the plague microbe, do not have such a major impact on the properties of *Y. pestis* populations.

The dynamic mutual interaction between resistance of the main host population and the population virulence of the plague causative agent can be represented in form of swing, where "A" is a position of dynamic stabilization (Figure 1). The oscillations of the system lead to epizootics, chronic courses of infection, and the peculiar characteristics of carrier hosts.

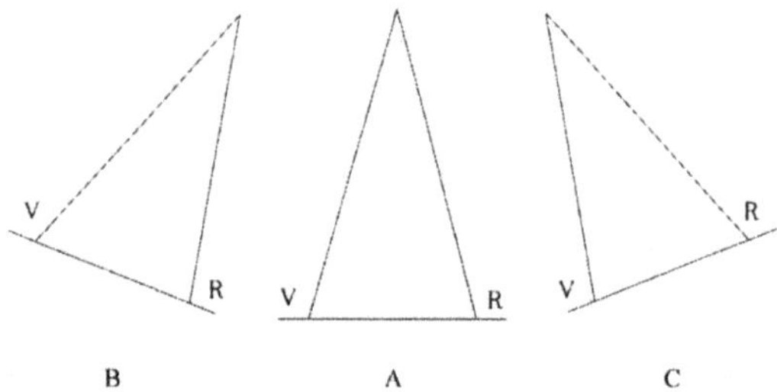

Figure 1. Diagram of mutual influence of resistance of main host population and *Y. pestis* virulence within natural focus. Note: R – is the level of resistance of main host population; V - is the level of *Y. pestis* population virulence.

In position "C" the level of host resistance is prevailing. It is possible that an increase in the proportion of immune individuals among the rodents, a decrease in the percentage of Pgm^+ and Ca- cells among the *Y. pestis* population, and an accumulation of these mutations in the bacterial population, will cause virulence levels to decrease. This situation, along with other factors, will lead to decreased epizootic activity in rodents. The situation "B" is possible when seasonal oscillations in resistance occur among rodents as a result of deteriorating living conditions, etc.

The virulence of *Y pestis* is determined by the integrated aggregate of its properties, which is represented as individual "bricks" on this slide, each of which contributes to the manifestation of the general phenomenon of virulence. Thus one can represent the population virulence as a "package" with a set of properties ("cassettes") lowering or raising the virulence (Figure 2). Each cassette represents the element of constitutive or functional virulence. Some of these characteristics can be functionally related to the extent to which a property is manifested, such as the percentage of Pgm^+ and Ca^- cells in strains, the level of auxotrophicity, etc. These characteristics can raise or lower the volume of a "package" resulting in possible gradations within each cassette.

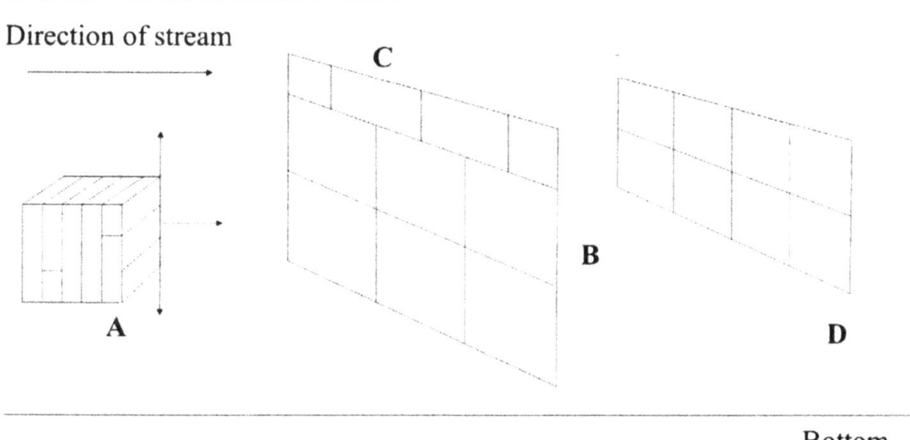

Figure 2. Diagram for selection of *Y. pestis* population within natural focus

Arbitrary designations: A - population virulence of *Y. pestis*; B - population resistance of main host; C – resistance of sensitive individuals of main host; D - population resistance of other hosts.

The effect of environmental factors are shown by liquid flow where the "cassette package" is placed within the flowstream; the latter has a certain floating ability and mass due to number and content of each cassette. The advantage of this graphical representation is that it reflects the dynamics and direction of selection. The third element of the diagram is a graphical representation of the population resistance of hosts in the form of nets with different sizes of cells; the higher the resistance the larger the cells in the net, which requires a correspondingly larger "package" to make become "trapped" within, or otherwise maintain itself, with the host "net". "Packages" that are too small and slip through some cell of the net represent bacteria that will be eliminated from the population (represented in diagram by size of "package"). The diagram shows the direction of selection, the capacity to eliminate the deviations from baseline (surface-bottom), and the involvement of bacteria and hosts populations having different virulence and resistance.

The survival of *Y. pestis* as a component within its specific ecological system of plague completely depends on two components of this system. In this system bacteria are connected with host and vector in one way, whereas transmission of bacteria between host and vector is two-way and provides a means for the stable circulation of *Y. pestis* in natural focus.

The enzootic process can be considered to be a planar model, involving numerous structural components of the plague ecological system, which overlap and interlink with each other (Figure 3).

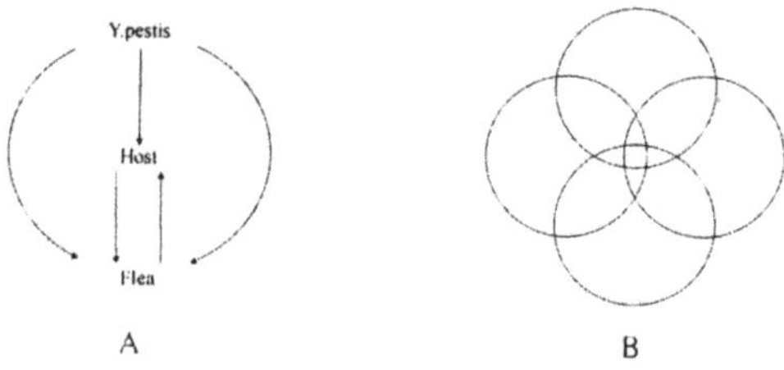

Figure 3. Scheme of *Y. pestis* circulation in natural plague foci

A plane diagram demonstrates this statement, while a principal plane model is multidimensional and complicated by the participation of minor and occasional hosts, and additional vectors. At the same time the rodent-flea biological system is self-sufficient system in that it is capable to exist outside the natural plague focus. This aspect is one of the unknown areas of natural focality of plague and other infections. It still is not clear what conditions are necessary for maintenance of the plague cycle in rodent populations and what factors are required for introduction of the pathogen into plague-free areas.

REFERENCES

Atshabar B.B., Isin J. M., Suleimenov B. M. Mutagenic activity of polymorphonuclear leukocytes on *Yersinia pestis*. J. Microbiology, Epidemiology, Immunology. Moscow - 1989. - No.4. - p.10-14.

Atshabar B.B. Plague biocenosis as ecological system, Saratov, 1999, Issue 79, p.36-44.

Chapter 64

Polymerase Chain Reaction Assays for the Presumptive Identification of *Yersinia pestis* Strains in Georgia

Lela BAKANIDZE, Ioseb VELIJANASHVILI, Merab KEKELIDZE, Levan BERIDZE, Ekaterine ZANGALADZE, Mariam ZAKALASHVILI, David TSERETELI and Paata IMNADZE
National Center for Disease Control of Georgia, Tbilisi, Georgia

1. INTRODUCTION

Plague has been a danger for the human society since the pre–Christian era. Events of fall 2001 made it clear to the whole world that particularly dangerous infectious agents can be easily used as biological weapons. Plague was mentioned for the first time in the Georgian manuscripts from the XI century. Later, in the XV century another Georgian manuscript – "Ustsoro Karabadini" ("Matchless Book of Medical Treatment") gave more detailed information on plague. Particularly, it described clinical manifestation of bubonic form of plague. In 1616, cases of plague were registered in western part of Georgia, and probably it was imported by catholic missionaries. In Georgian folklore plague is named "zhami" (in old Georgian it means "bad time").

2. NATURAL FOCI OF PLAGUE IN GEORGIA

There are two natural foci of plague in Georgia. One of the foci, the '*plain foothill region*', includes the territories of Dedoplistskaro, Signakhi, Sagarejo and Gardabani regions. Plague epizootics were identified in 1966 on Eldari Valley and in Karayazi Valley in 1968 – 1971. The main reservoir

The Genus Yersinia, Edited by Skurnik *et al.*
' ıwer Academic/Plenum Publishers, New York, 2003

of plague in these foci is *Meriones erythrourus (libicus)* and the main vectors, *Xenopsylla conformis* and *Ceratophyllus laeviceps*. Totally 83 *Y. pestis* strains were isolated: 30 strains from rodents and 53 strains from the vectors. Another focus, the '*high mountainous region*', includes the territories of Ninotsminda and Akhalkalaki regions on Javakheti plateau. Only one part of the focus is on the territory of Georgia, the focus is conventionally divided by state border between Georgia and Armenia. The first epizooty of plague in this focus was identified in 1979 and lasted until 1983. Epizooties were also registered in 1992 and 1997. The main reservoir in this focus was *Microtus arvalis* and the main vectors, *Callopsylla caspia, Nosopsillus consimilis*. From 39 *Y. pestis* strains isolated in this focus 5 strains were from rodents and 34 from vectors. All strains isolated in Georgia belong to the biotype Mediaevalis.

About fifty isolates of *Y. pestis* are kept in the microbial culture collection of the National Center for Disease Control of Georgia. NCDC is responsible for carrying out surveillance on particularly dangerous infections. Field teams go out every season for reconnoitering and gathering materials for laboratory investigation. Materials are taken from live rodents and rodent remains from locations of epizootics: parts of organs, spleen, liver, lungs, lymph nodes, bone marrow, brain, etc. Fleas and ticks are combed out, and collected from burrows. Material is investigated by microscopy, cultivation on selective media, serology and by infecting laboratory animals. Isolates are investigated by bacteriological, biochemical and biological methods.

3. IDENTIFICATION OF *Y. pestis* BY PCR ASSAYS

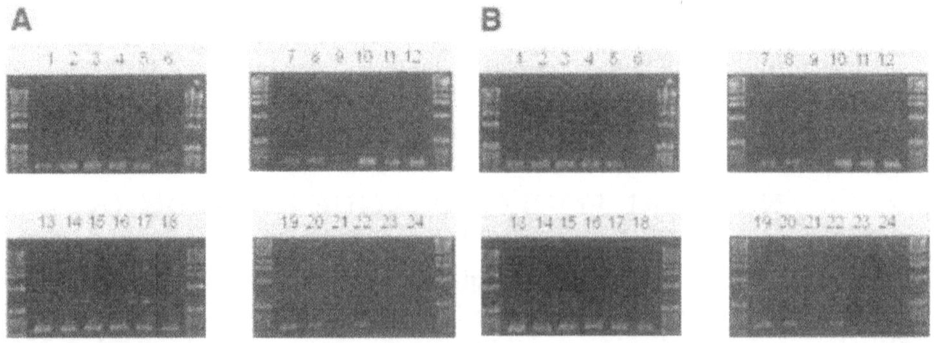

Figure 1. Identification of *Y. pestis* by PCR. Panel A. V antigen gene PCR; panel B, F1 antigen gene PCR. Lanes: 1-20 and 22 – *Y. pestis*, 21 and 23 – *Y. pseudotuberculosis*, 24 –*Y. enterocolitica.*

The existing laboratory tests are either indicative or take too long time (48 hours or longer), therefore, development of the rapid diagnostic tests for the identification of the plague bacillus is required We attempted to investigate field material by PCR and compare the results with those obtained using routine microbiological methods. Of note, these methods are time-consuming and expensive that is most important under our economic difficulties (when very often we have shortage of media).

In our earlier investigations we studied 24 strains of *Yersinia* by PCR for the presence of the genes for the V antigen, the F1 capsular antigen and the plasminogen-activator/coagulase. Altogether, 21 *Y. pestis* strains, isolated from the high-mountainous focus, one *Y. enterocolitica* strain and two *Y. pseudotuberculosis* strains were tested. All the *Y. pestis* strains were positive in the V antigen gene PCR (Figure 1).

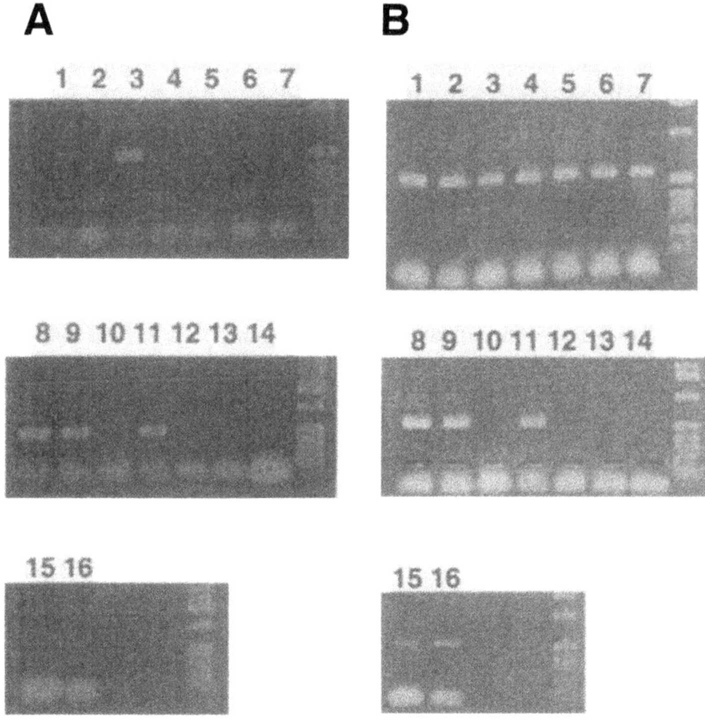

Figure 2. Identification of *Y. pestis* by PCR. Panel A. V antigen gene PCR. Panel B, *pla* gene PCR. Lanes: 1-2 Armenian strains, 3, *Y. pestis* from Azerbaijan; 4-7 and 10, Georgian strains of *Y. pestis*; 8-9, *Y. pestis* from Dagestan, 11, *Y. pestis* from Middle Asia; 12-16, suspension of fleas.

We investigated suspensions of fleas by PCR and, in two cases of five, the V antigen gene was amplified (Figure 2, lanes 15-16). As the field materials come usually to NCDC laboratories in huge amounts, it will be convenient to check them first by PCR prior to microbiological investigation. This will give us opportunity to save time and resources.

Strains of *Y. pestis* isolated from high-mountainous focus differ from strains of *Y. pestis* isolated from other plague natural foci by lower virulence. Strains with reduced virulence might be very interesting for further studies, as they can serve as starting point for developing plague vaccine.

REFERENCES

Brubaker, B. 1991. The V Antigen of *Yersiniae*: an overview. In: *Current Investigations of the Microbiology of Yersiniae* (T. Une and M. Tsubokura, eds.) Karger, Basel, pp.127-133.

Campbell, J., Lowe, J., Walz, S. and Essell J. 1993. Rapid and Specific Identification of *Yersinia pestis* by a Nested Polymerase Chain Reaction Procedure. J.Clin.Microbiol. 31: 758-759.

Chkheidze, G., Dzebisashvili, Iu., Nersesov, V., and Mukhadze, J. 1974. Natural Foci of Plague on the Territory of Georgia. In: J.Particularly Dangerous Infections in Caucasus, Stavropol 1: 70-72.

Filippov, A., Solodovnikov, N., Kookleva, L. and Prostenko O. 1990. Plasmid Content in *Yersinia pestis* Strains of Different Origin. FEMS Microbiol. Lett. 67: 45-48.

Neubauer, H., Meyer, H., Prior, J. *et al.*, 2000. A Combination of Different Polymerase Chain Reaction Assays for the presumptive Identification of *Yersinia pestis*. J.Vet. Med. B47: 573-580.

Chapter 65

Genetic (Sero)Typing of *Yersinia pseudotuberculosis*

Tatiana M. BOGDANOVICH[1], Elisabeth CARNIEL[3], Hiroshi FUKUSHIMA[4] and Mikael SKURNIK[1,2]
[1]Department of Medical Biochemistry and Molecular Biology, University of Turku and [2]Department of Bacteriology and Immunology, Haartman Institute, University of Helsinki, Finland; [3]Institute Pasteur, Paris, France; [4]The Shimane Prefectural Institute of Public Health and Environmental Science, Japan.

1. INTRODUCTION

Yersinia pseudotuberculosis strains are divided into serogroups (O:1-O:15) based mainly on antigenic differences in the lipopolysaccharide O-antigen (O-ag) (Tsubokura *et. al.*, 1995). However, conventional agglutination-based serotyping method is laborious, susceptible to interpretation errors and can not be applied to self-agglutinable strains.

Genes responsible for the O-ag biosynthesis are organized into a gene cluster between the *hem*H and *gsk* genes in the *Y. pseudotuberculosis* chromosome. The O-ag gene clusters of 21 reference strains of *Y. pseudotuberculosis* representing all known serotypes fall into distinct groups reflecting the differences in the chemical structure of the O-ags (Skurnik *et al.*, 2000, see also Figure 1 in Skurnik, 2003). Based on this information we developed a multiplex PÇR assay for genetic serotyping of *Y. pseudotuberculosis* as an alternative to the conventional serotyping.

2. METHODS AND RESULTS

Reference strains of *Y. pseudotuberculosis* were used to set up conditions for the multiplex PCR assay. A combination of O-ag gene cluster specific primers was selected to produce a unique PCR product pattern for each

The Genus Yersinia, Edited by Skurnik *et al.*
Kluwer Academic/Plenum Publishers, New York, 2003

reference strain. PCR conditions were optimized such that each PCR product amplified specifically without interference from the other amplifications running in the same tube.

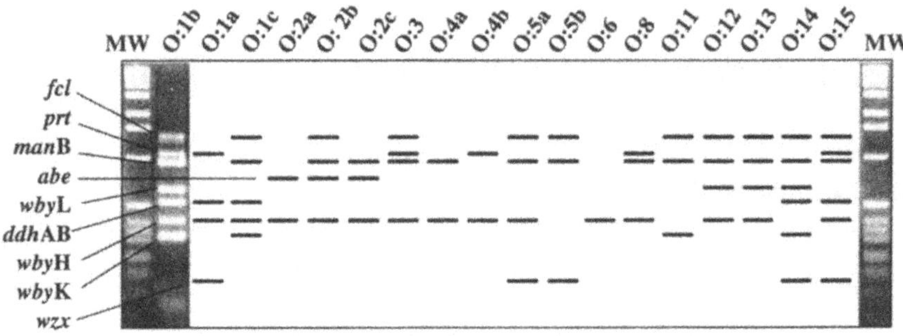

Figure 1. Multiplex PCR assay of the reference strains of *Y. pseudotuberculosis* (the actual PCR result is shown only for serotype O:1b, the others are given schematically). Serotypes are indicated above the lanes. Serotypes O:7, O:9 and O:10 were completely negative in the multiplex PCR and are not included in the figure. MW, molecular weight standard (1 kb ladder, Gibco BRL®; the 1.6 kb standard fragment migrates above the *fcl* PCR fragment).

Table 1. Genetic(sero)typing results discordant with conventional serotyping

No	Strain code	Source/Country	Serotype	Genotyping	Re-serotyping
1	CN2	House rat/China	O:1c	G:15a	Not done
2	No. 366	Human/Japan	O:2b	G:2a	Not done
3	PC94-72	Pig/Japan	O:4a	G:2c	Not done
4	R253	Mouse/Japan	O:6	G:12	Not done
5	MW387-1	River water/Japan	O:10	G:9	O:9
6	OK5608	Racoon dog/Japan	O:10	G:15a	O:15
7	MW900-3	River water/Japan	O:12	G:5c	Rough
8	MW896-2	River water/Japan	O:12	G:12a	O:12
9	N916	House rat/China	O:13	G:1b	Not done
10	N917	House rat/China	O:13	G:1a	O:13
11	J51	Rabbit/China	O:13	G:3	O:13
12	R626	Mole/Japan	Rough	G:9	Rough
13	PC95-219-1	Pig/Japan	Rough	G:4b	Rough
14	B55	Wild rat/China	Rough	G:4b	O:4b
15	B56	Wild rat/China	Rough	G:1a	Not done
16	B57	Wild rat/China	Rough	G:1a	Not done
17	No. 476	Pig/Italy	Rough	G:16	Not done
18	Pa8728	Human/Japan	Rough	G:3	Rough
19	RH805Ly	Mole/Japan	Rough	G:12	O:12
20	TE9343b	Racoon dog/Japan	Rough	G:4b	O:4b
21	NYP3	Well water/Japan	Rough	G:4a/8	Rough
22	SJ000101	Human/Korea	Rough	G:3	Rough
23	951-36	Human/Germany	Non-typable	G:4a/8	O:4a

No	Strain code	Source/Country	Serotype	Genotyping	Re-serotyping
24	955-36	Human/Germany	Non-typable	G:5c	O:1a

The resulting multiplex PCR consists of 9 pairs of primers and allows identification of 14 serotypes and 2 serogroups, the O:4a-O:8 and the O:12-O:13 groups (Figure 1). The remaining 3 serotypes (O:7, O:9, O:10) are negative in the multiplex PCR and require an additional double-PCR analysis, targeting *hem*H-*ddh*D and *wzz* genes.

The multiplex PCR was applied to evaluate 56 *Y. pseudotuberculosis* strains of known serotypes, 14 strains that had appeared rough, and 6 non-typable strains. Multiplex PCR results for 44 of the 56 serotyped strains were concordant with conventional serotyping. Most problems were seen with the strains of conventional serogroups O:12-O:13 (Table 1). While strain MW896-2 produced a new PCR pattern designated as G:12a, the O:13 strains were not of G:12-G:13 type and re-serotyping confirmed them to be O:13. Apparently our O:13 reference strain is actually O:12 and our scheme needs adjustment to genotype true O:13 strains, however, our results indicate that there is genetic heterogeneity within this group. Seven of the 20 rough and non-typable strains and one O:11 strain were negative in multiplex PCR and more thorough analyses confirmed that they were not *Y. pseudotuberculosis*. Eleven of the 20 strains were assigned to certain genetic (sero)type (Table 1), and some of these results were confirmed by conventional re-serotyping. And finally, one of the rough strains gave a completely new multiplex PCR pattern designated as G:16.

3. DISCUSSION AND CONCLUSIONS

Here we present a multiplex PCR assay with 9 sets of primers that can distinguish 18 out of 21 serotypes, including the most commonly circulating ones. The scheme requires additional double-PCR for serotypes O:7, O:9 and O:10. Our results proved that conventional serotyping is vulnerable to errors and that rough strains can be genetically (sero)typed. We detected heterogeneity within some serotypes and genetic (sero)typing will be useful to classification of the strains. Though the presented genetic typing system is not perfect, yet it represents a useful tool for rapid and reliable characterization of *Y. pseudotuberculosis* strains.

REFERENCES

Skurnik, M., Peippo, A., and Ervelä, E., 2000, Characterization of the O-antigen gene clusters of *Yersinia pseudotuberculosis* and the cryptic O-antigen gene cluster of *Yersinia pestis* showes that the plague bacillus is most closely related to and has evolved from *Y. pseudotuberculosis* serotype O:1b. *Mol. Microbiol.* 37: 316-330.

Skurnik, M., 2003, Molecular genetics, biochemistry and biological role of *Yersinia* lipopolysaccharide. In *The Genus Yersinia: entering the functional genomic era*, M. Skurnik, K. Granfors and J. A. Bengoechea, eds.: Kluwer Academic/Plenum Publishers, pp. 187-197.

Tsubokura, M., and Aleksic, C., 1995, A simplified antigenic scheme for serotyping of *Yersinia pseudotuberculosis*: phenotypic characterization of reference strains and preparation of O and H factor sera. *Contrib. Microb. Immunol.* 13: 99-105.

Chapter 66

Yersinia spp. in the Environment: Epidemiology and Virulence Characteristics

Deise P. FALCÃO[1], Ednéia F. CORRÊA[1] and Juliana P. FALCÃO[2].
[1]*Faculdade de Ciências Farmacêuticas-UNESP- Araraquara, SP,* [2]*Faculdade de Medicina-USP- Ribeirão Preto, SP.*

1. INTRODUCTION

Y. pestis, Y. pseudotuberculosis and some bio-serogroups of *Y. enterocolitica* have been shown to cause human diseases. On the other hand, different bio-serogroups of *Y. enterocolitica* and *Y. intermedia, Y. kristensenii, Y. frederiksenii, Y. bercovieri, Y. mollaretii ,Y. aldovae, Y. rohdei* and *Y. ruckeri* have not clearly demonstrated to cause human diseases; the majority of these species are found in the environment (Sulakvelidze, 2000). *Yersinia* spp. has been isolated from water and sewage in Brazil. The objectives of this work were to determine the species and the bio-serotypes of these strains, evaluate their drug resistance and some phenotypic and molecular virulence characteristics.

2. MATERIAL AND METHODS

A total of 145 strains of *Yersinia* spp. were studied. They were isolated from fresh and salt water and from sewage in the states of São Paulo (SP) and Rio de Janeiro (RJ), Brazil.

The *Yersinia* isolates were fully characterized for species, biotypes and serotypes (Aleksic and Bockmühl, 1999)

All the isolates were tested for drug resistance using the disc diffusion technique (Bauer *et al.*, 1966). The following antimicrobial drugs were used: amikacin (Ami), chloramphenicol (Clo), cefoxitin (Cfo), cephalotin (Cfl),

The Genus Yersinia, Edited by Skurnik *et al.*
uwer Academic/Plenum Publishers, New York, 2003

cefotaxime (Ctx) gentamicin (Gen), kanamicin (Kn), imipenen (Ipm), tobramycin (Tob), cefazolin (Cfz), ampicillin (Amp), sulfamethoxazole-trimethropin (Sut) and tetracycline (Tet).

The isolates were tested for phenotypic virulence characteristics such as, temperature-dependent autoagglutination; calcium growth at 37°C; Congo Red absorption; pyrazinamidase production; salicin fermentation and esculin hydrolysis (Farmer *et al.*,1992). Detection of the genes *inv, ail, yst* and *vir*F were performed by the PCR technique. Primers for the amplification of the *inv, ail, yst* and *virF* genes were used as described by Rasmussen *et al.*, (1994), Nakajima *et al.*, (1992), Ibrahim *et al.*, (1997) and Wren and Tabaqchali (1990), respectively.

3. RESULTS

The 145 *Yersinia* isolates were classified as *Y. enterocolitica* (67 strains), *Y. intermedia* (64 strains), *Y. frederiksenii* (10 strains), *Y. kristensenii* (3 strains) and *Yersinia* non-typable (2 strains). All the environmental *Yersinia* strains isolated in Brazil were found only in two states: São Paulo and Rio de Janeiro. The majority of the isolates (90%) were classified as *Y. enterocolitica* and *Y. intermedia*

Some of the bio-serotypes of *Y. enterocolitica* found in the environment in Brazil, like Ye 2/O:5,27 and Ye 3/O:5,27 are related to human diseases while others such as Ye 1A/O:5 , 1A/O:10, 1A/O:16 and 1A/ O:27 are not (Alęksic and Bockmühl, 1999).

All the *Yersinia* isolates were sensitive to amikacin, chloramphenicol, cefotaxime and to sulfametaxozol-trimetropin. Despite not being considered pathogenic, 70% of *Y. intermedia,* 80% of *Y. frederiksenii* and 100% of the non-typable *Yersinia* strains showed resistance to three or more drugs while only 18% of the environmental *Y. enterocolitica* presented multiple drug resistance. The three *Y. kristensenii* strains were resistant only to cephalotin.

Four isolates of *Y. enterocolitica* biotype 3 and serogroup O:5,27 were virulent according to the phenotypic virulence tests used. These four strains presented the genes *inv*, *ail* and *vir*F and two of them also the gene *yst.* They were isolated from ocean water (3 strains) and one of from sewage. The two strains positive for the gene *yst* were isolated from the ocean water. We would like to point out that the four positive strains for virulence markers did not present multiple drug resistance.

4. CONCLUSION

The occurrence of some virulence markers in four *Y. enterocolitica* strains of biotype 3, serogroup O:5,27, isolated from water and sewage shows that the environment can be responsible for human infection by *Y. enterocolitica* in Brazil.

ACKNOWLEDGEMENTS

We would like to thank FAPESP for financial support.

REFERENCES

Aleksic, S.; Bockemühl, J. 1999. *Yersinia* and other *enterobacteriaceae*. In:Murray et al.(ED.). Manual of Clinical Microbiology 7ed.Washington, D.C.:American Society for Microbiology, pp.483-496.

Bauer, A.W., Kirby, W. M, Sherris, J. C. and Turck, M. 1966. Antibiotic susceptibility testing by a standardized single disc method. Am. J. Clin. Pathol.,45 : 493-496.

Farmer III, J.J., Carter, G.P., Miller, V.L., Falkow, S.and Wachsmuth, I.K. 1992. Pyrazinamidase, CR-MOX agar, salicin fermentation-esculin hydrolysis and D-xylose fermentation for identifying pathogenic serotypes of *Yersinia enterocolitica*. J.Clin. Microbiol. , 30: 2589-2594.

Ibrahim,A. , Liesack, W. Griffiths , M.,W. and Robins-Browenw,R.M. 1997. Development of a highly specific assay for rapid identification of pathogenic strains of *Yersinia enterocolitica* based on PCR amplification of *Yersinia* heat-stable enterotoxin gene (yst). J.Clin. Microbiol.,.35 :1636-1638.

Nakajima, H., Inoue, M., Mori, T., Itoh, K.I., Arakawa, A.E. and Watanabe,H. 1992. Detection and identification of *Yersinia pseudotuberculosis* and pathogenic *Yersinia enterocolitica* by an improved polymerase chain reaction method. J. Clin. Microbiol., 30 : 2484-2486.

Rasmussen, H.N., Rasmussen, O.F., Andersen, J.K. and Olsen, J.E. , 1994, Specific detection of pathogenic *Yersinia enterocolitica* by two-step PCR using hot-start and DMSO. Mol. Cel. Probes, 8: 99-108.

Sulakvelidze, A. 2000. *Yersiniae* other than *Y. enterocolitica, Y. pseudotuberculosis* and *Y. pestis*: the ignored species. Microbes and Infection, 2: 497-513.

Wren, B.W. and Tabaqchali S. 1990. Detection of pathogenic *Yersinia enterocolitica* by the polymerase chain reaction. Lancet., 336: 693

Chapter 67

Molecular Characterization of *Yersinia enterocolitica* 1A Strains Isolated from Buenos Aires Sewage Water

Mirtha E. FLOCCARI[1], Heinrich K.J. NEUBAUER[2], Stella M. GÓMEZ[1], Czilla LODRI[2] and Jose L. PARADA[1]
[1]*Facultad de Ciencias Exactas y Naturales. Universidad de Buenos Aires. Argentina;*
[2]*Institute of Microbiology German Federal Armed Forces, Neuherbergstr. 11, 80937 München, Germany*

1. INTRODUCTION

Yersinia enterocolitica is an important enteric pathogen which has well-defined virulence determinants that allow the bacteria to become established in their hosts and overcome host defence. The infection is food borne and causes gastrointestinal syndromes with a tendency to spread into liver, kidneys, spleen and lung. Inflammatory acute enteritis, occasionally bloody, with fever and diarrhoea is the most frequent syndrome seen, especially in children. The appendicitis like syndrome, mesenteric lymphadenitis and terminal ileitis can also be included in the diagnosis. In immune suppressed individuals septicaemia can also be developed. Direct inocculation through blood transfusion may occur.

Y. enterocolitica has been found in both, aquatic and terrestrial environment and in warm and cold blooded animals, implying a contamination risk for dairy and meat products or vegetables.

The species consists of a heterogeneous group of strains which includes several bio-serotypes. On biochemical basis, *Y. enterocolitica* strains are divided into six biotypes. Each biotype comprises several serotypes based on lipopolysaccharide O antigens, with different virulent potentials.

Strains of biogroups 1B, 2, 3, 4 and 5 are generally accepted as virulent and have well established virulence characteristics, namely a plasmid of 70

The Genus Yersinia, Edited by Skurnik *et al.*
'luwer Academic/Plenum Publishers, New York, 2003

Kb (pYV) and chromosomal markers such as *ail*, *myf*, *ystA*, and *inv*. Biogroup 1A strains lack some of those characteristics and are considered to be non pathogenic. However, strains of this biogroup have been isolated from several cases of intestinal infections. A relevant association of diarrhea with the precence of biogroup 1A strains could be supposed suggesting that an additional mechanism of pathogenicity could be present in these strains.

The objective of this study was to characterize five *Y. enterocolitica* biotype 1A strains, isolated previously from Buenos Aires City sewage water by detecting chromosomal *ail*, *inv* and *yst*, plasmid encoded *yadA*, V-antigen gene, and *virF* virulence genes by PCR.

2. MATERIAL AND METHODS

Strains: BA1 serotype 0:5; BA2 serotype 0:7,8; BA3 serotype 0:5; BA4 serotype 0:25,35; BA5 serotype 0:6 (Floccari and Peso, 1984); Y11: Europen origen DSMZ 13030 (Neubauer *et al.,* 2000); Y286: American origen ATCC 9160 (Neubauer *et al.,* 2000).

DNA preparation: DNA preparation was done as described by Neubauer *et al.,* (2000).

Primers: chromosomal genes. 16S rRNA gene: 238 bp; SP1 and SP3 (Neubauer *et al.,* 2000). *yst*: 208 bp, pr1a and pr1b (Ibrahim *et al.,* 1997). *inv*: 570 bp, Inv1 and Inv2 (Rasmussen *et al.,* 1993). *ail*: 170 bp, Ail1 and Ail2 (Nakajima *et al.,* 1992). Plasmid genes. *virF*: 591 bp, virF1 and virF2 (Bhaduri and Cotrell, 1998). *yadA*: 191 bp, yadA1 and yadA2 (Blais & Phillipe, 1995). V-antigen gene: 524 bp, V1 and V2: (Price *et al.,* 1989)

Sequencing: Products of the amplification of the 16S rRNA gene and the inv were purified by the QIAquick PCR Purification Kit (QIAGEN, Germany) and sequenced using ABI PRISM sequencer.

3. RESULTS

There was no amplification of DNA from strains BA1-5 applying PCRs targeting genes *yst*, *ail*, *virF* and *yadA* or the V- antigen gene. Figure 1 shows the amplification of DNA of strains BA2, 4 and 5 in the *inv* PCR assay. The sequence analysis of these bands did not indicate the functionality of the gene. Sequencing the amplicons of the 16S rRNA gene proved that all Argentinian strains tested belonged to the European type of *Y. enteroclitica*, termed *Y. enterocolitica palearctica* (Neubauer *et al.,* 2000).

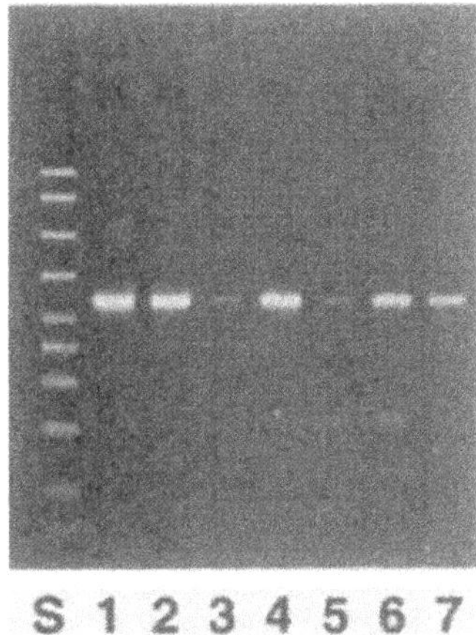

Figure 1. Amplification of *Y. enterocolitica* DNA in the *inv* PCR-assay. Lanes: 1. Y11; 2. Y286; 3.BA1; 4. BA2; 5. BA3; 6.BA4; 7. BA5; S. Ladder (bp): 2.000, 1.500, 1.000, 700, 500, 400, 300, 200, 100 bp

4. CONCLUSIONS

None of the plasmid encoded genes *virF*, *yadA* and the V-antigen gene were present in strains isolated from Buenos Aires sewage water suggesting that the pYV is absent. This was confirmed by plasmid isolation (data not shown). Chromosomally encoded *ail* and *yst* were absent in all five strains, but *inv* was detected in strains BA2, 4 and 5. Sequence analysis of the amplicons obtained did not show evidence that the gene might be functional. Interestingly, the 16S rRNA gene sequence of all 5 Argentinian strains was specific for the European subspecies *Y. enterocolitica palearctica* pointing to the fact that *Y. enterocolitica* was imported into Argentina with animals, food, feed or even humans and that it was possible for theses strains to get accustomed to the adverse 'life conditions' on a new continent. It can be concluded that this might also be possible for enteropathogenic *yersiniae*. However, invasion of tissue culture cells will be carried out with strains BA1-5 to analyze their real invasive potential.

REFERENCES

Bhaduri, S. and Cottrell, B., 1998, A simplified sample preparation method from various foods for PCR detection of pathogenic *Yersinia enterocolitica*: a possible model for other food pathogens. *Mol. Cell. Probes*, 12: 79-83.

Blais, B.W. and Phillipe, L.M., 1995, Comparative analysis of *YadA*, and *ail* polymerase chain reaction methods for virulent *Yersinia enterocolitica. Food Control*, 6: 211-214.

Floccari, M.E. and Pęso O.A., 1984, *Yersinia enterocolitica* and related species in the sewage discharge from the city of Buenos Aires: use of Schiemann medium in their isolation. *Rev. Argent. Microbiol.*, 16: 57-66.

Ibrahim, A., Liesack, W., Griffiths, M.W., and Robins-Brown, R.M., 1997, Development of a highly specific assay for rapid identification of pathogenic strains of *Yersinia enterocolitica* based on PCR amplification of the *Yersinia* heat-stable enterotoxin gene (*yst*). *J. Clin. Microbiol.* 35: 1636-1638.

Nakajima, H., Inoue, M., Mori, T., Itoh, K.-I., Arakawa, E, and Watanabe, H., 1992, Detection and identification of *Yersinia pseudotuberculosis* and pathogenic *Yersinia enterocolitica* by an improved Polymerase Chain Reaction method. *J. Clin. Microbiol.* 30: 2484-2486.

Neubauer, H., Hensel, A., Aleksic, S., Finke, E.-J., and Meyer H., 2000, *Yersinia enterocolitica* 16S rRNA gene types belong to the same genospecies but form three different homology clusters. *Int. J. Med. Microbiol.*, 290: 61-64.

Price, S.B., Leung K.Y., Barve, S.S. and Straley, S.C., 1989, Molecular analysis of lcrGVH, the V antigen operon of *Yersinia pestis. J Bacteriol.,* 171: 5646-5653.

Rasmussen, H.N., Rasmussen, O.F., Andersen, J.K., and Olsen J.E., 1993, Specific detection of pathogenic *Yersinia enterocolitica* by two-step PCR using hot-start and DSMO. *Mol. Cell. Probes,* 8: 99-108.

Chapter 68

Molecular Typing of *Yersinia* Strains by Pulsed-Field Gel Electrophoresis and RAPD-PCR.

Laura FRANZIN and Daniela CABODI
Infectious Diseases Unit, University of Turin, Corso Svizzera 164, 10149 Turin, Italy

1. INTRODUCTION

Yersinia enterocolitica 4/0:3 and 2/0:9 strains are the most common pathogenic bioserotypes isolated in Italy. In epidemiological studies it is necessary to differentiate species into types in order to investigate the circulation and the prevalence of pathogenic types and identify reservoir of infection and association between clinical cases. Molecular typing methods are useful to differentiate subtypes, lineages and clonal lines. Pulsed-field gel electrophoresis (PFGE) and random amplified polymorphic DNA PCR (RAPD-PCR) have been successfully used by some authors in epidemiological surveys of *Y. enterocolitica* (Fredriksson-Ahomaa *et al.*, 1999; Gray *et al.*, 2001; Iteman *et al.*, 1991; Hosaka *et al.*, 1997; Najdenski *et al.*, 1994; Odinot *et al.*, 1995; Rasmussen *et al.*, 1994; Shayegani *et al.*, 1995). The aim of the study was the typing of human *Yersinia* strains isolated in our laboratory by PFGE and by RAPD-PCR for epidemiological purposes.

2. MATERIALS AND METHODS

Twenty strains of *Y. enterocolitica* isolated in our laboratory from 15 symptomatic patients (10 males and 5 females; 13 children and 2 adults) with enterocolitis (13 subjects), arthralgia (1 adult) and acute appendicitis (1 child) were studied. Eleven *Y. enterocolitica* 4/0:3 strains and 2 *Y. enterocolitica* 2/0:9 were typed by PFGE and by RAPD-PCR; 7 strains of *Y. enterocolitica* 1/10,34 were also typed by RAPD-PCR.

The Genus Yersinia, Edited by Skurnik *et al.*
wer Academic/Plenum Publishers, New York, 2003

For PFGE a standardized suspension of 24 hour-old culture of *Yersinia* on Peptone water was concentrated by centrifugation and the pellet mixed with equal volumes of molted 1.5% low melting point agarose. The plugs were treated with digestion solution containing proteinase K (2 mg/ml) and then with restriction enzyme *Not* I (20 U) and *Xba*I (20 U). Electrophoresis for cleaved DNAs with *Not*I was performed on 1% agarose gel in 0.5% Tris-borate-EDTA buffer with CHEF-DR III at +6 V/cm, 120° switch angle and 14°C, with four electrophoresis conditions (pulse time: linear switch time ramp 5-70 sec for 24 h; 3-10 sec for 15 h and 15-20 sec for 9 h; 5-15 sec for 18 h; 5-15 sec for 22 h) and with *Xba*I with pulse time 15 sec for 22 h. Genomic fragments were stained with ethidium bromide and were photographed under UV illumination. The results were interpreted following Tenover criteria (Tenover *et al.,* 1995).

For RAPD-PCR, a standardized suspension of 48 hour-old culture of *Yersinia* on Tryptic Soy agar was treated with 10% Chelex (Gomez-Lus *et al.,* 1993) and amplified with two primers (primer 2 and 5, Pharmacia), using Ready-To-Go RAPD Analysis Beads (Pharmacia). The amplification products were analyzed by electrophoresis on polyacrylamide gel (CleanGel Pharmacia and silver staining) with GenePhor System.

3. RESULTS

The interpretation of PFGE patterns was difficult because of the large number of fragments very closely spaced. Two electrophoresis conditions (3-10 sec for 15 h and 15-20 sec for 9 h; 5-15 sec for 22 h) for *Not*I gave the best results. Two completely different PFGE patterns were obtained by *Not*I between *Y. enterocolitica* 0:3 and O:9, while similar patterns were shown by *Y. enterocolitica* O:3 strains, except one. Patterns profiles obtained by *Xba*I were more difficult to interpret than those of *Not*I. RAPD-PCR with two primers showed different patterns for *Y. enterocolitica* O:3 and O:9, confirming the results of PFGE. Two *Y. enterocolitica* 2/0:9 strains, isolated from mother and child, shared the same profile with all the methods used. RAPD-PCR confirmed the identity of two *Y. enterocolitica* 1/10,34 strains isolated from stool and intestinal mucosa of a child with enterocolitis.

4. CONCLUSIONS

Many techniques have been used in the molecular epidemiology of *Y. enterocolitica* such as restriction enzyme analysis of plasmids and chromosomes, ribotyping, PFGE and RAPD-PCR.

PFGE has been successfully employed in characterization of *Y. enterocolitica* isolates. Profiles of pathogenic serotypes are different from non pathogenic strains. High similarity is however observed in *Y. enterocolitica* bioserotype 4/0:3 (Najdenski *et al.*, 1994; Shayegani *et al.*, 1995). RAPD-PCR fingerprinting offers some advantages over other methods: it is rapid, it requires minimal labor and is reliable technique using standardized reagents (Fredriksson-Ahomaa *et al.*, 1999).

Different profiles between *Y. enterocolitica* O:3 and O:9 were found in our study, confirming previous observations (Najdenski *et al.*, 1994; Shayegani *et al.*, 1995). Interpretation of PFGE patterns of *Y. enterocolitica* is however difficult. Strains *Y. enterocolitica* 2/O:9 isolated from mother and child shared the same profile, confirming a common source of infection (consumption of raw sausage). Strains of *Y. enterocolitica* 1/10,34 isolated from different samples of a child were the same with both methods.

We conclude that PFGE and RAPD-PCR are useful tool to type clinical isolates of *Yersinia* and to recognized epidemiological related strains. Best results are however obtained by the combined used of the two techniques.

ACKNOWLEDGEMENTS

We thank Specchio dei tempi - La Stampa Foundation, Italy for financial support.

REFERENCES

Fredriksson-Ahomaa, M., Autio, T., and Korkeala, H., 1999, Efficient subtyping of *Yersinia enterocolitica* bioserotype 4/O:3 with pulsed-field gel electrophoresis. *Lett. Appl. Microbiol.* 29, 308-312.

Gomez-Lus, P., Fields, B., Benson, R.F. *et al.*, 1993, Comparison of arbitrarily primed polymerase chain reaction, ribotyping, and monoclonal antibody analysis for subtyping *Legionella pneumophila* serogroup 1. *J. Clin. Microbiol.* 31,1940-1942.

Gray, J.T., WaKabongo, M., Campos, F.E., Diallo, A.A., Tyndal, C., and Tucker, C.A., 2001, Recognition of *Yersinia enterocolitica* multiple strain infection in twin infants using PCR-based fingerprinting. *J. Appl. Microbiol.* 90, 358-364.

Iteman, I., Baril, C., Saint Girons, I., and Carniel. E., 1991, Pulse field electrophoresis of the chromosome of the pathogenic Yersiniae. *Contrib. Microbiol. Immunol.* 12, 198-202.

Hosaka, S., Uchiyama, M., Ishikawa, M., *et al.*, 1997, *Yersinia enterocolitica* serotype O:8 septicemia in an otherwise healthy adult: analysis of chromosome DNA pattern by pulsed-field gel electrophoresis. *J. Clin. Microbiol.* 35, 3346-3347.

Najdenski, H., Iteman, I,. and Carniel, E., 1994, Efficient subtyping of pathogenic *Yersinia enterocolitica* strains by pulsed-field gel electrophoresis. *J. Clin. Microbiol.* 32, 2913-2920.

Odinot, P.T., Meis, J.F.G.M., van den Hurk, P.J.J.C., *et al.,* 1995, PCR-based DNA fingerprinting discriminates between different biotypes of *Yersinia enterocolitica. Contr. Microbiol. Immunol.* 13, 93-98.

Rasmussen, H.N., Olsen, J.E., and Rasmussen, O.F., 1994, RAPD analysis of *Yersinia enterocolitica. Lett. Appl. Microbiol.* 19, 359-362.

Shayegani, M., Maupin, P.S., and Waring, A., 1995, Prevalence and molecular typing of two pathogenic serogroups of *Yersinia enterocolitica* in New York State. *Contr. Microbiol. Immunol.* 13, 33-38.

Tenover, F., Arbeit, R.D., Goering, R.V., *et al.,* 1995, Interpreting chromosomal DNA restriction patterns produced by pulsed-field gel electrophoresis: criteria for bacterial strain typing. *J. Clin. Microbiol.* 33, 2233-2239.

Chapter 69

Bacteriocin Susceptibility of Clinical *Yersinia* Strains

Laura FRANZIN, Daniela CABODI, and Cristina FANTINO
Infectious Diseases Unit, University of Turin, Corso Svizzera 164, 10149 Turin, Italy.

1. INTRODUCTION

Bacteriocins are defined as antagonistic proteins or protein complexes synthesized by a variety of Gram-positive and Gram-negative bacterial strains, that show bactericidal activity against species closely related to the producer bacterium. Production of bacteriocins or bacteriocin-like substances has been reported from *Y. enterocolitica*, *Y. pestis*, *Y. pseudotuberculosis*, *Y. frederiksenii*, *Y. kristensenii* and *Y. intermedia* (Bottone *et al.*, 1979). Pathogenic serogroups of *Y. enterocolitica* may be inactivated by bacteriocins from *Y. kristensenii* (Toora *et al.*, 1994) and other *Yersinia* spp. (Cafferkey *et al.*, 1989; Strauch *et al.*, 2001). Bacteriocins have found widespread application in epidemiological studies as specific marker of many bacteria including *Yersinia* (Toora, 1995).

The purpose of the study was to evaluate bacteriocin susceptibility of *Yersinia* strains belonging to different serogroups which were isolated in our laboratory from symptomatic patients.

2. MATERIALS AND METHODS

Three bacteriocin producer strains (Y. *enterocolitica* 1/5/Xz [N.8], Y. *enterocolitica* 1/7,8,19 [N.19] and Y. *enterocolitica* 1/AA [N.32]), previously recognized in our Laboratory (Franzin *et al.*, 1998), were tested against 128 clinical strains (81 *Y. enterocolitica*, 40 *Y. frederiksenii*, 5 *Y. intermedia* and 2 *Y. kristensenii)* isolated from 93 patients (116 isolates from stool, 9 from appendix and 3 from intestinal mucosa). The strains were all

The Genus Yersinia, Edited by Skurnik *et al.*
'luwer Academic/Plenum Publishers, New York, 2003

isolated in our laboratory and typed at Institut Pasteur, Paris (Prof. H.H. Mollaret and Dr. E. Carniel). The bacteriocin production was detected by the colony diffusion technique (Toora *et al.,* 1994). A loopful culture of the producer strain, grown in Tryptic Soy Broth at 25°C for 24 h, was stubbed on Tryptic Soy Agar plate incubated at 25°C for 24 h. The colony was exposed to chloroform vapour for 10 minutes. The assay plate was overlaid with 10 mL of 0.8% Tryptic Soy Agar seeded with 0.1 mL of indicator strains (grown in Tryptic Soy Broth at 25°C for 24 h) and incubated at 25°C for 18h. The plate was observed for clear zone of inhibition around the colony of the producer due to the bacteriocin production. The surface-spotting technique was used to study some strains. To evaluate the proteinaceous nature of bacteriocin, the test was performed after treating bacteriocin solution (obtained by centrifugation from broth culture) with trypsin (bovine pancreas type III, Sigma) at final concentration of 500 μg/mL for 60 minutes at 37°C.

The plasmid analysis and *in vitro* virulence markers (pyrazinamidase activity, salicin fermentation, esculin hydrolysis, calcium dependency, autoagglutination, congo red uptake) (Wauters *et al.,* 1987) were also performed.

3. RESULTS AND CONCLUSIONS

A total of 79 (61.7%) strains were susceptible at least to one bacteriocin producer strains: 17 (13.3%) to first producer, 3 (2.3%) to the second and 68 (53.1%) to the third.

Table 1. Results of bacteriocin susceptibility of pathogenic *Y. enterocolitica* (Y.e.) strains from different samples

Strains	*n*	Origin	Producer 8	Producer 19	Producer 32
Y.e. 4/3/IX$_a$ Arab- Malt-	1	Appendix	+	+	+
Y.e. 4/3/VIII	1	Stool	+	+	+
Y.e. 4/3/VIII	2	Stool	+	-	+
Y.e. 4/3/VIII	2	Stool	+	-	-
Y.e. 4/3/VIII	2	Stool	-	-	+
Y.e. 4/3/VIII	1	Stool	-	-	-
Y.e. 2/9	1	Stool	+	-	+
Y.e. 2/9	1	Stool	-	-	+

Susceptibility of 81 *Y. enterocolitica* strains to the three producers were respectively: 13 (16%), 3 (3.7%) and 39 (48.1%). 10 of 11 *Y. enterocolitica* pathogenic strains (serogroups O:3 and O:9) were susceptible at least to one bacteriocin: 6 to the first producer, 2 to the second and 8 to the third producer (Table 1). 27 of 40 (67.5%) strains of *Y. frederiksenii* were susceptible to the first producer, 2 to the second and 8 to the third producer. No correlation with plasmids was observed. The bacteriocin activity was completely lost after trypsin treatment.

Bacteriocins production was described from many Gram-negative bacteria, both of clinical and food origin. Bacteriocin-like antagonism by strains of *Yersinia* spp. against other *Yersinia* species has been reported. (Toora *et al.*, 1994).

Antagonisms between bacteria are very important, for example inhibition of foodborne microorganisms by the lactic acid bacteria in food. Very little of the physico-chemical aspects and genetics of the bacteriocin production has been studied; however, the findings show that this bacteriocin is proteinaceous in nature, as complete loss of its activity was observed in the presence of proteolytic enzyme (trypsin) (Toora *et al.*, 1994). Recently, a phage tail-like bacteriocin of *Yersinia* was characterized; this enterocoliticin does not contain nucleic acids, shows contraction upon contact with susceptible bacteria, has channel-forming activity and shows inhibition against enteropathogenic strains of *Y. enterocolitica* of serogroup O:3, O:5,27 and O:9 (Strauch *et al.*, 2001)

In this study an elevated percentage (61.7%) of clinical strains were susceptible to the bacteriocin produced by three strains recognized in our laboratory. The third producer was more active. Stronger inhibition was observed against *Y. enterocolitica* pathogenic strains, followed by *Y. frederiksenii*. Bacteriocin antagonism may play an important role in human infections caused by pathogenic serogroups of *Y. enterocolitica*, but the mechanism should be further evaluated.

REFERENCES

Bottone, E.J., Sandhu, K.K., and Pisano, M.A., 1979, *Yersinia intermedia*: temperature-dependent bacteriocin production. J. Clin. Microbiol. 10: 433-436.

Cafferkey, M.T., McClean, K., and Drumm, M.E., 1989, Production of bacteriocin-like antagonism by clinical isolates of *Yersinia enterocolitica*. J. Clin. Microbiol. 27: 677-680.

Franzin, L., Vicelli, S., Cabodi, D., and Gioannini, P., 1998, Bacteriocin-like substances production of *Yersinia strains*. Nederland Tijdschrift voor Medische Microbiologie 6 (suppl 2), p.24.

Strauch, E., Kaspar, H., Schaudinn, C., Dersch, P., *et al.*, 2001, Characterization of enterocoliticin, a phage tail-like bacteriocin, and its effect on pathogenic *Yersinia enterocolitica* strains. Appl. Environ. Microbiol. 67: 5634-5642.

Toora, S., Budu-Amoako, E., Ablett, R.F., and Smith, J., 1994, Inhibition and inactivation of pathogenic serogroups of *Yersinia enterocolitica* by a bacteriocin produced by *Yersinia kristensenii*. Lett. Appl. Microbiol. 19: 40-43.

Toora, S., 1995, Application of *Yersinia kristensenii* bacteriocin as a specific marker for the rapid identification of suspected isolates of *Yersinia enterocolitica*. Lett. Appl. Microbiol. 20: 171-174.

Wauters, G., Kandolo, K., and Janssens, M., 1987, Revised biogrouping scheme of *Yersinia enterocolitica*. Contr. Microbiol. Immunol. 9: 14-21.

Chapter 70

Molecular Epidemiology of *Yersinia pseudotuberculosis*

Hiroshi FUKUSHIMA
The Shimane Prefectural Institute of Public Health and Environmental Science. 582-1 Nishihamasada, Matsue, Shimane 690-0122, Japan

Yersinia pseudotuberculosis has a wide distribution in most countries with cold climates and is recognized as an important causal agent of sporadic and epidemic human enteric disease. The pathogenicity of *Y. pseudotuberculosis* depends on the presence of 70-kb virulence plasmid (pYV) which is essential for virulence. Additionally, *Y. pseudotuberculosis* produces a novel superantigenic toxin designated YPMa (*Y. pseudotuberculosis*-derived mitogen), YPMb or YPMc and has a pathogenicity island termed HPI (high-pathogenicity island) or R-HPI (a right-hand part of the HPI with truncation in its left-hand part) which are encoded by the chromosome. *Y. pseudotuberculosis* has been classified into serotypes O1 to O15, the serotypes of which O1 to O5 have been isolated in Europe and the Far East and almost all are pathogenic, while serotypes O6 to O14 have been isolated only from wild animals and environments in the Far East but never from clinical samples. There are numerous reports studying the virulence factors of *Y. pseudotuberculosis*, however, comparisons of the prevalence of these virulence factors in the wild *Y. pseudotuberculosis* strains have never been documented. We investigated the distribution of virulence factors; pYV, YPMs and HPI among 2,235 wild *Y. pseudotuberculosis* strains of various serotypes isolated from patients, domestic and wildlife animals and natural environments all over the world (Fukushima *et al.*, 2001). Analysis of the distribution of these virulence factors allowed for differentiation of species *Y. pseudotuberculosis* into six subgroups thus reflecting the geographical spread of two main clones (Figure 1): the YPMa$^+$ and HPI$^-$ Far Eastern systemic pathogenic type belonging to serotypes O1b, 2a, 2b, 2c, 3, 4a, 4b, 5a, 5b, 6, 10 and 15 and the YPMs$^-$ and HPI$^+$ European gastroenteric pathogenic type belonging to

The Genus Yersinia, Edited by Skurnik *et al.*
¹luwer Academic/Plenum Publishers, New York, 2003

serotypes O1a and 1b. The YPMa$^+$ and HPI$^+$ pathogenic type belonging to serotypes O1b, 3, 5a, 5b and 15 and the YPMb$^+$ and HPI$^-$ non-pathogenic type belonging to non-melibiose-fermenting serotypes O1b, 5a, 5b, 6, 7, 9, 10, 11 and 12 were prevalent in the Far East. The YPMc$^+$ and R-HPI$^+$ European low-pathogenic type belonging to non-melibiose-fermenting serotype O3 and the YPMs$^-$ and HPI$^-$ pathogenic type belonging to 15 serotypes were found to be prevalent all over the world. This new information is useful for a better understanding of the evolution and spread of *Y. pseudotuberculosis.*

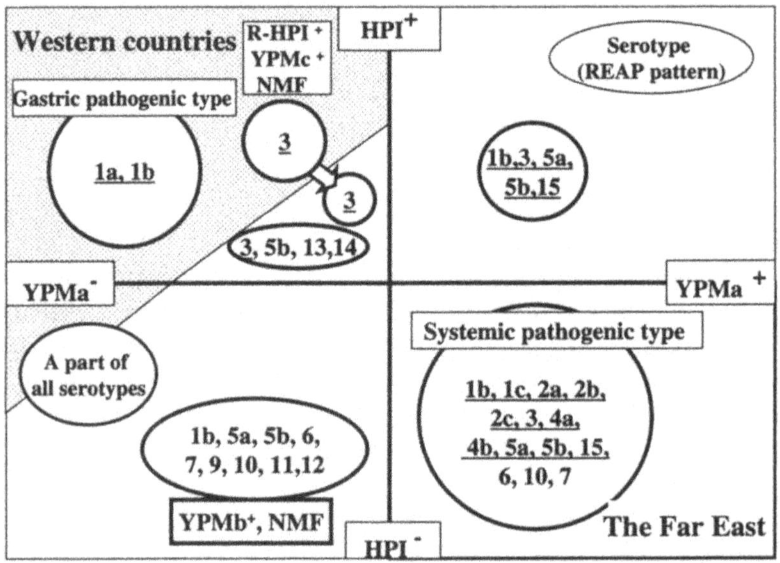

Figure 2. Geographical distribution of *Y. pseudotuberculosis* serotypes. Serotypes with underlining: clinical isolates, pYV+, serotypes without underlining: animal and environment isolates, pYV+ or pYV-

REFERENCE

Fukushima, H., Y. Matsuda, R. Seki, M. Tsubokura, N. Takeda, F.N. Shubin, I-K. Paik, and X.B. Zheng. 2001. Geographical heterogeneity between Far East and Western countries in prevalence of the virulence plasmid, the superantigen *Yersinia pseudotuberculosis*-derived mitogen and the high-pathogenicity island among *Yersinia pseudotuberculosis* strains. J. Clin. Microbiol. 39:3541-3547.

Chapter 71

Growth of *Yersinia enterocolitica* in Inegol Meatballs

Ugur GUNSEN
Food Control and Central Research Institute, 16036, Bursa, TURKEY

1. INTRODUCTION

Refrigeration technique has been accepted as a unique factor in controlling foodborne pathogens. Some psychrophilic pathogens such as *Y. enterocolitica* may survive and grow in foods during refrigerated storage (Aulisio *et al.*, 1983 and Berrang *et al.*, 1989).

This study was performed in order to investigate the bacterial flora and *Y. enterocolitica* growth in vacuum packed and refrigerated Inegol meatballs, a special kind of meatball produced in Inegol Town of Bursa Province of Turkey. Nowadays these meatballs are widely sold as ready made products.

2. MATERIAL AND METHODS

The strain of *Y. enterocolitica* 300290500009Y UN 2814 classe 6.2 was obtained from Institut Pasteur, Center National de Reference des *Yersinia*, Paris, France. Stock culture was propagated in Brain Heart Infusion (BHI) agar and incubated at 30 °C. One loopfull was streaked on Celfsulodin Irgasan Novobiocin (CIN) agar and incubated at 27–30 °C for 24 h. The levels of *Y. enterocolitica* were determined per ml of BHI broth for the detection of the initial inoculum level of Inegol meatballs.

Inegol meatballs used as research material were collected daily from the producing restaurant and transported to the laboratory under chilled conditions. Before the *Y. enterocolitica* inoculations, *Y. enterocolitica* could not be isolated from the Inegol meatballs using the FDA methods (Weagant *et al.*, 1995).

The Genus Yersinia, Edited by Skurnik *et al.*
Kluwer Academic/Plenum Publishers, New York, 2003

Inegol meatballs were inoculated with two different *Y. enterocolitica* levels were 3.4 x 10^3 cfu g^{-1} and 2.7 x 10^5 cfu g^{-1} and then vacuum packed and refrigerated at +4 °C for 7 days in the first group and –18 °C for 90 days in the second group. The levels of *Y. enterocolitica* were determined on 0, 1, 2, 4 and 7 days of the storage in the first group and on 0, 15, 30 and 90 days of the storage in the second group. Further, both of the groups were analysed for mesophylic aerobic bacteria, psychrophilic bacteria, coliform bacteria, staphylococcus and micrococcus, enterococcus, lactobacilli, yeasts and molds (ICMSF, 1982). pH values were determined by using Metrohm 744 pH meter and a_w values were detected by using Durotherm a_w – Wert-Messer during the storage.

Values from the replicate trials were used for statistical analyses. Data were analyzed by ANOVA (analyses of variance) with the general linear model procedure of the MINITAB statistical package. Data reported were average of duplicate samples and replicate trials.

3. RESULTS

Tables 1 and 2 show the growth of *Y. enterocolitica* in vacuum packaged Inegol meatballs, refrigerated at different temperatures. Tables 3 and 4 show the changes of bacterial flora in vacuum packaged Inegol meatballs after refrigeration at different temperatures.

Table 1. Growth of *Y. enterocolitica* in vacuum packaged Inegol meatballs, refrigerated at +4°C for 7 days (cfu g^{-1}).

Groups	Days				
	0	1	2	4	7
1	3.4 x 10^3	4.6 x 10^3	8.5 x 10^3	3.1 x 10^4	8.4 x 10^4
2	2.7 x 10^5	5.9 x 10^5	8.9 x 10^5	4.4 x 10^6	1.2 x 10^7

Table 2. Growth of *Y. enterocolitica* in vacuum packaged Inegol meatballs, refrigerated at –18°C for 90 days (cfu g^{-1})

Groups	Days			
	0	15	30	90
1	3.4 x 10^3	3.7 x 10^4	8.6 x 10^4	2.5 x 10^5
2	2.7 x 10^5	4.5 x 10^6	1.3 x 10^7	5.9 x 10^7

During the storage at +4°C for 7 days, pH values were 7.4 in the beginning, 7.9 on day 1 and 6.7 on day 7. During the storage at –18°C for 90 days, pH values were 7.4 in the beginning, 7.6 on day 1 and 6.4 on day 90. a_w values changed between 0.96 and 0.98 in both of the groups during the storage.

4. CONCLUSION

The results showed that vacuum packaging and refrigeration of Inegol meatballs at + 4°C for 7 days and –18°C for 90 days did not prevent the growth of *Y. enterocolitica* and there were significant differences in the bacterial counts in both of the groups ($p < 0.01$). The presence of a competitive bacteria flora did not influence the growth of *Y. enterocolitica* at any of the storage temperatures. The high levels of Lactobacilli and psychrophilic bacteria could not prevent the growth of *Y. enterocolitica* at any of the storage temperatures.

Table 3. The changes of bacterial flora in vacuum packaged Inegol meatballs, refrigerated at + 4°C for 7 days (cfu g^{-1})

Bacterial flora	Days				
	0	1	2	4	7
Meso. Aerobic bacteria	1.8×10^4	7.3×10^4	8.1×10^4	1.7×10^4	1.1×10^4
Psychrophilic bacteria	2.3×10^6	2.1×10^7	4.7×10^7	8.9×10^7	4.4×10^8
Coliform bacteria	4.1×10^2	5.6×10^2	6.1×10^2	3.9×10^2	2.2×10^2
Staph.- micrococcus	1.4×10^5	2.5×10^5	2.9×10^5	1.8×10^5	2.7×10^4
Enterococci	5.3×10^4	3.7×10^4	5.6×10^4	2.8×10^4	6.5×10^4
Lactobacilli	9.2×10^6	6.2×10^6	7.5×10^6	9.3×10^6	1.7×10^7
Yeasts and moulds	4.5×10^4	1.7×10^4	3.3×10^4	2.4×10^4	3.8×10^4

Table 4. The changes of bacterial flora in vacuum packaged Inegol meatballs, frozen at –18°C for 90 days (cfu g^{-1})

Bacterial flora	Days			
	0	15	30	90
Meso. Aerobic bacteria	1.8×10^4	9.3×10^3	7.5×10^3	2.8×10^3
Psychrophilic bacteria	2.3×10^6	2.4×10^8	9.2×10^8	2.3×10^9
Coliform bacteria	4.1×10^2	6.7×10^2	4.6×10^2	3.8×10^2
Staph.-micrococcus	1.4×10^5	5.8×10^3	2.7×10^4	9.6×10^4
Enterococci	5.3×10^4	6.5×10^4	4.6×10^4	1.7×10^4
Lactobacilli	9.2×10^6	8.7×10^6	4.8×10^6	1.3×10^7
Yeasts and moulds	4.5×10^4	2.6×10^4	5.2×10^4	7.2×10^4

CONCLUSIONS

Our results demonstrate that more importance should be given to the hygienic conditions during the production of the Inegol meatballs and that efficient grilling process should be applied when preparing the meatballs.

NOTES

This project was supported by Turkish Republic of Ministry of Agriculture and Rural Affairs.

ACKNOWLEDGEMENTS

I would like to special thanks to Elisabeth CARNIEL, MD PhD, Director of the National Reference Laboratory and WHO Collaborating Center for Yersinia, Institute Pasteur, and to the staff of Microbiology and Mycotoxin Laboratory in our Institute.

REFERENCES

Aulisio, C.C.G., Stanfield, J.T., Weagant, S.D. and Hill, W.E., 1983, Yersiniosis associated with tofu consumption: serological, biochemical and pathogenicity studies of *Yersinia enterocolitica* isolates. J. Food Prot. 46: 226-230.

Berrang, M.E., Bracket, R.E., Beuchat, L.R., 1989, Growth of *Listeria monocytogenes* on fresh vegetables stored under a controlled atmosphere. J.Food Prot. 52: 702-705.

Weagant, S.D., Feng, P., Stainfield, J.T., 1995, *Yersinia enterocolitica* and *Yersinia pseudotuberculosis*. F.D.A. Bacterial Analytical Manual. 8th ed. A.O.A.C. International, Gaithersburg, U.S.A. Chapter 8 (8.01-8.13).

Chapter 72

Molecular Genetic Typing of *Yersinia enterocolitica* Serovar O:8 Isolated in Japan

Hideki HAYASHIDANI[1], Yuki ISHIYAMA[1], Tomomitsu A. OKATANI[1], Shin-Ichiro YOSHIDA[2], Motoko ISHIKAWA[2], Yukio KATO[3], Yoshimitsu OHTOMO[4], Masaaki SAITO[5], Tomoko HORISAKA[1], Ken-Ichi KANEKO[1], and Masuo OGAWA[2]

[1]Tokyo University of Agriculture and Technology, 3-5-8 Saiwai-cho, Fuchu, Tokyo183-8509, [2]Japan Food Research Laboratories, Tama Laboratory, 6-11-10 Nagayama, Tama, Tokyo 206-0025, [3]Azabu University, Fuchinobe, Sagamihara, Kanagawa 229-8501, [4]Aomori Prefectural Institution of Public Health and Environment, 1-1-1 Higashi-Tsukurimichi, Aomori 030-8566, [5]Adult Diseases Examination Center, Hirosaki Medical Association, 2-7-1 Noda, Hirosaki, Aomori 036-8045, Japan

1. INTRODUCTION

Yersinia enterocolitica serovar O:8 is known to be an important human enteric pathogen (Munnich *et al.*, 2001). In 1989, this pathogen was first isolated from small rodents living in the wild in mountainous areas in Japan (Iinuma *et al.*, 1992). In 1990, an isolate of the serovar was isolated from a human patient in Aomori Prefecture, Japan (Ichinohe *et al.*,1991). Following that, human patients infected with serovar O:8 have been sporadically reported in the northern area of Honshu Island in Japan (Hayashidani *et al*, 1995).

In the present study, we examined the genetic relationship of *Y. enterocolitica* serovar O:8 isolated in Japan using molecular typing methods.

2. MATERIALS AND METHODS

A total of 22 O:8 strains isolated from human patients (9), wild rodents (9), pigs (2) and environmental sources (2) in Japan, and WA strain originated from USA was used (Table1). These isolates were examined by means of pulse-field gel electrophoresis (PFGE), ribotyping using The Riboprinter system (Qualicon™), and restriction endonuclease analysis of virulence plasmid (REAP).

PFGE electrophoresis was performed for 24 h at 12 mA at 200V with pulse times of 1 to 25s using by Gene Navigator (Pharmacia Biotech). Ribotyping was performed using The Riboprinter system according to the manufacturer's instruction for gram-negative bacteria. REAP was performed with the enzyme *Eco*RI and *Bam*HI as described by Nesbakken et al.(1987), with 1.2% agarose gel electrophoresis.

Table 1. Typing results for *Y. enterocolitica* O:8 isolated in Japan using PFGE, ribotyping and REAP

No	Strain	Region	Source	PFGE pattern	Ribo-pattern	REAP pattern	Geno-type[a]
1	YE87069	Aomori	Patient	B	R3	P2	IV
2	YE89023	Aomori	Patient	A	R4	P1	II
3	YE93009	Aomori	Patient	A	R1	P1	I
4	YE93017	Aomori	Patient	A	R1	P1	I
5	YE93019	Aomori	Patient	A	R1	P1	I
6	APCC Y9314	Aomori	Patient	B	R1	P2	III
7	APCC Y9415	Aomori	Patient	D	R2	P3	VII
8	NY9306005	Aomori	Wild rodent	B	R1	P2	III
9	NY9306089	Aomori	Wild rodent	A	R1	P1	I
10	NY9406043	Aomori	Wild rodent	A	R1	P1	I
11	NY9406064	Aomori	Wild rodent	A	R1	P1	I
12	NY9406085	Aomori	Wild rodent	A	R1	P1	I
13	NY9504002	Aomori	Wild rodent	C	R1	P1	V
14	NY9406007	Aomori	Stream water	B	R1	P2	III
15	NY9407007	Aomori	Stream water	A	R1	P1	I
16	Towada33	Aomori	Pig	B	R1	P2	III
17	Towada45	Aomori	Pig	A	R1	P1	I
18	APCC Y9433	Akita	Patient	A	R1	P1	I
19	YE9809001	Yamagata	Wild rodent	D	R2	P3	VII
20	NY891001	Nigata	Wild rodent	D	R1	P3	VI
21	NY891004	Nigata	Wild rodent	D	R1	P3	VI
22	NY9809001	Gifu	Patient	D	R1	NT[b]	VI
23	WA	USA	Patient	E	R5	P4	VIII

[a] Genotype was produced by combining the results obtained using PFGE with *Not*I and ribotyping

[b] This strain lacks virulence plasmid

3. RESULTS AND DISCUSSION

Twenty-two strains isolated in Japan were grouped into four pulsotypes by *Not*I, into four ribotypes by *Eco*RI and into three REAP types by *Eco*RI and *Bam*HI endonucleases. Although the number of types generated by PFGE and ribotyping were the same, their differentiation patterns differed. By combining the results obtained using PFGE and ribotyping, seven (I-VII) different genotype were distinguished (Table1). Types I and III from Aomori Prefecture were commonly distributed in human patients, wild rodents, pigs, and environmental sources. The type obtained by means of a combination of PFGE and ribotyping was different in the wild rodent isolates in terms of regional groups. WA strain originated from USA was grouped into genotype VIII which differed from the Japanese isolates.

These results suggest that PFGE and ribotyping using The Riboprinter should be used together for epidemiological analysis of *Y. enterocolitica* serovar O:8 infection.

REFERENCES

Hayashidani,H., Ohtomo,Y., Toyokawa,.Y., Saito,M., Kaneko,K., Kosuge,J., Kato,M., Ogawa,M., and Kapperud,G., 1995, Potential source of sporadic human infection with *Yersinia enterocolitica* serovar O:8 in Aomori prefecture, Japan. *J. Clin. Microbiol.* 33: 1253-1257.

Iinuma,Y., Hayashidani,H., Kaneko,K., Ogawa,M., and Hamasaki, S., 1992, Isolation of *Yersinia enterocolitica* serovar O8 from free-living small rodents in Japan. *J. Clin. Microbiol.* 30: 240-242.

Ichinohe.H., Yoshioka.M., Fukushima,H., Kaneko,S., and Maruyama,T., 1991, First isolation of *Yersinia enterocolitica* serotype O:8 in Japan. *J.Clin.Microbiol.* 29:846-847.

Munnich,S.A., Smith,M.J., Weagant,S.D., and Feng,P, 2001, *Yersinia*, In *Foodborne Disease Handbook Vol1:Bacterial pathogens* (Y.H.Hui, M.D.Pierson, and J.R.Gorham, eds.), Marcel Dekker, New York, pp.471-514.

Nesbakken,T., Kapperud,G., Sørum,H., and Dommarsnes,K., 1987, Structural variability of 40-50 Mdal virulence plasmid from *Yersinia enterocolitica. Acta Pathol. Microbiol.Immunol.Scand.Sect.*B.95:167-173

Chapter 73

Prevalence and Characterisation of *yadA*-Positive *Yersinia enterocolitica* in Pig Tonsils in 1995 and 1999

Tiina KORTE[1], Maria FREDRIKSSON-AHOMAA[1,2] and Hannu KORKEALA[1]
[1]Department of Food and Environmental Hygiene, University of Helsinki, Finland; [2]Institute of Hygiene and Technology of Food of Animal Origin, Ludwig-Maximilians-University, Munich, Germany

1. INTRODUCTION

Yersinia enterocolitica is an important food-borne pathogen in Finland. The most common bioserotype is 4/O:3. The mean prevalence of *yadA*-positive *Y. enterocolitica* in pig tonsils was 37% in 1995 when samples were collected from 9 slaughterhouses (Fredriksson-Ahomaa *et al.,* 2000). Two common genotypes were obtained with PFGE using *Not*I and *Xba*I enzymes. Genetic diversity of bioserotype 4/O:3 is limited. However, with *Not*I, *Apa*I and *Xho*I enzymes this bioserotype can efficiently be subtyped (Fredriksson-Ahomaa *et al.,* 1999). The aim of this work was to study the prevalence of *yadA*-positive *Y. enterocolitica* in pig tonsils and to compare the genotypes found in 1995 between the genotypes found in 1999.

2. MATERIALS AND METHODS

Tonsils of finishing pigs were studied in four slaughterhouses in 1995 and 1999. The slaughterhouses are located in various parts of Finland. A total of 83 and 120 samples were collected in 1995 and 1999, respectively. The prevalence was determined by culture method using selective enrichment in MRB, and PCR method detecting *yadA* on the virulence

The Genus Yersinia, Edited by Skurnik *et al.*
Kluwer Academic/Plenum Publishers, New York, 2003

plasmid. The recovered *Y. enterocolitica* isolates of bioserotype 4/O:3 were characterised with PFGE using *Not*I, *Apa*I and *Xho*I enzymes.

3. RESULTS

The mean prevalence of *yadA*-positive *Y. enterocolitica* was 30% and 65% in 1995 and 1999, respectively. The prevalence varied from 13% to 42% between the slaughterhouses in 1995 and between 50% and 87% in 1999 (Table 1). Bioserotype 4/O:3 was the only pathogenic bioserotype found with the culture method.

Table 1. Prevalence of *yadA*-positive *Yersinia enterocolitica* in pig tonsils

	1995		1999	
Slaughter-house	No. of samples	Detection rate	No. of samples	Detection rate
I	21	33%	30	87%
II	20	35%	30	63%
III	23	13%	30	50%
IV	19	42%	30	60%
Total	83	30%	120	65%

A total of 24 and 74 *Y. enterocolitica* 4/O:3 strains were obtained in 1995 and 1999, respectively. In all, 17 genotypes were found in 1995 and 28 genotypes in 1999. Six of the genotypes (GT2, GT3, GT4, GT7, GT8 and GT11) were found both in 1995 and 1999 (Table 2). Two genotypes, GT3 and GT8, were the most common types in 1995 and 1999. In one slaughterhouse, all the three genotypes (GT2, GT4 and GT8) found in 1995 were also found in 1999.

Table 2. Genotypes of *Yersinia enterocolitica* 4/O:3 strains found in pig tonsils

	1995		1999	
Slaughter-house	No. of strains	Genotypes	No. of strains	Genotypes
I	9	**GT2, GT3**,GT6,**GT7**, GT9, GT11, GT12, GT13, GT14	31	**GT2, GT3, GT8, GT11**, GT19, GT21, GT25, GT29, GT31, GT33, GT37, GT38, GT40
II	4	GT1, **GT3**, GT16	15	**GT3**, GT26, GT34, GT39
III	4	**GT2, GT4, GT8**	15	**GT2, GT4, GT8**, GT19, GT22, GT23, GT27, GT28, GT30, GT35
IV	7	**GT3**, GT5, **GT8**, GT10, GT15, GT17	13	**GT7, GT8**, GT20, GT24, GT32, GT36

4. CONCLUSIONS

Prevalence of pathogenic *Y. enterocolitica* in pig tonsils has increased significantly in the four slaughterhouses during 1995-99. Some of the genotypes were shown to be stable and were found from same slaughterhouses in 1995 and 1999. Two same genotypes were dominating in 1995 and 1999.

REFERENCES

Fredriksson-Ahomaa, M., Autio, T., and Korkeala, H., 1999, Efficient subtyping of *Yersinia enterocolitica* bioserotype 4/O:3 with pulsed-field gel electrophoresis. Lett. Appl. Microbiol. 29: 308-312.

Fredriksson-Ahomaa, M., Björkroth, J., Hielm, S., and Korkeala, H., 2000, Prevalence and characterisation of pathogenic *Yersinia enterocolitica* in pig tonsils from different slaughterhouses. Food Microbiol. 17: 93-101.

Chapter 74

Yersinia pseudotuberculosis in Pigs and Pig Houses in Finland

Riikka LAUKKANEN[1], Taina NISKANEN[1], Maria FREDRIKSSON-AHOMAA[1,2] and Hannu KORKEALA[1]
[1]*Department of Food and Environmental Hygiene, Faculty of Veterinary Medicine, University of Helsinki, Helsinki, Finland;* [2]*Institute of Hygiene and Technology of Food of Animal Origin, Ludwig-Maximilians-University, Munich, Germany*

1. INTRODUCTION

Yersinia pseudotuberculosis causes sporadic enteric infections and sometimes outbreaks in humans. Pigs and wild animals are considered to be carriers of *Y. pseudotuberculosis* and contaminated pork is one possible cause of yersiniosis in humans.

2. MATERIALS AND METHODS

A total of 161 pooled fecal and 116 environmental (floor, trough and air) samples were collected from 8 pre-selected pig houses in Finland during summer 2000. Each pooled sample contained feces from 5 different pigs. *Y. pseudotuberculosis* was previously isolated from tonsil samples of pigs coming from these farm (Niskanen *et al.*, 2001).

Samples were plated directly and after three weeks' cold enrichment in peptone sorbitol bile salt broth on cefsulodin irgasan novobiosin agar plates. Typical bull's eye colonies, which were urease positive, were identified using API 20E (BioMérieux®, France). *Y. pseudotuberculosis* strains were serotyped with O:1-O:6 antisera (Denka Seiken®, Japan). The virulence plasmid of the isolates was determined using PCR-assay targeting the *virF* gene (Nakajima *et al.*, 1992).

The Genus Yersinia, Edited by Skurnik *et al.*
Kluwer Academic/Plenum Publishers, New York, 2003

3. RESULTS

Y. pseudotuberculosis O:3 was isolated from 15 (9%) fecal samples and from 1 (1%) environmental (floor) sample (Table 1). *Y. pseudotuberculosis* was found from 4 out of 8 pig houses investigated. All isolates were melibiose negative and O:3 was the only serotype found. All isolates investigated harboured *virF* gene.

Table 1. Y. pseudotuberculosis in environment and in pooled fecal samples

Origin of samples	No. of samples	No. of positive samples	
		n	%
Environment	116	1	1
Floors	37	1	3
Troughs	41	0	0
Air	38	0	0
Feces	161	15	9
Fattening pigs	107	15	13
Others[a]	17	0	0

[a]piglets, sows and boars

4. CONCLUSIONS

In this study *Y. pseudotuberculosis* was found in 9% of fecal samples studied. In previous reports, *Y. pseudotuberculosis* has been isolated from 0,6 to 3% of fecal samples (Zen-Yoji *et al.*, 1974; Tsubokura *et al.*, 1976; Weber ja Knapp, 1981; Fukushima *et al.*, 1989; Fukushima *et al.*, 1990; Inoue, 1991; Fain-Binda, 1999). The higher occurrence in this study is possibly due to pre-selection of pig farms. This indicates that *Y. pseudotuberculosis* could be a farm-related problem. Serotype O:3 is the most common type found in pigs. Also the latest human outbreaks in Finland have been caused by serotype O:3. No previous information is found about the occurrence of *Y. pseudotuberculosis* in the pig house environment. Low amount of *Y. pseudotuberculosis* in environment and a relatively insensitive culture method might be one of the reasons why *Y. pseudotuberculosis* was found in only one environmental sample. The one positive environmental sample could indicate some role of the pigs' immediate environment in the infection process. Further investigation is nevertheless needed. Also the possible connection between pork and yersiniosis in humans needs further investigation.

REFERENCES

Fain-Binda, J.C., Gaia, O., Cornale, J., Comba, E., Francois, S., Drab, S., and Fain-Binda, V., 1999, Baja prevalencia de *Yersinia pseudotuberculosis* en especies pecuarias de Argentina. *Vet. Argent.* **157**: 493-500.

Fukushima, H., Marayama, K., Omori, I., Ito, K., and Iorihara, M., 1989, Role of the contaminated skin of pigs in fecal *Yersinia* contamination of pig carcasses at slaughter. *Fleishwirtschaft* **69**: 369-372.

Fukushima, H., Maruyama, K., Omori, I., Ito, K., and Iorihara, M., 1990, Contamination of pigs with yersinia at the slaughterhouse. *Fleishwirtschaft* **70**: 1300-1302.

Inoue, M., Nakashima, H., Mori, T., Sakazaki, R., Tamura, K., and Tsubokura, M., 1991, *Yersinia pseudotuberculosis* infection in the mountain area. In *Current Investigations of the Microbiology of Yersiniae* (Une, T., Maruyama, T. & Tsubokura, M. eds.) Karger, Basel, pp. 307-310.

Nakajima, H., Inoue, M., Mori, T., Itoh, K.-I., Arakawa, E., and Watanabe, H., 1992, Detectionand identification of *Yersinia pseudotuberculosis* and pathogenic *Yersinia enterocolitica* by an improved polymerase chain reaction method. *J. Clin. Microbiol.* **30**: 2484-2486.

Niskanen, T., Fredriksson-Ahomaa, M., and Korkeala, H., 2002, *Yersinia pseudotuberculosis* with limited genetic diversity is a common finding in tonsils of fattening pigs. *J. Food Protect.* **65**: 540-5.

Tsubokura, M., Otsuki, K., Fukuda, T., Kubota, M., Imamura, M., and Itagi, K., 1976, Studies on *Yersinia pseudotuberculosis* IV. Isolation of *Y. pseudotuberculosis* from healthy swine. *Jap. J. Vet. Sci.* **38**: 549-552.

Weber, A. and Knapp, W., 1981, Nachweis von *Yersinia enterocolitica* und *Yersinia pseudotuberculosis* in Kotproben gesunder Schlachtschweine in Abhängigkeit von der Jahrzeit. *Zentralbl. Vet. Med. B* **28**: 407-413.

Zen-Yoji, H., Sakai, S., Maruyama, T. and Yanagawa, Y., 1974, Isolation of *Yersinia enterocolitica* and *Yersinia pseudotuberculosis* from swine, cattle and rats at an abattoir. *Japan. J. Microbiol.* **18**: 103-105.

Chapter 75

Evaluation of Pulsed-Field Gel Electrophoresis (PFGE) for *Yersinia enterocolitica* Molecular Epidemiology Investigations

Marina MARRANZANO, Maria A. CONIGLIO and Luisa MAURO
Department of Hygiene and Public Health, University of Catania, via S. Sofia 87 – 95123, Catania, Italy

1. INTRODUCTION

Yersinia enterocolitica is an important enteric pathogen associated with a variety of human and animal diseases. It is primarily transmitted by foods, although contaminated water and blood transfusions have been involved in several cases.

Among the virulent *Y. enterocolitica*, serogroups O:3, O:9 and O:5,27 are pathogenic in humans and are not lethal to mice, while serogroups O:8; O:4,32; O:13,18; O:20 and O:21 are pathogenic in humans and lethal to mice.

There is a close correlation between the biotypes, serotypes and phage types of the strains, making it impossible to distinguish isolates of the same serotype with the classical phenotypic markers. In fact, serotyping, biotyping and phage-typing, currently used as primary tools for the identification of clinically significant strains of *Yersinia*, are not suitable for epidemiological studies because they do not allow to investigate relationships among strains; so they remain of limited use due to its low discriminative capabilities.

Among the molecular methods, conventional agarose and polyacrylamide gel electrophoresis have been the main methods of separating DNA fragments smaller than 20 kbp, while DNA fragments larger than 20 kbp lose molecular weight size resolution.

On the contrary, pulsed-field gel electrophoresis (PFGE) is a well-established technique capable of separating and resolving large fragments

The Genus Yersinia, Edited by Skurnik *et al.*
'uwer Academic/Plenum Publishers, New York, 2003

(up to 2 Mbp) of bacterial DNA generated by rare cutting restriction endonucleases. This system is based on the idea that molecules in an agarose gel will orient themselves in the direction of an applied electric field. If the direction of the electric field is changed abruptly, the DNA molecules will disentangle themselves from the gel matrix and begin to move along the reoriented field vector. Contour clamped homogeneous electric field (CHEF) gel electrophoresis is a particular PFGE in which the electric field is distributed along the contour of a hexagonal array of 24 electrodes. The two opposite sides of the hexagon are activated alternately, for the optimum 120° reorientation angle. Under directional switching of the electric field, the DNA fragments change direction in the gel. The migration rate of DNA molecules through the agarose gel is dependent on pulse time, voltage, and run time; the separation is due to the fact that the time each fragment takes to move in a new direction is proportional to its molecular weight.

In the present study CHEF technique was employed to verify whether the infections caused by *Y. enterocolitica* isolated in Sicily (south Italy) during the last few years, were determined by single strains or not and to verify the existence of correlations among the pathogenic strains and the environmental strains isolated in Sicily during the same period.

2. MATERIALS AND METHODS

A total of 23 *Y. enterocolitica* strains isolated in Sicily during the last few years were selected for this study: 12 strains belonging to pathogenic serogroups (6 to biotype 4/O:3 and 6 to biotype 2/O:9) isolated from diarrhoeal faeces and 11 strains belonging to environmental serogroups (O:7,8; O:6; O:5; O:4,32-10,16; O:19,8; O:30 and NAG), isolated from fresh waters and from vegetables.

For analysis by CHEF, a single colony of each strain was inoculated into 10 ml of Luria broth and incubated for 10 hr at 37°C and at 27°C overnight. Next cells were harvested, washed and suspended in low melting point agarose plus 2%. Agarose plugs containing cells of *Y. enterocolitica*, were then incubated in a proteinase K solution (50 mM Tris, 50 mM EDTA and 2 mg/ml of proteinase K), at 50°C overnight.

After the extraction in low melting point agarose plugs the bacterial DNA was digested with the endonuclease *Not*I (Life Technologies) as recommended by the manufacturer.

High-molecular-weight restriction fragments were resolved by CHEF (CHEF-DR II; BioRad Laboratories; Richmond, CA) pulsed-field system. An electrophoretic regimen of 200V for 21 hr at a temperature of 15°C and a switching time from 1 to 26 seconds was employed to separate fragments.

*Hind*III-cleaved DNA-λ. (New England Biolabs) was used as molecular weight marker. The gels were stained in ethidium bromide solution and photographed.

3. RESULTS AND CONCLUSIONS

By visual inspection under UV light only few similarities were observed among serogroups O:3 (range 700-900 kb) and among serogroups O:9 (range 500-870 kb). Among the environmental strains, completely different patterns were observed for serogroups O:7,8. On the contrary, remarkable similarities were observed in the range of 291 kb and of 921.5 kb for serogroups O:6. Among the 3 NAG strains only two showed similar banding patterns while the third one showed only few similar bands (range 582-921.5 kb). Remarkable similarities were observed in the range 291-873 kb among serogroups O:30 and O:5. By comparing banding patterns, the existence of many differences was noticed mainly among the pathogenic strains.

This may indicate the existence of different subgroups among *Y. enterocolitica* strains, even when they are serologically identified as belonging to the same serogroups. In conclusion, we think that CHEF technique is probably more suitable than other genotypic methods for molecular epidemiological studies of *Y. enterocolitica.*

REFERENCES

Blackwood, R.A., Rode, C.K., Pierson, C.L., and Bloch, C.A., 1997, Pulsed-field gel electrophoresis genomic fingerprinting of hospital *Escherichia coli* bacteraemia isolates. *J Med Microbiol* **46**: 506-510.

Buchrieser, C., Buchrieser, O., Kristl, A., Kaspar, C.W., 1994, Clamped homogeneous electric field gel-electrophoresis of DNA restriction fragments for comparing genomic variations among strains of *Yersinia enterocolitica* and *Yersinia* spp. *Zbl Bakt* **281:** 457-470

Curran, R., Hardie, K.R. and Towner, K.J., 1994, Analysis by pulsed-field gel electrophoresis of insertion mutations in the transferring-binding system of *Haemophilus influenzae* type b. *J Med Microbiol* **41**: 120-126.

Wei, M.Q., Fu Wang, and Grubb, W.B., 1992, Use of contour-clamped homogeneous electric field electrophoresis (CHEF) to type methicillin-resistant *Staphylococcus aureus. J Microbiol* **36**: 172-176

Weller, T.M.A., MacKenzie, F.M., and Forbes, K.J., 1997, Molecular epidemiology of a large outbreak of multiresistant *Klebsiella pneumoniae. J Med Microbiol* **46**: 921-926

Olsen, J.E., Skov, M.N., Threfall, E.J., and Brown, D.J., 1994, Clonal lines of *Salmonella enterica* serotype Enteritidis documented by IS200-, ribo-, pulsed-field gel electrophoresis and RFLP typing. *J Med Microbiol* **40**: 15-22

Chapter 76

F1-Negative Natural *Y. pestis* Strains

Tatyana V. MEKA-MECHENKO
M. Aikimbayev's Kazakh Scientific Center for Quarantine and Zoonotic Diseases, 14, Kapalskaya str., Almaty, Kazakhstan

1. BACKGROUND

An ability to synthesize the F1 capsular antigen is a steady genetic feature of *Yersinia pestis,* the plague pathogen. As a rule, fully virulent plague strains synthesize F1 intensively and F1 confers to the plague microbe resistance to phagocytosis by polymorphonuclear leukocytes. This feature is typical for the majority of strains isolated on the territory of the Kazakhstan natural plague foci. Nevertheless sometimes F1-negative *Y. pestis* strains are isolated.

2. METHODS

In this work we report our data on F1-negative *Y. pestis* strains. We investigated 48 *Y. pestis* strains from 4 natural plague foci of Kazakhstan including 34 strains from the Central Asian desert natural foci, 4 strains from the Volgo-Ural sandy natural foci, 5 strains from Tien-Shanian and 5 strains from the Near-Caspian steppe natural foci. The plague strains were isolated from 13 various sources (different kinds of mammals and ectoparazites). The isolation of the *Y. pestis* strains took place during several years and were stored in a strain collection of alive cultures at KSCQZD. Strains were investigated by standard bacteriological and molecular methods in the laboratories of KSCQZD and CDC (Fort Collins, Colorado, USA).

'he Genus Yersinia, Edited by Skurnik *et al.*
wer Academic/Plenum Publishers, New York, 2003

Table 1. Details of the F1-negative *Y. pestis* strains

Isolate #	Geographic location	Source of the strain
1	Near-Balkhash autonomous focus of Central Asian desert natural plague focus	Blood of *Meriones meridianus*
2	Near-Balkhash autonomous focus of Central Asian desert natural plague focus	Nonidentified fleas combed from *Rhombomys opimus*
3	Near-Balkhash autonomous focus of Central Asian desert natural plague focus	Blood of *Rhombomys opimus*
4	Near-Balkhash autonomous focus of Central Asian desert natural plague focus	Liver of *Rhombomys opimus*
5	Sarydzas autonomous focus of Tien - Shanian natural plague focus	*Ixodes* ticks combed from *Marmota baibacina*

3. RESULTS

No F1-antigen could be detected in Western Blot analysis from 5 of the 48 investigated strains. Four of the strains were isolated from the Near-Balkhash autonomous focus of the Central Asian desert natural plague focus, and one strain from the Sarydzas autonomous focus of the Tien-Shanian natural plague focus (Table 1). In immunofluorescence analysis 4 of the 5 F1-negative strains were totally negative while one strain (#4) reacted weakly (Table 2). Only one strain (#1) of the F1-negative strains (isolated from the Near-Balkhash autonomous focus) contained the three *Y. pestis* plasmids (110, 70 and 9.5 kb) and it was also positive in a Multiplex PCR for the genes encoding for Pst, Rep and F1. Two other strains from the same focus were missing the 110 kb plasmid and were negative for the F1 encoding gene in PCR. One strain from Near-Balkhash autonomous focus contained only the 70 kb plasmid and it was also negative for the F1 encoding gene in PCR (Table 2).

Table 2. Molecular characteristics *Y. pestis* isolates

## strains	FA	Protein profile	F1 in Western blot	Plasmids	PCR
Control strains:					
Y. pestis A1122	POS	Yes	Yes	110, 19,	Pst, F1
Y. pseudotuberculosis	NEG	Yes	No	none	None
Y. pestis EV76	POS weak	Yes	Yes	110, 70,	Rep, F1
Kazakh isolates:					
Y.pestis 1	NEG	Yes	No	110, 70, 9,5	Pst, Rep, F1
Y.pestis 2	NEG	Yes	No	70, 9,5	Pst, Rep
Y.pestis 3	NEG	Yes	No	70, 95	Pst, Rep
Y.pestis 4	POS weak	Yes	No	70	Pst, Rep
Y.pestis 5	NEG	Yes	No	70	Pst, Rep

Most of the *Y. pestis* strains from Central Asian desert natural focus belong to the biotype Medievalis since they ferment glycerol and do not have ability of nitrate reduction. The F1 negative *Y. pestis* strain #4 from Near-Balkhash autonomous focus of Central Asian desert natural plague focus belongs to the biotype Antiqua. The F1-negative strain #5 from Sarydzhas autonomous focus of Tien-Shanian natural plague focus had only one plasmid (70 kb) and was also negative in the F1 gene PCR. This strain was positive in the nitrate reduction and glycerol fermentation tests placing also it to biotype Antiqua.

4. CONCLUSION

Our results showed that F1-negative *Y. pestis* strains with changed plasmid profile and genetic profile are isolated alongside with F1-positive strains from the natural foci of plague of Kazakhstan. These findings require further studies that will be carried out in collaboration with the CDC laboratory in Fort Collins, Colorado, USA, within the ISTC project framework.

ACKNOWLEDGEMENTS

We are very grateful to our foreign colleagues Dr. May C. Chu, Mr. Leon G. Carter and Mr. Brook M. Yockey, and to our colleagues at KSCQZD.

Chapter 77

Occurence of *Yersinia pseudotuberculosis* in Iceberg Lettuce and Environment

Taina NISKANEN[1], Maria FREDRIKSSON-AHOMAA[1, 2] and Hannu KORKEALA[1]

[1]Department of Food and Environmental Hygiene, Faculty of Veterinary Medicine, University of Helsinki, Finland; [2]Institute of Hygiene and Technology of Food of Animal Origin, Ludwig-Maximilians-University, Munich, Germany

1. INTRODUCTION

Six outbreaks of *Yersinia pseudotuberculosis* serotype O:3 have occurred between 1997 and 2001 in Finland. After the outbreak in 1998 a large population-based case-control study was done and a domestic iceberg lettuce was implicated as a vehicle of the outbreak (Nuorti *et al.*, 2002). The source of the infection was traced to three salad farms in Ahvenanmaa Island. An investigation was carried out to study the occurrence of *Y. pseudotuberculosis* in iceberg lettuce, soil, and water samples in these farms.

2. MATERIALS AND METHODS

Altogether 265 iceberg lettuce, soil, and water samples were investigated. From these 128 were iceberg lettuce samples (Table 1). A 10-g of soil and salad sample was transferred into 90 ml of phosphate-buffered saline (PBS) with 1.0% mannitol and 0.15% bile salts. Water samples were passed trough membrane 0,45 μm filters and placed into 90 ml of PBS. The samples were studied with culture methods including direct plating and 7, 14 and 21 days cold enrichment combined with KOH treatment. *Y. pseudotuberculosis* isolates were identified using API 20E, and further serotyped with slide agglutination using O:1 - O:6 antisera. The pathogenicity was determined by

The Genus Yersinia, Edited by Skurnik *et al.*
Kluwer Academic/Plenum Publishers, New York, 2003

PCR targeting the *virF* gene located on the virulence plasmid (Nakajima *et al.*, 1992).

3. RESULTS

A total of four samples were positive for *Y. pseudotuberculosis. Y. pseudotuberculosis* was isolated from one soil and water sample, and two iceberg lettuce samples. All positive samples were obtained from the same farm. One strain recovered from iceberg lettuce was serotype O:2 and *virF* positive when studied by PCR. However, the other *Y. pseudotuberculosis* strains were not agglutinated with O:1-O:6 antisera and were *virF* negative.

Table 1. Yersinia pseudotuberculosis isolated in environmental samples on Ahvenanmaa Island

Origin of samples	No. of samples	No. of positive samples	
		n	%
Soil	51	1	1.7
Sludge	21	0	0
Water	39	1	2.5
Waterpipe	22	0	0
Compost	4	0	0
Lettuce	128	2	1.6
Total	265	4	1.5

4. CONCLUSIONS

The *virF*-positive *Y. pseudotuberculosis* serotype O:2 strain was isolated from food in Finland for the first time. *Y. pseudotuberculosis* is considered to be a food-borne pathogen, but it has rarely been isolated from foods. Iceberg lettuce are grown on open fields and irrigated with water from lakes, rivers, and ponds on Ahvenanmaa Island. *Y. pseudotuberculosis* is frequently isolated from many domestic and wild animals (Fukushima and Gomyoda, 1991; Sanford *et al.,* 1995; Niskanen *et al.,* 2001). There is a large population of deers and hares on Ahvenanmaa Island. These animals can be asymptomatic carriers and reservoirs of *Y. pseudotuberculosis* (Griffin *et al.,* 1992). Wildlife had access to lettuce fields and the ponds used to collect irrigation water. It is possible, that wild animals contaminate water, soil, and lettuce on the fields with their feces containing *Y. pseudotuberculosis* (Inoue *et al.,* 1991; Fukushima *et al.,* 1995). Contamination of iceberg lettuce was likely resulted from use of irrigation water contaminated with animal feces. Also nontypable *Y. pseudotuberculosis* strains without the virulence plasmid were recovered from environmental samples and a lettuce sample. Non-pathogenic *Y. pseudotuberculosis* serotypes O:7-O:14 have been previously recovered in environment (Nagano *et al.,* 1996). Lettuce produced on open fields irrigated with water from natural water systems may contain pathogenic microbes, and should be concern as a possible source of *Y. pseudotuberculosis* infections.

ACKNOWLEDGEMENTS

We thank the following persons for their assistance in the investigation: Mikael Gruner, Mikael Karring and Jaana Nieminen.

REFERENCES

Fukushima, H., and Gomyoda, M., 1991, Intestinal carriage of *Yersinia pseudotuberculosis* by wild birds and mammals in Japan. Appl. Environ. Microbiol. **57**: 1152-1155.

Fukushima, H., Gomyoda, M., Tsubokura, M., and Aleksic, S., 1995, Isolation of *Yersinia pseudotuberculosis* from river waters in Japan and Germany using direct KOH and HeLa cell treatment. Zbl. Bakt. **282**: 40-49.

Griffin, J. F. T., Thomson, A. J., Cross, J. P., Buchan, G. S., Mackintosh, C. G., and Brown, R. D., 1992, The impact of domestication on red deer immunity and disease resistance. The biology of deer. **14**: 120-125.

Inoue, M., Nakashima, H., Mori, T., Sakazaki, R., Tamura, K., and Tsubokura, M., 1991, *Yersinia pseudotuberculosis* infection in the mountain area.In: Une, T., Maruyama, T., Tsubokura, M. (editors). Currrent Investigations of the Microbiology of Yersiniae, Karger, Basel. pp. 307-310.

Nagano, T., Kiyohara, K., Tsubokura, M., and Otsuki, K., 1997, Identification of pathogenic strains within serogroups of *Yersinia pseudotuberculosis* and the presence of non-pathogenic strains isolated from animals and the environment. J. Vet. Med. Sci. **59**: 153-158.

Nakajima, H., Inoue, M., Mori, T., Itoh, K.-I., Arakawa, E., and Watanabe, H., 1992, Detectionand identification of *Yersinia pseudotuberculosis* and pathogenic *Yersinia enterocolitica* by an improved polymerase chain reaction method. J. Clin. Microbiol. **30**: 2484-2486.

Niskanen, T., Fredriksson-Ahomaa, M., and Korkeala, H., 2002, *Yersinia pseudotuberculosis* with limited genetic diversity is a common finding in tonsils of fattening pigs. J. Food Prot. 3: 540-545.

Nuorti, J. P., Niskanen, T., Hallanvuo, S., Mikkola, J., Kela, E., Lyytikäinen, O., Hatakka, M., Korkeala, H., Siitonen, A., and Ruutu, P., 2002, A nationwide outbreak of *Yersinia pseudotuberculosis* O:3 infections from iceberg lettuce in Finland. Submitted.

Sanford, S. E., 1995, Outbreaks of yersiniosis caused by Yersinia pseudotuberculosis in farmed cervids. J Vet Diagn Invest. **7**: 78-81.

Picture 24. Lidiya Sayapina, Valentina Fedorova, Igor Smirnov and Sergei Balakhonov at the City Reception.

Chapter 78

A Rapid Method for the Detection of Enteropathogenic *Yersinia* in Routine Diagnostics of Yersiniosis and Pseudotuberculosis

Igor V. SMIRNOV, Vitali V. TSARKOV and Anton V. YAKUSHEV
Pavlov's State Medical University, Ryazan, RUSSIA

1. INTRODUCTION

Epidemiological surveillance of the presence of enteropathogenic *Yersinia* in food stuffs, in environment, in animals and in humans needs an easy and cheap procedure for routine work. CIN agar is widely used for this purpose, but colonies of pathogenic and nonpathogenic *Yersinia enterocolitica* need further differentiation by a pure culture serotype/biotype determination. Moreover growth of *Y. pseudotuberculosis* strains may be inhibited on CIN agar. In this work we evaluated the usefulness of selective culture combined with determination of pYV-associated antigens as a routine diagnostic approach for the identification of enteropathogenic *Yersinia*.

2. METHODS

Forty six strains were investigated: 21 were *Y. enterocolitica* O3 (19 pYV+ and 2 pYV-); 3 were *Y. enterocolitica* O9 (2 pYV+ and 1 pYV-); 3 were *Y. pseudotuberculosis* O1 (2 pYV+ and 1 pYV-); and also 19 strains of other genera: *Enterobacter, Escherichia, Klebsiella, Proteus, Providencia, Salmonella, Serratia, Enterococcus, Staphylococcus* and *Pseudomonas*. All the nonreference strains were of human origin. Tryptone Soya Agar (M290, HiMedia Laboratories Ltd., India) was a general purpose medium and 4

The Genus Yersinia, Edited by Skurnik *et al.*
Kluwer Academic/Plenum Publishers, New York, 2003

agars were used as selective media for enteropathogenic *Yersinia*: CIN 1, CIN 2, I-agar and SBTS. Yersinia Selective Agar Base with addition of cefsulodine-irgasan-novobiocine supplement (M843 and FD034, respectively, HiMedia) was marked as CIN 1. It is a common medium prepared according to Schiemann (1979) formulation. Self-prepared agar base with the supplement FD034 was marked as CIN 2. I-agar is the same medium but it contains irgasan (0,004 g/l) instead of Yersinia Selective Supplement (FD034). Finally SBTS is an agar medium which contains bovine bile as the main selective ingredient. Last medium is widely used in Russian laboratories for routine diagnostics of yersiniosis. Buffered peptone water was used as a cold enrichment medium for *Yersinia* in swabs from swine tonsils.

Immune absorbed rabbit serum with agglutinins to pYV-associated antigens (SVY) was used for determination of virulent (pYV+) *Yersinia* cultures by slide agglutination. The procedure for the preparation and detailed characteristics of this serum were published by us earlier (Smirnov and Gorochov, 1990).

Young cultures of reference strains grown at optimal temperature were diluted in saline and 300 CFU were spread on agar surface of the medium. All media were incubated at 32°C aerobically for 48 h. The mean diameter of colonies, quantity and morphology of colonies, growth time and other characteristics were monitored. At least 20 colonies from duplicate 90 mm Petri dishes were included in a morphometric trial. Appearance of typical *Yersinia* colonies ("red bulls-eye") was recorded by hourly inspection. The obtained colonies were reidentified using Gram stain, common biochemical tests and SVY-agglutination.

The different media were used to test 43 tonsil samples from healthy pigs. Swabs were enriched in buffered peptone water at +4°C during 7 days. Inoculation of plate media was on 3rd, 5th and 7th day and the plates were incubated at 32°C for 48 h. We also tested in a similar way 12 mixtures containing *Y. enterocolitica* O3 pYV+ (100 CFU/ml) artificially mixed with *E. coli, E. cloacae, E. faecalis, P. aeruginosa, P. vulgaris, S. aureus, S. marcescens* and *Salmonella enterica* serotype Typhimurium (1000 CFU/ml).

3. RESULTS AND DISCUSSION

All the 22 pYV+ *Yersinia* strains grew in typical colonies on modified CIN agar (I-agar) and the slide agglutination of the colonies with SVY was positive within 24 h (Figure 1).

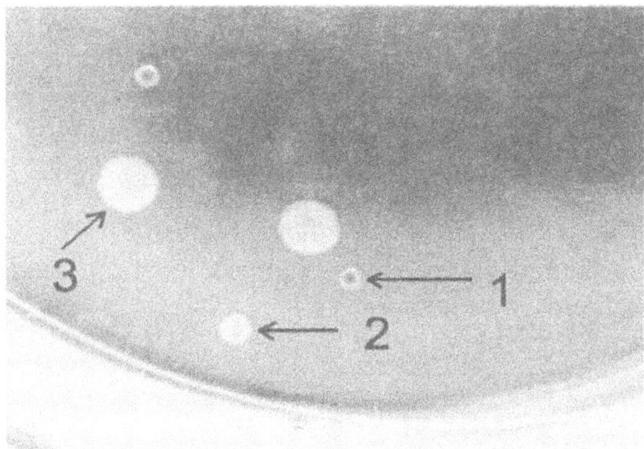

Figure 1. Colonies of *Y. enterocolitica* O3 pYV+ (1), *P. aeruginosa* (2) and *S. marcescens* (3) grown on I-agar at 32°C

pYV- *Yersinia* grew as typical colonies, but they were SVY negative during 48 h. *P. aeruginosa* and *S. marcescens* strains also grew on this agar, but their colonies were not similar to *Yersinia* colonies and the result of agglutination with SVY was also negative. Growth of other bacteria was mainly inhibited in these conditions (Figs. 1 and 2). I-agar functioned as a good selective medium in identification of enteropathogenic *Yersinia* (Figure 2). The mean diameters of the colonies were 0,6-0,8 mm and 1,1-1,2 mm after 24 h and 48 h of incubation, respectively. Growth time for *Yersinia* colonies was 22,7-24,5 h. The cultures of *Yersinia* grown on I-agar had typical morphology in Gram-staining. They did not demonstrate any unusual biochemical or antigenic features including results of the SVY-agglutination.

No enteropathogenic *Yersinia* were detected in the 43 tonsil samples of healthy swine during the above-mentioned period of enrichment. At the same time pYV+ *Y. enterocolitica* O3 bacteria were detected in all of the 12 artificially contaminated samples. The recovery depended in the medium used.

In our experiments the exclusion of cefsulodin and novobiocin from the CIN agar resulted in useful changes in selectivity of this medium. Cultures of *Y. pseudotuberculosis* O1 grew on I-agar as typical colonies without any inhibition. Colonies of some other bacteria that could grow on I-agar could be easily differentiated from enteropathogenic *Yersinia* by their colony morphology and negative result of the SVY-agglutination test.

In this comparative study of the 4 selective media the results suggested that the usage of I-agar (modified CIN agar) and SVY-agglutination allows determination of pYV+ *Yersinia*. Further testing of I-agar combined with the SVY-agglutination test using clinical and environmental specimens will be needed to assess its usefulness in detection of enteropathogenic *Yersinia*.

This approach shows promise as a rapid procedure for the routine diagnostics of yersiniosis and pseudotuberculosis.

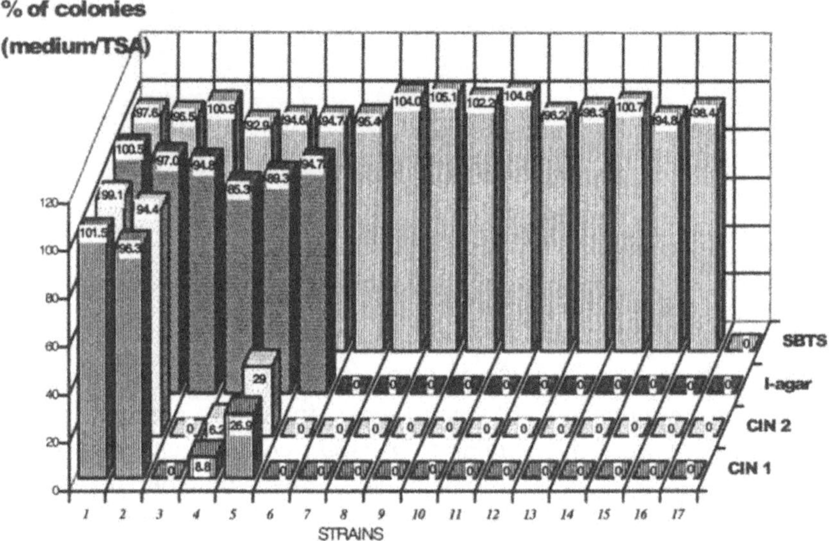

Figure 2. Ability of different strains to grow on the four *Yersinia* selective media compared to that on TSA. Strains: 1, *Y. enterocolitica* O3; 2, *Y. enterocolitica* O9; 3, *Y. pseudotuberculosis* O1; 4, *S. marcescens*; 5, *S. liquefaciens*; 6, *P. aeruginosa*; 7, *S.* serotype Typhimurium; 8, *P. mirabilis*; 9, *P. vulgaris*; 10, *E. cloacae*; 11, *E. faecalis*; 12, *E. coli*; 13, *K. pneumoniae*; 14, *K. oxytoca*; 15, *P. stuartii*; 16, *P. rettgeri*; 17, *S. aureus*.

ACKNOWLEDGEMENTS

We thank prof. G. Y. Tseneva and prof. L. S. Strachounsky who kindly provided us clinical and reference strains for this work and also Dr. G. M. Warke and Dr. Sanjay Singh for their excellent technical assistance.

REFERENCES

Schiemann, D.A., 1979, Synthesis of a selective agar medium for *Yersinia enterocolitica*. Can. J. Microbiol. 25: 1298-1304.

Smirnov, I.V., and Gorokhov, V. I., 1990, A diagnostic serum with antibodies to virulent Yersinia [Diagnosticheskaya syvorotka s antitelamy k virulentnym yersiniyam]. Laboratornoe delo 10: 66-68.

Chapter 79

Molecular Epidemiological Characterization of *Yersinia pseudotuberculosis* Circulating in Different Geographic Areas of the Russian Federation

Ekaterina A. VOSKRESSENSKAYA[1], Valery T. KLIMOV[2], GalinaY. TSENEVA[1], Elisabeth CARNIEL[3], Jeannine FOULON[3] and Margarita V. CHESNOKOVA[2]
[1]St. Petersburg Pasteur Institute; [2]Plague Institute of Siberia and Far East, Irkutsk; [3]Pasteur Institute, Paris

1. INTRODUCTION

Infections due to *Y. pseudotuberculosis* are rather common in Russia and they are manifested as outbreaks or sporadic cases. Contaminated vegetables are the major source for spreading *Y. pseudotuberculosis* and different kinds of rodents serve as the reservoir.

Virulence of *Y. pseudotuberculosis* is dependent on chromosomal and plasmid encoded genes. *Y. pseudotuberculosis* infections are manifested in great variability of clinical symptoms. Modern molecular epidemiological methods are needed to elucidate the significance of *Y. pseudotuberculosis* as an etiological agent and for the monitoring of the infections.

The aim of our work was to characterize genotypic variants of *Y. pseudotuberculosis* serotype O:1 circulating in different geographic zones of Russia.

2. TASKS

- molecular typing (ribotyping) of different isolates.
- determination of geographic distribution of the different ribotypes.

The Genus Yersinia, Edited by Skurnik *et al.*
·wer Academic/Plenum Publishers, New York, 2003

- comparative analysis of strain polymorphism between the Russian strains and strains from other geographic locations.

3. RESULTS

A total of 167 isolates from three regions of Russia (North–West, Siberia, Far East), isolated between 1981 and 2000, including strains from 14 outbreaks and from sporadic cases of human, rodent, environment and food origin were studied (Figure 1).

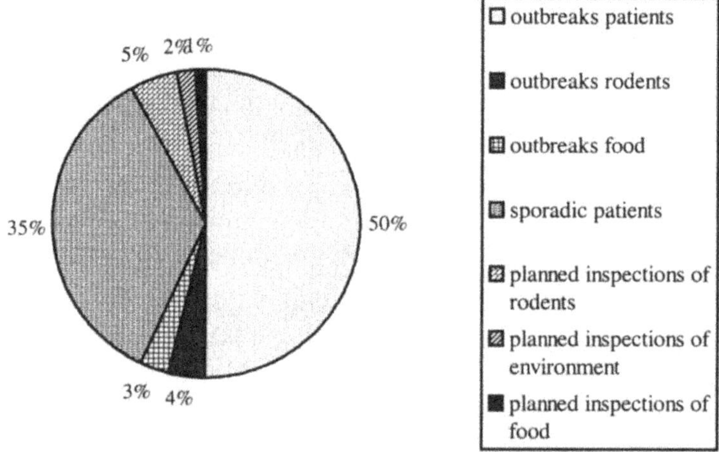

Figure 1. Sources of *Y. pseudotuberculosis* strains isolated in the different regions of Russia.

For comparison, 28 strains of *Y. pseudotuberculosis* serotype 1 from other countries (mostly from Western Europe, and also in Canada, Tunisia, Australia and New Zealand) and different sources were studied (Table 1).

Our results demonstrated that 7 different ribotypes are circulating in Russia (Figure 2). Most strains belonged to the ribotype 1' that prevailed in all three Russian regions studied. In addition, region-specific ribotypes were identified: ribotype 2' in North-West, ribotypes 5' and 7', in Far East, and ribotype 6', in Siberia.

Ribotype 1' strains were identified most frequently, and during the entire surveillance period they were encountered every year. They originated from different sources: human outbreak and sporadic cases, rodents, food, and environment. At the same time, ribotype 2' and 7' strains were isolated only from rodents and environment; no link was found between these ribotypes and disease development in humans.

Three ribotypes (2', 3', 4') were identified both among Russian strains and those from other countries (Table 1). On the other hand, ribotype 1' strains were found only among Russian strains and can therefore be considered as specific for Russia.

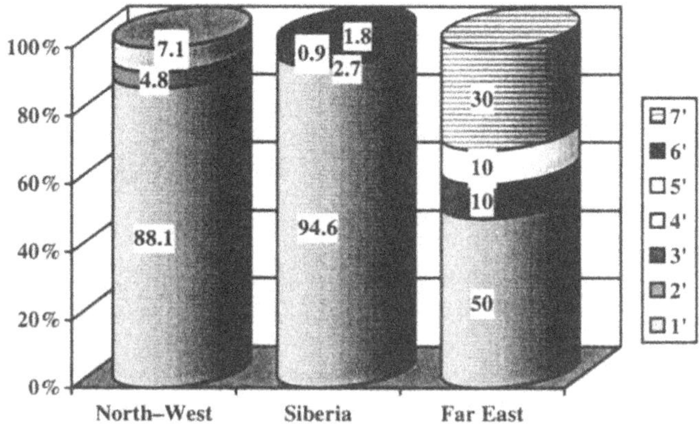

Figure 2. Distribution of ribotypes of *Y. pseudotuberculosis* O:1 in different geographic areas of Russian Federation

Table 1. Geographical distribution and frequency of different ribotypes of *Y. pseudotuberculosis* O:1 strains from the Pasteur Institute strain collection, Paris.

Ribotype	2'	3'	4'	8'	9'	10'	23'
Place of isolation	France	France, Tunisia Czech Republic	Italy, Canada, Tunisia, Yugoslavia, Romania, Germany, France, Great Britain, Switzerland	France, New Zealand, Australia, Netherlands	Switzerland	Yugoslavia	France
Frequency, %	3.6	11.0	43.0	28.0	3.8	3.6	7.0

4. CONCLUSIONS

Ribotyping was used to genotype for the first time *Y. pseudotuberculosis* serotype O:1 strains isolated from different geographic areas of Russia. The results demonstrated that majority of the strains belong to (Russian) ribotype 1'.

PART VI

DISEASES, VACCINES AND DIAGNOSTICS

Picture 25. Lela Bakanidze, David Tsereteli and Marina Darsavelidze.

Chapter 80

Second and Third Generation Plague Vaccines

Richard W. TITBALL and E. Diane WILLIAMSON
Defence Science and Technology Laboratory, Porton Down, Salisbury, Wiltshire SP4 0JQ, UK

1. INTRODUCTION

For many years it has been recognised that there is a risk of contracting plague as a consequence of handling the bacterium in the laboratory. In addition, and because plague is endemic in some parts of the world, it is recognised that individuals in these areas may be at risk of contracting the disease, especially during periods of plague activity (Leary and Titball, 1998). More recently, the threat to human health posed by *Yersinia pestis* used as a weapon by terrorists has also been recognised. Against this background there has been a long-standing need for vaccines and anti-infective pre-treatments and therapies against plague. Vaccines have been used in at risk populations, such as workers in laboratories handling *Y. pestis* and the cases of human plague which do occur nowadays are usually treated with antibiotics. However, these approaches to plague prophylaxis and therapy have several significant disadvantages. It is now generally accepted that the existing killed whole cells vaccines offer minimal protection against pneumonic plague (Russell *et al.*, 1995; Titball *et al.*,1999; Titball and Williamson, 2001). In addition, these vaccines required a course of vaccinations over a period of 6 months and carry a significant risk of transient and some times severe side effects (Titball *et al.*, 1999). Although most strains of *Y. pestis* are susceptible to a range of commonly available antibiotics, multiply antibiotic resistant strains of the bacterium, which caused disease in humans, have recently been reported (Galimand *et al.*, 1997). Clearly there is a need for both improved vaccines and improved antibiotics for the prevention and treatment of plague.

'he *Genus Yersinia*, Edited by Skurnik *et al.*
wer Academic/Plenum Publishers, New York, 2003

Previous workers have considered two approaches to the generation of an improved vaccine against plague, either the construction of a live attenuated mutant or the identification of components of the bacterium (sub-units) which can be formulated to provide an effective vaccine. Of these approaches, the most feasible appears to be the development of a sub-unit vaccine. This paper reviews progress in developing and licensing a sub-unit vaccine against plague.

2. IDENTIFICATION OF PROTECTIVE SUB-UNITS

A number of components of the cell wall of *Y. pestis* have been identified as candidate sub-units. Some of these components have been isolated from the bacterium whilst others have been produced in *Escherichia coli* using recombinant DNA technology. Of the components evaluated in mice, only immunisation with the V-antigen (a component of the type III system) or the F1-antigen (which forms a polypeptide capsule on the cell surface) has been shown to induce a good protective response which protects against challenge with *Y. pestis* (Table 1). The F1- and V-antigens have therefore been identified as suitable components for the development of a sub-unit vaccine against plague.

Table 1. Sub-units evaluated as vaccines for plague

Sub-unit	Immunogenic	Protective efficacy (bubonic/peumonic model)	Reference
Pla	Y	Not tested	Easterbrook *et al.*, 1995
pH6 antigen	Y	bubonic – not protective	Payne *et al.*, unpubllished
LPS	Y	bubonic – not protective	Prior *et al.*, 2001
F1 antigen	Y	bubonic & pneumonic – protective	Williamson *et al.*, 1995; Simpson *et al.*, 1990
YopD	Y	bubonic – partially protective	Andrews *et al.*, 1999
YopH	Y	bubonic – not protective	Andrews *et al.*, 1999
YopE	Y	bubonic – not protective	Andrews *et al.*, 1999; Leary *et al.*, 1999
YopN	Y	bubonic – not protective	Andrews *et al.*, 1999; Leary *et al.*, 1999
YopK	Y	bubonic – not protective	Andrews *et al.*, 1999; Leary *et al.*, 1999
YopM	Y	bubonic – not protective	Andrews *et al.*, 1999; Nemeth and Straley 1997
Ypk A	Y	bubonic – delayed time to death	Andrews *et al.*, 1999
V antigen	Y	bubonic & pneumonic – protective	Williamson *et al.*, 1995; Leary *et al.*, 1995; Hill *et al.*, 1997; Anderson *et al.*, 1996

3. EXPRESSION SYSTEMS

Whilst F1-antigen can be isolated in relatively high yield from cultures of *Y. pestis*, this procedure carries some risk to human health, even if partially attenuated strains of *Y. pestis* are used. However, the isolation of V-antigen from *Y. pestis* has proven to be notoriously difficult, both because the yield is low and because the protein appears to be quite susceptible to proteolysis. For these reasons, it is preferable to express both these antigens from recombinant *E. coli* which also allows vaccine production to be carried out in a conventional pharmaceutical manufacturing plant. High-level expression of the genes encoding F1- and V-antigens (*caf1* and *lcrV* respectively) in *E. coli* has proven to be relatively straightforward, possibly reflecting the close phylogenetic relationship between *Y. pestis* and *E. coli* and the similar codon usage in these species. The biosynthesis of F1-antigen in *Y. pestis* is encoded by the *caf* operon, which directs not only the expression of the F1-subunit but also its export and assembly on the cell surface. The transfer of the entire *caf* operon of *Y. pestis* into *Salmonella typhimurium* resulted in the biosynthesis and export of the recombinant F1-antigen (rF1) onto the surface of the bacterium (Titball *et al.,* 1997). Similar results have been obtained on expression of the *caf* operon in *E. coli* (Miller *et al.,* 1998). Harvesting of the exported rF1-antigen, which is only loosely attached to the cell surface, has proven to be relatively straightforward (Miller *et al.,*1998). The purified rF1-antigen forms a high molecular weight polymer (Miller *et al.,*1998; Tito *et al.,* 2001) and there is some evidence that the rF1 monomers might assemble into a fibrillar type structure on the cell surface (Tito *et al.,* 2001). The polymeric form of the rF1-antigen is thought to induce an optimal protective immune response after immunisation (Miller *et al.,* 1998).

The expression of the V-antigen at high levels in *E. coli* has proven to be quite straightforward if the protein is fused with a carrier such as glutathione-s-transferase (GST). Initial studies with this expression system resulted in the purification of a GST-V-antigen fusion protein which was then treated with the site-specific protease, factor X_a, to release the recombinant V-antigen (rV) moiety (Leary *et al.,* 1995). However, factor X_a is of animal origin (the protease is derived from snake venom) and V-antigen produced in this way is more suited to research use. For the production of V-antigen suitable for inclusion in a human vaccine, a modified GST expression system (based on vector pGEX-6P-2) was used in which the rV-antigen moiety was cleaved from the GST carrier using PreScission protease™ (Carr *et al.,* 2000). The yield of purified rV-antigen produced using this procedure at laboratory scale was reported to be 31 mg/l (Carr *et al.,* 2000).

4. PROTECTION AGAINST PARENTERAL AND INHALATION CHALLENGE

The immunisation of mice with rF1 and rV-antigens, given with alhydrogel adjuvant, is able to induce a protective immune response against a subcutaneous (s.c.) challenge (i.e. mimicking bubonic plague) or an inhalation (inh) challenge (i.e. mimicking pneumonic plague) with *Y. pestis* (Table 2). At all of the challenge doses used, complete protection of immunised animals was demonstrated. The immunisation schedule used in most of these experiments involved two doses of 10μg each of rF1- and rV-antigens. Protection against a s.c. challenge after mice received 2 doses of 5μg of rF1- and rV-antigens has also been demonstrated (Williamson *et al.*, 1999) and represents the minimum protective dose in the mouse, since, reducing the dose of rF1- and rV-antigens to 1μg of each sub-unit resulted in some loss of protection at the highest challenge dose tested (10^7 colony forming units; cfu). A single dose of 10μg of each of these antigens was also able to induce a high level of protection against an airborne challenge with 10^6 cfu of *Y. pestis* (approximately 10^4 LD_{50} doses; Williamson *et al.*, 2001).

Table 2. Protection afforded by recombinant F1- and V-antigens against challenge with *Y. pestis* strain GB

Treatment	Challenge dose (cfu)	Survivors / total after challenge		
		Mouse (s.c.) challenge [a]	Mouse (inh) challenge [a]	Guinea pig (s.c.) challenge [b]
rF1 + rV	$4x10^4$	n.d.	6/6	n.d.
rF1 + rV	10^5	5/5	n.d.	6/6
rF1 + rV	10^6	n.d.	6/6	5/6
rF1 + rV	10^7	6/6	n.d.	3/6
none	10^4	0/6	0/6	0/6

[a] immunised with 10 μg each of recombinant F1- and V-antigens; [b] immunised with 50 μg each of recombinant F1- and V-antigens

Equally importantly, an analysis of spleen, liver and lung samples 12-14 days post s.c. or inhalation challenge failed to reveal the presence of *Y. pestis* (Williamson *et al.*, 1995; Williamson *et al.*, 1997; Williamson *et al.*, 2001). These findings indicate that the response induced by immunisation with these antigens is able to clear the *Y. pestis* challenge fully.

Most of the protection studies carried out to date have involved challenge with a typical *Y. pestis* biovar *orientalis* strain (strain GB). However, there is good reason to believe that this vaccine will also provide protection against other strains of *Y. pestis*. Immunisation of mice with rV-antigen has been shown to provide protection against a *Y. pestis* strain which was isolated from a pneumonic case of plague (strain CO92; Anderson *et al.*, 1996). Mice immunised with rV-antigen were also protected against challenge with a naturally occurring but virulent F1-capsule-negative strain of *Y. pestis* (strain Java9; Anderson *et al.*, 1996).

5. PROTECTION IN DIFFERENT ANIMAL MODELS OF DISEASE

A range of animal models have been described which can be used to assess the efficacy of plague vaccines. The murine model of disease is generally accepted to provide a meaningful indication of the efficacy of plague vaccines. However, mice are susceptible to the effects of the murine exotoxin (Montie and Montie, 1971). Guinea pigs have been proposed as an alternative model for assessing the efficacy of plague vaccines (Von Metz *et al.*, 1971). Guinea pigs immunised with the rF1 and rV vaccine develop IgG1 and IgG2a antibody to both of the immunising antigens (Jones *et al.*, submitted). In addition, immunised guinea pigs were solidly protected against a subcutaneous challenge of 10^5 cfu of *Y. pestis*. At higher challenge doses some of the immunised animals were protected from death (Table 2) and those animals which died after challenge showed a significantly increased time to death (13 days±1.25 days) compared with control animals (6.8 days±0.15 days).

6. VACCINE FORMULATION

Most of the immunisation studies to date have used rF1- and rV-antigens formulated in alhydrogel adjuvant. This adjuvant has previously been approved for use in vaccines intended for man and is therefore ideally suited to the development of a next generation plague vaccine. The quantities of rF1- and rV-antigens which are required for the development of an optimal protective immune response have been investigated (Williamson *et al.*, 1999) and on the basis of these studies the current formulation of the vaccine for use in humans contains equal quantities of both proteins. Although a single dose of this vaccine has been shown to induce good protective responses (Williamson *et al.*, 2001), the proposed immunisation schedule in humans will involve 2 doses of vaccine being given on days 1 and 21.

7. CORRELATES OF PROTECTION

The murine model of protection has previously been approved by the U.S. Public Health Service for the testing of plague vaccines. The efficacy of the killed whole cells vaccine was determined by measuring the ability of sera from immunized mice, guinea pigs, monkeys, or humans to protect mice against *Y. pestis*. Serum was injected intravenously into groups of 10 mice (0.5 mL serum per mouse), and the mice were then challenged subcutaneously with 100 LD_{50} of *Y. pestis* (Meyer and Foster 1948, Bartelloni *et al.*, 1973). The mouse protection index (MPI) was expressed as the percentage mortality of the group of mice (over 14 days) divided by the average time to death. MPI values of 10 or less were considered to be indicative of protection (Meyer, 1970).

Antibody plays a major role in the protection afforded by the rF1 and rV vaccine. Serum taken from mice immunised with this vaccine was able to protect naïve SCID/Beige mice against a s.c. or an inhalation challenge with *Y. pestis* (Green *et al.*, 1999). In addition, the passive transfer of antisera from immunised guinea pigs to naive mice provided protection against a s.c. challenge with 1000cfu of *Y. pestis* (Jones *et al.*, submitted). In mice, the titre of IgG1 subclass antibody to the F1 and V-antigens has been correlated with protection against disease. More recently, a competitive ELISA which exploits the recognition of V antigen by a neutralising monoclonal antibody which has previously been shown to protect mice against challenge with *Y. pestis* (Hill *et al.*, 1997), has allowed an estimation of titre of neutralising antibody and this may be more predictive of protection. Although antibody raised in the vaccinee against rF1 and rV does appear to play a key role in protection it is likely cellular immunity and the ability to mount a pro-inflammatory response to clear the bacteria are also essential to resolve the infection completely.

The precise mechanism by which antibody provides protection against plague is not fully clarified at this time. However, it is reported that V-antigen can be secreted from the bacterium, and that this extracellular protein is able to elicit immunomodulatory affects on the host by down-regulating the production of TNFα and interferon-γ (Nakajima *et al.*, 1995). Therefore, it is possible that antibody to V-antigen induced after immunisation is able to block this immunomodulatory activity (Nakajima *et al.*, 1995). However, V-antigen is also thought to play a key role in the translocation of Yops via the type III system (Pettersson *et al.*,1999) and antibody against V-antigen is reported to protect macrophages from type III-mediated cytotoxicity and to enhance phagocytosis of *Y. pestis* (Weeks *et al.*, 2002). The mechanism by which antibody to the F1-antigen provides protection against plague is less certain. However, in view of the proposed function of the F1-antigen capsule in preventing phagocytosis (Du *et al.*, 2002) it is possible that antibody to this protein is able to opsonise bacteria and therefore to promote phagocytosis and killing.

8. NON-INVASIVE DELIVERY SYSTEMS

Whilst the proposed second generation vaccines against plague would require the injected delivery of recombinant F1 and V-antigens, there are a number of reports of the use of non-invasive delivery systems for this vaccine. The oral immunisation of mice with a recombinant *aroA* mutant of *S. typhimurium*, which encoded the *caf* operon, resulted in the induction of high levels of protective immunity against plague (Titball *et al.*,1997). More recently the immunisation of mice with an *aroA* mutant of *S. typhimurium* expressing an F1-V fusion protein has been reported to induce a response against both antigens, and the induction of protective immunity (Leary *et al.*,1997) However, such recombinant live vaccines do not appear to have been further developed. An alternative approach to the non-invasive delivery of the vaccine involves the microencapsulation of the rF1 and rV proteins in poly L-lactide (PLLA) microspheres. This approach appears to offer good potential for the non-invasive delivery of the vaccine. A PLLA microsphere formulation suitable for nasal or possibly inhalational administration has been derived which is fully protective against an inhalational challenge with *Y. pestis* after only two immunising doses in the mouse model (Eyles *et al.*, 2000, 2001).

9. CONCLUSION

A defined sub-unit vaccine against plague now seems to be feasible. The further development of this vaccine will be dependent on clinical trials with this vaccine. In parallel, there is good reason to believe that next-generation vaccines, will be given orally or intranasally.

REFERENCES

Anderson, G.W.Jr., Leary, S.E.C., Williamson, E.D., Titball, R.W., Welkos, S.L., Worsham, P.L. and Friedlander, A.M., 1996, Recombinant V antigen protects mice against pneumonic and bubonic caused by F1-capsule-positive and -negative strains of *Yersinia pestis*. *Infect. Immun.* **64:** 4580-4585.

Andrews, G.P., Strachan, S.T., Benner, G.E., Sample, A.K., Anderson, G.W., Adamovicz, J.J., Welkos, S.L., Pullen, J.K., Friedlander, A.M., 1999, Protective efficacy of recombinant *Yersinia* outer proteins against bubonic plague caused by encapsulated and nonencapsulated *Yersinia pestis*. *Infect. Immun.* **67:** 1533-1537.

Bartelloni, P.J., Marshall, J.D., Cavanaugh, D.C., 1973, Clinical and serological responses to plague vaccine. *Mil. Med.* **138:** 720–722.

Carr, S., Miller, J., Leary, S.E.C., Bennett, A.M., Ho, A and Williamson, E.D., 1999, Expression of a recombinant form of the V antigen of *Yersinia pestis* using three different expression systems. *Vaccine* **18:** 153-159.

Du, Y., Rosqvist, R., Forsberg, A., 2002, Role of fraction 1 antigen of *Yersinia pestis* in inhibition of phagocytosis. *Infect. Immun.* **70:** 1453-1460.

Easterbrook, T.J., Reddin, K., Robinson, A., Modi, N., 1995, Studies on the immunogenicity of the Pla protein from *Yersinia pestis*. *Contrib. Microbiol. Immunol.* **13:** 214-215.

Eyles, J.E., Williamson, E.D., Spiers, I.D., Stagg, A.J., Jones, S.M. and Alpar, H.O., 2000, Generation of protective immune responses to plague by mucosal administration of microsphere coencapsulated recombinant sub-units. *J. Cont. Rel.* **63:** 191-200.

Eyles, J.E., Bramwell, V.W., Williamson, E.D. and Alpar, H.O., 2001, Microsphere translocation and immunopotentiation in systemic tissues following intranasal administration. *Vaccine* **19:** 4732-4742.

Galimand, M., Guiyoule, A., Gerbaud, G., Rasoamanana, B., Chanteau, S., Carniel, E., Courvalin, P., 1997, Multidrug resistance in *Yersinia pestis* mediated by a transferable plasmid. *N. Engl. J. Med.* **337:** 677-680.

Green, M., Rogers, D., Russell, P., Stagg, A.J., Bell, D.L., Eley, S.M., Titball, R.W., Williamson, E.D., 1999, The SCID/Beige mouse as a model to investigate protection against *Yersinia pestis*. *FEMS Immunol. Med. Microbiol.* **23:** 107-113.

Hill, J., Leary, S.E.C., Griffin, K.F., Williamson, E.D. and Titball, R.W., 1997, Regions of *Yersinia pestis* V antigen that contribute to protection against plague identified by passive and active immunisation. *Infect. Immun.* **65:** 4476-4482.

Jones, S.M., Griffin, K.F., Hodgson, I. and Williamson, E.D., 2002, Protective efficacy of a fully recombinant plague vaccine in the guinea pig. *Vaccine*, submitted.

Leary, S.E.C., Williamson, E.D., Griffin, K.F., Russell, P., Eley, S.M and Titball, R.W., 1995, Active immunisation with recombinant V antigen from *Yersinia pestis* protects mice against plague. *Infect. Immun.* **63:** 2854-2858.

Leary, S.E.C., Griffin, K.F., Garmory, H.S., Williamson, E.D. and Titball, R.W., 1997, Expression of an F1/V fusion protein in attenuated *Salmonella typhimurium* and protection of mice against plague. *Microb. Pathog.* **23:** 167-179.

Leary, S.E.C. and Titball, R.W., 1998, Plague. *Brit. Med. Bull.* **54:** 625-633.

Leary, S.E.C., Griffin, K.F., Galyov, E., Hewer, J., Williamson, E.D., Hölmström, A., Forsberg, A. and Titball, R.W., 1999, *Yersinia* outer proteins (YOPS) E,K and N are antigenic but non-protective compared to V antigen, in a murine model of bubonic plague. *Microb. Pathog.* **26:** 159-169.

Meyer, K.F., 1970, Effectiveness of live or killed plague vaccines in man. *Bull. World Health Organ.* **42:** 653–666.

Meyer, K.F., and Foster, L.E., 1948, Measurement of protective serum antibodies in human volunteers inoculated with plague prophylactics. *Stanford Med. Bull.* **6:** 75-79.

Miller, J., Williamson, E.D., Lakey, J.H., Pearce, M.J., Jones, S.M., and Titball, R.W., 1998, Macromolecular organisation of recombinant *Yersinia pestis* F1 antigen and the effect of structure on immunogenicity. *FEMS Immunol. Med. Microbiol.* **21:** 213-221.

Montie, T.C., and Montie, D.B., 1971, Protein toxins of *Pasturella pestis*. Subunit composition and acid binding. *Biochemistry* **10:** 2094–2100.

Nakajima, R., Motin, V.L., and Brubaker, R.R., 1995, Suppression of cytokines in mice by protein A-V antigen fusion peptide and restoration of synthesis by active immunization. *Infect. Immun.* **63:** 3021-3029.

Nemeth, J., and Straley, S.C., 1997, Effect of *Yersinia pestis* YopM on experimental plague. *Infect. Immun.* **65:** 924-930.

Payne, D.W., Oyston, P.C.F., and Williamson, E.D., Immunisation with pH6 antigen in the mouse is not protective (unpublished data).

Pettersson, J., Holmström, A., Hill, J., Leary, S., Frithz-Lindsten, E., von Euler-Matell, A., Carlsson, E., Titball, R.W., Forsberg, A. and Wolf-Watz, H., 1999, The V-antigen of *Yersinia* is surface exposed before target cell contact and involved in virulence protein translocation. *Mol. Microbiol.* **32:** 961-976.

Prior, J., Hitchin, P.G., Wiliamson, E.D., Reason, A.J., Morris, H.R., Dell, A., Wren, B. and Titball, R.W., 2001, Characterisation of the lipopolysaccharide of *Yersinia pestis*. *Microbial. Pathog.* **30:** 49-57.

Russell, P, Eley, S.M., Hibbs, S.E., Manchee, R.J., Stagg, A.J. and Titball, R.W., 1995, A comparison of plague vaccine, USP and EV76 vaccine induced protection against *Yersinia pestis* in a murine model. *Vaccine* **13:** 1551-1556.

Simpson, W.J., Thomas, R.E., and Schwan, T.G., 1990, Recombinant capsular antigen (fraction 1) from *Yersinia pestis* induces a protective antibody response in BALB/c mice. *Am. J. Trop. Med. Hyg.* **43:** 389-396.

Titball, R.W., Howells, A.M., Oyston, P.C.F., Williamson, E.D., 1997, Expression of the *Yersinia pestis* capsular antigen (F1 antigen) on the surface of an *aroA* mutant of *Salmonella typhimurium* induces high level protection against plague. *Infect. Immun.* **65:** 1926-1930.

Titball, R.W., Eley, S., Williamson, E.D., and Dennis, D.T., 1999, Plague. In *Vaccines* (S.A. Plotkin and W.A. Orenstein, eds.), W.B. Saunders Company, Philadelphia, pp.734-742.

Titball, R.W. and Williamson, E.D., 2001, Vaccination against bubonic and pneumonic plague. *Vaccine* **19:** 4175-4184.

Tito, M.A., Miller, J., Griffin, K.F., Williamson, E.D., Titball, R.W. and Robinson, C.V., 2001, Macromolecular organisation of the *Yersinia pestis* capsular F1-antigen: insights from time of flight mass spectrometry. *Prot. Sci.* **10:** 2408-2413.

Von Metz, E., Eisler, D.M., Hottle, G.A., 1971, Immunogenicity of plague vaccines in mice and guinea pigs. *Appl. Microbiol.* **22:** 84–88.

Weeks, S., Hill, J., Friedlander, A., Welkos, S., 2002, Anti-V antigen antibody protects macrophages from *Yersinia pestis* -induced cell death and promotes phagocytosis. *Microb. Pathog.* **32:** 227-237.

Williamson, E.D., Eley, S.M., Griffin, K.F., Green, M., Russell, P., Leary, S.E.C., Oyston, P.C.F., Easterbrook, T., Reddin, K.M., Robinson, A. and Titball, R.W., 1995, A new improved sub-unit vaccine for plague: the basis of protection. *FEMS Immunol. Med. Microbiol.* **12:** 223-230.

Williamson, E.D., Eley, S.M., Stagg, A.J., Green, M., Russell, P., Titball, R.W., 1997, A sub-unit vaccine elicits IgG in serum, spleen cell cultures and bronchial washings and protects immunised animals against pneumonic plague. *Vaccine* **15:** 1079-1084.

Williamson, E.D., Vesey, P.M., Gillhespy, K,J., Eley, S.M., Green, M. and Titball, R.W., 1999, An IgG_1 titre to the F1 and V antigens correlates with protection against plague in the mouse model. *Clin. Exp. Immunol.* **116:** 107-114.

Williamson, E.D., Eley, S.M. Stagg, A.J., Green, M., Russell, P and Titball, R.W., 2001, A single dose sub-unit vaccine protects against pneumonic plague. *Vaccine* **19:** 566-571.

Chapter 81

Yersinia Outer Protein E, YopE

A Versatile Type III Effector Molecule for Cytosolic Targeting of Heterologous Antigens by Attenuated Salmonella

Holger RÜSSMANN

Max von Pettenkofer-Institut für Hygiene und Medizinische Mikrobiologie, Ludwig Maximilians Universität München, Pettenkoferstrasse 9a, 80336 München, Germany

1. ABSTRACT

Many Gram-negative pathogens evade the host's immune response by utilizing a specialized protein secretion machinery, known as type III secretion system (TTSS). Virulence factors such as the *Yersinia* outer protein E (YopE) are delivered directly into the cytosol of target cells in a TTSS-dependent fashion. This unique translocation mechanism can be used by attenuated *Salmonella* carrier vaccines for the delivery of heterologous antigens fused to YopE into the MHC class I-restricted antigen processing pathway. In orally immunized mice, this novel vaccination strategy results in the induction of pronounced peptide-specific cytotoxic CD8 T cell responses.

2. TTSS IN *Salmonella* AND *Yersinia*

In the past ten years, highly conserved type III secretion systems (TTSS) encoded by pathogenicity islands have been identified in many Gram-negative bacterial plant and human pathogens (Hueck, 1998). Probably the most fascinating aspect of TTSS is the fact that virulence factors are not only secreted from the bacterial cytoplasm but also delivered directly to the inside of the eukaryotic host cell, therefore effectively working as a "molecular syringe" (Kubori *et al.*, 1998; Galán, 1998).

The Genus Yersinia, Edited by Skurnik *et al.*
wer Academic/Plenum Publishers, New York, 2003

The life cycles of enteropathogenic *Salmonella* and *Yersinia* species differ significantly from each other. *Yersinia* employs type III effector proteins to destroy key functions of immune cells (Cornelis *et al.*, 1998). After binding to the surface of macrophages, at least six effector proteins (*Yersinia* outer proteins, Yops) are translocated in a type III-dependent fashion into the cytosol of the eukaryotic host cell (Lee, 1997) mediating the ability to resist phagocytosis (Rosqvist *et al.*, 1988; Black *et al.*, 1997), to trigger apoptosis (Ruckdeschel *et al.*, 1997) and to suppress TNF-α and IFN-γ release (Autenrieth *et al.*, 1992). The translocation process leads to extracellular survival and proliferation of *Yersinia* in the infected host (Simonet *et al.*, 1990). In contrast to this survival strategy, *Salmonella* uses type III effector proteins encoded in the *Salmonella* Pathogenicity Island 1 (SPI1) to actively invade eukaryotic cells (Galán, 1996). Once inside the host cell, *Salmonella* species reside in membrane-bound vacuoles (macropinosomes) during their entire intracellular life cycle (Alpuche-Aranda *et al.*, 1994). From this endosomal compartment, *Salmonella* continues to translocate type III effector proteins into the cytoplasm of eukaryotic cells (Galán, 1998).

3. TTSS-MEDIATED TRANSLOCATION OF HETEROLOGOUS ANTIGENS BY *Salmonella*

Virtually all viruses and many bacterial pathogens invade mammalian host cells, thereby potentially escaping complement- and antibody-mediated defense mechanisms. In many cases these intracellular infectious agents induce strong antigen-specific CD8 T cell responses that might contribute to rapid clearance of the pathogens. CD8 T cells detect peptides bound to MHC class I molecules. Most peptides presented by MHC class I molecules derive from endogenous proteins degraded by proteasomes in the cytosol of the host cell. The resulting peptides are transported by TAP (transporter associated with antigen processing) molecules into the endoplasmic reticulum, where they are loaded onto newly synthesized MHC class I molecules. Thus, CD8 T cells have been implicated in defense against cytosolic microorganisms.

To elicit suitable CD8 T cell responses, a live vaccine vector must be capable of delivering pathogen-derived antigenic peptides in the vaccinated individual in a way that they are made available for binding to MHC class I molecules of appropriate antigen-presenting cells (APC) (Germain, 1994; Germain, 1995). Endosomal-bound *Salmonella* is not an ideal vaccine carrier to deliver heterologous proteins to the cytosol of host cells. In fact, secretion of foreign proteins into the macropinosome by *Salmonella* induces peptide-specific CD4 T cell priming (Wick *et al.*, 1994). We have developed a new vaccination strategy by using the TTSS to inject heterologous antigens fused to type III effector proteins directly into the cytosol of APC for efficient induction of MHC class I-restricted immune responses.

4. A *Yersinia* TYPE III CARRIER PROTEIN FOR ANTIGEN DELIVERY BY *Salmonella*

A versatile antigen delivery system should be capable of targeting large protein fragments derived from diverse pathogens. Thus, we were interested to identify a type III effector protein that could be used in *Salmonella* for this purpose.

Among different bacterial species, many components of type III secretion systems reveal functional conservation probably due to the fact that shared type III genes were recruited by horizontal transfer during evolution (Hueck, 1998). One of the best studied type III effector protein is the 25-kDa *Yersinia* outer protein E (YopE). During the interaction of *Yersinia* species with professional phagocytes, YopE translocation disturbs eukaryotic cytoskeleton dynamics and inhibits phagocytosis. YopE is a GTPase-activating protein that is active towards G proteins from the Rho family (Black *et al.*, 2000, Von Pawel-Rammingen *et al.*, 2000). Cytosolic delivery of YopE into host cells was first observed using immunofluorescence microscopy (Rosqvist *et al.*, 1991; Rosqvist *et al.*, 1994). Alternative techniques to demonstrate YopE translocation were based on reporter enzyme such as a calmodulin-dependent adenylate cyclase, a neomycin phosphotransferase, and the green fluorescent protein fused to various residues of YopE (Sory *et al.*, 1994; Sory *et al.*, 1995; Schesser *et al.*, 1996; Jacobi *et al.*, 1998; Lee *et al.*, 1998). These studies led to the characterization of N-terminal signals required for type III-dependent secretion and translocation of YopE. The minimal sequence shown to be sufficient for secretion of YopE was found to comprise 11-15 amino acids, whereas the minimal domain required for translocation of YopE across the eukaryotic cell membrane was reduced to 50 residues.

We have made use of YopE in *Yersinia enterocolitica* for the injection of a heterologous antigen into the cytosol of infected host cells and to deliver foreign antigenic peptides to the MHC class I-restricted antigen presentation pathway (Rüssmann *et al.*, 2000). The p60 protein of *Listeria monocytogenes*, an intracellular bacterium, was used as a model antigen to construct various hybrid YopE proteins (Bubert *et al.*, 1992). Two immunodominant p60 nonamer peptides ($p60_{217\text{-}225}$ and $p60_{449\text{-}457}$) are presented to CD8 T cells by MHC class I molecules and it has been shown that CD8 T cells specific for $p60_{217-225}$ can transfer immunity against *L. monocytogenes* to naive mice (Pamer, 1994; Harty *et al.,* 1995; Vijh *et al.,* Sijts *et al.,* 1996; 1997; Harty *et al.,* 1999). To assess the ability of YopE to secrete and translocate large foreign protein fragments, the C-terminal 354 amino acids of p60 were fused to various N-terminal parts of YopE (Rüssmann *et al.,* 2000). Immunoblot analysis of *Yersinia*-infected epithelial or macrophage-like tissue culture cells revealed that hybrid $YopE_{1-18}$-$p60_{130-484}$ lacking the YopE translocation domain was secreted into the culture supernatant but was not translocated into the host cell cytosol. In contrast, chimeric $YopE_{1-138}$-$p60_{130-484}$ containing both the secretion and translocation domain of YopE was translocated into the cytosol of infected cells in a type III-dependent fashion. In further experiments, the ability of *Y. enterocolitica* was studied to deliver the C-terminal amino acids 130-484 containing the antigenic nonamer peptides 217-225 and 449-457 of the p60 molecule to the MHC class I-restricted antigen presenting pathway (Rüssmann *et al.,* 2000). APC infected with *Yersinia* expressing and translocating $YopE_{1-138}$-$p60_{130-484}$ stimulated specifically both $p60_{217-225}$- and $p60_{449-457}$-specific T cells. CD8 T cell stimulation was strictly dependent on the cytosolic delivery of the C-terminal portion of p60 by the *Yersinia* type III apparatus, as APC infected with an attenuated *Y. enterocolitica* strain secreting but not translocating $YopE_{1-18}$-$p60_{130-484}$ did not stimulate p60-specific T cells.

In a previous study, it has been shown that full-length wild-type YopE can be secreted and translocated by *S. typhimurium* in a type III-dependent manner (Rosqvist *et al.,* 1995). This observation on one hand and our promising experiences with YopE as a carrier molecule in *Yersinia* on the other hand prompted our laboratory to investigate the possible employment of YopE for the delivery of heterologous antigens by attenuated *Salmonella.* The above mentioned defined secretion and translocation domains of YopE were fused to the immunodominant T cell antigens listeriolysin O (LLO) and p60 of *L. monocytogenes* (Pamer *et al.,* 1991). In vitro experiments showed that *S. typhimurium* allows secretion and translocation of large hybrid YopE proteins in a type III-dependent fashion. Translocation and cytosolic delivery of these chimeric proteins into host cells, but not secretion into endosomal macropinosomes, led to efficient MHC class I-restricted antigen presentation of listerial nonamer peptides (Rüssmann *et al.,* 2001). As determined by ELISPOT assay, mice orally vaccinated with a single dose of attenuated *S. typhimurium* expressing either translocated YopE/LLO or YopE/p60 proteins revealed high numbers of IFN-γ-producing cells reactive with LLO_{91-99} or $p60_{217-225}$, respectively (Rüssmann *et al.,* 2001). However, these recombinant *Salmonella* strains did not elicit the same protective ability as an immunizing sublethal dose of wild-type *L. monocytogenes*, which naturally displays a variety of listerial peptides from different antigens to

CD8 T cells of the vaccinated host. Therefore, polyvalent antigen delivery by bacterial live carrier vaccines is a desirable feature to elicit efficient protection against infectious agents. In a recent study, we demonstrate that type III-mediated concomitant translocation of YopE/LLO and YopE/p60 by a *Salmonella* vaccine strain significantly surpassed single cytosolic antigen delivery in the protection ability against *Listeria* (Igwe *et al.,* 2002). In summary, these findings emphasize the versatility of YopE for heterologous antigen delivery purposes in *Salmonella.* Use of TTSS will expand the efficiency of *Salmonella*-based live vaccines to induce MHC class I-restricted immune and antigen-specific CD8 T cell responses.

REFERENCES

Alpuche-Aranda, C. M., Racoosin, E. L., Swanson, J. A., Miller, S. I.: *Salmonella* stimulate macrophage macropinocytosis and persist within spacious phagosomes. J. Exp. Med. 179, 601-608 (1994).

Autenrieth, I. B., Heesemann, J.: In vivo neutralization of tumor necrosis factor-alpha and interferon-gamma abrogates resistance to *Yersinia enterocolitica* infection in mice. Med. Microbiol. Immunol. Berl. 181, 333-338 (1992).

Black, D. S., Bliska, J. B.: Identification of p130Cas as a substrate of *Yersinia* YopH (Yop51), a bacterial protein tyrosine phosphatase that translocates into mammalian cells and targets focal adhesions. EMBO J. 16, 2730-2744 (1997).

Black, D. S., Bliska, J. B.: The RhoGAP activity of the *Yersinia pseudotuberculosis* cytotoxin YopE is required for antiphagocytic function and virulence. Mol. Microbiol. 37, 515-527 (2000).

Bubert, A., Kuhn, M., Goebel, W., Köhler, S.: Structural and functional properties of the p60 protein from different Listeria species. J. Bacteriol. 174, 8166-8171 (1992).

Cornelis, G. R., Boland, A., Boyd, A. P., Geuijen, C., Iriarte, M., Neyt, C., Sory, M.-P., Stanier, I.: The virulence plasmid of *Yersini*a, an antihost genome. Microbiol. Mol. Rev. 62, 1315-1352 (1998).

Galán, J. E.: Molecular genetic bases of *Salmonella* entry into host cells. Mol. Microbiol. 20, 263-271 (1996).

Galán, J. E.: Interactions of *Salmonella* with host cells: Encounters of the closest kind. Proc. Natl. Acad. Sci. USA. 95, 14006-14008 (1998).

Germain, R. N. 1994.: MHC-dependent antigen processing and peptide presentation: providing ligands for T lymphocyte avtivation. Cell. 76, 287-299 (1994).

Germain, R. N.: The biochemistry and cell biology of antigen presentation by MHC class I and II molecules: implications for development of combination vaccines. Ann. NY Acad. Sci. 754, 114-125 (1995).

Harty, J. T., Pamer, E. G.: CD8 T lymphocytes specific for the secreted p60 antigen protect against *Listeria monocytogenes* infection. J. Immunol. 154, 4642-4650 (1995).

Harty, J. T., Bevan, M. J.: Responses of CD8+ T cells to intracellular bacteria. Curr. Opin. Immunol. 11, 89-93 (1999).

Hueck, C. J. 1998.: Type III protein secretion systems in bacterial pathogens of animals and plants. Microbiol. Mol. Biol. Rev. 62, 379-433 (1998).

Igwe, E. I., Geginat, G., Rüssmann, H.: Concomitant cytosolic delivery of two immunodominant listerial antigens by *Salmonella enterica* serovar Typhimurium confers superior protection against murine listeriosis. Infect. Immun. 70: 7114-7119 (2002).

Jacobi, C. A., Roggenkamp, A., Rakin, A., Zumbihl, R., Leitritz, L., Heesemann, J.: In vitro and in vivo expression studies of YopE from *Yersinia enterocolitica* using the gfp reporter gene. Mol. Microbiol. 30, 865-882 (1998).

Kubori, T., Matsushima, Y., Nakamura, D., Uralil, J., Lara-Tejero, M., Sukhan, A., Galán, J. E., Aizawa, S. I.: Supramolecular structure of the *Salmonella typhimurium* type III protein secretion system. Science. 280, 602-605 (1998).

Lee, C. A.: Type III secretion systems: machines to deliver bacterial proteins into eukaryotic cells. Trends Microbiol. 5, 148-156 (1997).

Lee, V. T., Anderson, D. M. Schneewind, O.: Targeting of *Yersinia* Yop proteins into the cytosol of HeLa cells: one-step translocation of YopE across bacterial and eukaryotic membranes is dependent on SycE chaperone. Mol. Microbiol. 28, 593-601 (1998).

Pamer, E. G., Harty, J. T., Bevan, M. J.: Precise prediction of a dominant class I-restricted epitope of *Listeria monocytogenes*. Nature. 353, 852-855 (1991).

Pamer, E. G.: Direct sequence identification and kinetic analysis of an MHC class I-restricted *Listeria monocytogenes* CTL epitope. J. Immunol. 152, 686-694 (1994).

Rosqvist, R., Bolin, I., Wolf-Watz, H.: Inhibition of phagocytosis in *Yersinia pseudotuberculosis*: a virulence plasmid-encoded ability involving the Yop2b protein. Infect. Immun. 56, 2139-2143 (1988).

Rosqvist, R., Forsberg, A., Wolf-Watz, H.: Intracellular targeting of the *Yersinia* YopE cytotoxin in mammalian cells induces actin microfilament disruption. Infect. Immun. 59, 4562-4569 (1991).

Rosqvist, R., Magnusson, K. E., Wolf-Watz, H.: Target cell contact triggers expression and polarized transfer of *Yersinia* YopE cytotoxin into mammalian cells. EMBO J. 13, 964-972 (1994).

Rosqvist, R., Hakansson, S., Forsberg, A., Wolf-Watz, H.: Functional conservation of the secretion and translocation machinery for virulence proteins of Yersiniae, Salmonellae and Shigellae. EMBO J. 14, 4187-4195 (1995).

Ruckdeschel, K., Roggenkamp, A., Lafont, V., Mangeat, P., Heesemann, J., Rouot, B.: Interaction of *Yersinia enterocolitica* with macrophages leads to macrophage cell death through apoptosis. Infect. Immun. 65, 4813-4821 (1997).

Rüssmann, H., Weissmüller, A., Geginat, G., Igwe, E. I., Roggenkamp, A., Bubert, A., Goebel, W., Hof, H., Heesemann, J.: *Yersinia enterocolitica*-mediated translocation of defined fusiom proteins to the cytosol of mammalian cells results in MHC class I-restricted antigen presentation. Eur. J. Immunol. 30, 1375-1384 (2000).

Rüssmann, H., Igwe, E.I., Sauer, J., Hardt, W.-D., Bubert, A., Geginat, G.: Protection against murine listeriosis by oral vaccination with recombinant *Salmonella* expressing hybrid *Yersinia* type III proteins. J. Immunol. 167, 357-365 (2001).

Schesser, K., Frithz-Lindsten, E., Wolf-Watz, H.: Delineation and mutational analysis of the *Yersinia pseudotuberculosis* YopE domains which mediate translocation across bacterial and eukaryotic cellular membranes. J. Bacteriol. 178, 7227-7233 (1996).

Sijts, A. J. A. M., Neisig, A., Neefjes, J., Pamer, E. G.: Two *Listeria monocytogenes* CTL epitopes are processed from the same antigen with different efficiencies. J. Immunol. 156, 683-692 (1996).

Simonet, M., Richard, S., Berche, P.: Electron microscopic evidence for in vivo extracellular localization of *Yersinia pseudotuberculosis* harboring the pYV plasmid. Infect. Immun. 58, 841-845 (1990).

Sory, M.-P., Cornelis, G. R.: Translocation of a hybrid YopE-adenylate cyclase from *Yersinia* enterocolitica into HeLa cells. Mol. Microbiol. 14, 583-594 (1994).

Sory, M.-P., Boland, A., Lambermont, I., Cornelis, G. R. 1995.: Identification of the YopE and YopH domains required for secretion and internalization into the cytosol of macrophages, using the cyaA gene fusion approach. Proc. Natl. Acad. Sci. USA. 92, 1998-2002 (1995).

Vijh, S., Pamer, E. G.: Immunodominant and subdominant CTL responses to *Listeria monocytogenes* infection. J. Immunol. 158, 3366-3371 (1997).

Von Pawel-Rammingen, Telepnev, U., M. V., Schmidt, G., Aktories, K., Wolf-Watz, H. Rosqvist, R.: GAP activity of the *Yersinia* YopE cytotoxin specifically targets the Rho pathway: a mechanism for disruption of actin microfilament structure. Mol. Microbiol. 36, 737-748 (2000).

Wick, M. J., Harding, C. V., Normark, S. J., Pfeifer, J. D.: Parameters that influence the processing efficiency of antigenic epitopes expressed in *Salmonella typhimurium*. Infect. Immun. 62, 4542-4548 (1994).

Chapter 82

Immunological Characterisation of Sub-Units of the *Yersinia* Type III Secretion Apparatus

James HILL[1], Cindy D. UNDERWOOD[1], Lena SUNDBERG[2], Hanna ÅSTRÖM[2], Sophie E.C. LEARY[1], Åke FORSBERG[2] and Richard W. TITBALL[1]

[1]Chemical and Biological Sciences, Dstl, Porton Down, Wiltshire SP4 0JQ, UK; [2]Swedish Defence Research Agency, FOI, NBC-Defence, 901 82 Umeå, Sweden

1. INTRODUCTION

Yersinia pestis, the causative agent of plague, is an important human pathogen. *Y. pestis* shares a number of virulence factors with *Yersinia pseudotuberculosis* and *Yersinia enterocolitica*, including a plasmid-encoded type III secretion (TTS) system. TTS is a contact-dependent virulence mechanism that results in the direct injection of a set of anti-host molecules into eukaryotic cells (Cornelis, 1998, Hueck, 1998). The system consists of regulators of secretion, Yersinia secretion complex (Ysc) proteins, translocator proteins, and effector proteins (Yops) that promote resistance to phagocytosis and induce cytotoxixity. Bacteria deficient in plasmid-encoded TTS are highly attenuated in animal models. A chromosomally encoded TTS system has recently been identified but virulence studies suggest that this system is less important for *Y. enterocolitica* infection than the plasmid-encoded TTS system. A key component of the TTS system is V antigen (LcrV). A number of roles have been proposed for LcrV, for example it is thought to interact intracellularly with LcrG (another TTS component) in order to regulate TTS (Nilles *et al*, 1997). LcrV appears to localise to the *Yersinia* surface prior to cell contact (Pettersson *et al*, 1999). This might explain why LcrV is an effective vaccine in plague infection models. LcrV is currently being evaluated in clinical trials, along with F1 antigen, as a plague sub-unit vaccine. It is thought that protection is derived either from the

The Genus Yersinia, Edited by Skurnik *et al.*
Kluwer Academic/Plenum Publishers, New York, 2003

neutralisation of LcrV's anti-inflammatory properties (Nedialkov *et al,* 1997, Schmidt, Rollinghoff *et al* 1999, Sing *et al*, 2002, Welkos *et al*, 1998), or by blocking of TTS (Pettersson *et al*, 1999, Weeks *et al*, 2002), or through a combination of both. The amino acid sequences of LcrV homologues in *Y. pseudotuberculosis* and *Y. enterocolitica* are similar (96 % and 93-95 % respectively) to that of *Y. pestis*, however there are differences in the protective properties between strains (Roggenkamp *et al*, 1997, Schmidt, Schaffelhofer *et al*, 1999).

2. RESULTS AND DISCUSSION

In this study we asked whether components of the *Yersinia* TTS system, other than LcrV, are effective as plague vaccines. We looked at membrane-associated or surface exposed components of the *Yersinia* secretion complex as protective antigens.

Table 1. TTS complex proteins are immunogenic, but do not protect mice against plague.

	RIBI		Alhydrogel	
	Titre	TTD±SEM (days)	Titre	TTD±SEM (days)
PBS	-	5.2±0.3	-	4.9±0.3
GST	83 520	5.4±0.6	98 880	5.6±0.4
GST-YscO	432 000	5.3±0.4	477 440	5.0±1.1
GST-YscP	1 646 080	6.1±0.4	1 495 040	4.4±0.4
GST-YscF	124 160	6.0±0.9	79 200	5.2±0.3
GST-TyeA	195 840	6.0±0.4	378 880	6.1±0.6
His-YscJ	211 200	5.2±0.8	161 920	5.5±0.4
His-VirG	2 115	5.7±0.6	6 300	5.4±0.7

Mice were immunised with recombinant YscJ, VirG, YscO, YscP, YscF or TyeA protein, with Alhydrogel or Ribi as an adjuvant. ELISA data indicated the presence of high levels of IgG in all groups except VirG-immunised mice, where relatively low IgG levels were detected. None of the proteins appeared to extend the time to death after challenge with 50 MLD of *Y. pestis* GB (Table 1).

This study indicated that although none of the TTS proteins used were able to confer immunity to challenge by *Y. pestis*, the majority of them were able to trigger a noticeable immune response. This may be significant; it suggests that if the immune response were to be higher, protection may result. It is possible that repetition of the study with an increased dose of the TTS proteins, or perhaps by variation of the type and preparation of adjuvants used, may allow this to happen.

It is also possible that, assuming any protection achieved would be antibody mediated, it may be dependent on the correct conformation of protective epitopes. While the TTS proteins used in this study are immunogenic, they are also recombinant and adjuvanted and may therefore differ slightly in conformation to their native forms.

Further work is therefore required to explore these theories and to test whether TTS proteins could make effective plague vaccines.

REFERENCES

Cornelis, G. R. 1998. The *Yersinia* deadly kiss. Journal of Bacteriology 180:5495-5504.

Hueck, C. J. 1998. Type III protein secretion systems in bacterial pathogens of animals and plants. Microbiology and Molecular Biology Reviews 62:379

Nedialkov, Y. A., V. L. Motin, and R. R. Brubaker. 1997. Resistance to lipopolysaccharide mediated by the *Yersinia pestis* V antigen-polyhistidine fusion peptide: Amplification of interleukin-10. Infection and Immunity 65:1196-1203.

Nilles, M. L., A. W. Williams, E. Skrzypek, and S. C. Straley. 1997. *Yersinia pestis* LcrV forms a stable complex with LcrG and may have a secretion-related regulatory role in the low-Ca^{2+} response. Journal of Bacteriology 179:1307-1316.

Pettersson, J., A. Holmstrom, J. Hill, S. Leary, E. Frithz-Lindsten, A. von Euler-Matell, E. Carlsson, R. Titball, A. Forsberg, and H. Wolf-Watz. 1999. The V-antigen of *Yersinia* is surface exposed before target cell contact and involved in virulence protein translocation. Molecular Microbiology 32:961-976.

Roggenkamp, A., A. M. Geiger, L. Leitritz, A. Kessler, and J. Heesemann. 1997. Passive immunity to infection with *Yersinia* spp mediated by anti-recombinant V antigen is dependent on polymorphism of V antigen. Infection and Immunity 65:446-451.

Schmidt, A., M. Rollinghoff, and H. U. Beuscher. 1999. Suppression of TNF by V antigen of *Yersinia* spp. involves activated T cells. European Journal of Immunology 29:1149-1157.

Schmidt, A., S. Schaffelhofer, K. Muller, M. Rollinghoff, and H. U. Beuscher. 1999. Analysis of the *Yersinia enterocolitica* O:8 V antigen for cross protectivity. Microbial Pathogenesis 26:221-233.

Sing, A., A. Roggenkamp, A. M. Geiger, and J. Heesemann. 2002. *Yersinia enterocolitica* evasion of the host innate immune response by V antigen-induced IL-10 production of macrophages is abrogated in IL-10-deficient mice. Journal of Immunology 168:1315-1321.

Weeks, S., J. Hill, A. Friedlander, and S. Welkos. 2002. Anti-V antigen antibody protects macrophages from *Yersinia pestis*-induced cell death and promotes phagocytosis. Microbial Pathogenesis 32:227-237.

Welkos, S., A. Friedlander, D. McDowell, J. Weeks, and S. Tobery. 1998. V antigen of *Yersinia pestis* inhibits neutrophil chemotaxis. Microbial Pathogenesis 24:185-196.

Picture 26. Sergei Balakhonov, Rima Shaikhutdinova and Svetlana Dentovskaya in a Turku café.

Picture 27. Rima Shaikhutdinova, Svetlana Dentovskaya and Andrei Anisimov at the Symposium dinner

Chapter 83

A Recombinant Prototrophic *Yersinia pestis* Strain Over-Produces F1 Antigen with Enhanced Serological Activity

Svetlana V. DENTOVSKAYA, Rima Z. SHAIKHUTDINOVA, and Andrei P. ANISIMOV
State Research Center for Applied Microbiology, Obolensk, Moscow Region, Russia

1. INTRODUCTION

Immunization with the capsular F1 antigen induces protection against lethal plague challenge (Meyer *et al.*, 1974; Simpson *et al.*, 1990). Our recent studies have shown that *Y. pestis* cells expressing the recombinant pFSK3 plasmid produce F1 antigen with 10^2-10^4-fold enhanced ability to react with antibody to F1 when compared to the amount of F1 obtained from the wild type *Y. pestis*. Moreover, the recombinant strain possessed 30-fold improved protection when compared with the Russian commercial live vaccine (EV line NIIEG) (Anisimov *et al.*, 1995). Over-production of F1 along with its enhanced potency is of obvious for formulation of a subunit plague vaccine.

2. ENGINEERING AND STUDYING *Yersinia pestis* STRAIN OVERPRODUCING F1 ANTIGEN WITH ENHANCED SEROLOGICAL ACTIVITY

In this study plasmid pFSK3 was transferred into the prototrophic, plasmid-less derivative of *Y. pestis* vaccine strain EV, EV11M. Bacteria were grown for 24 h at 28 °C or 37 °C in LB broth. The results of determinations of the total-protein concentrations (Lowry *et al.*, 1951) in the culture supernatants are shown in the Table 1.

The Genus Yersinia, Edited by Skurnik *et al.*
Kluwer Academic/Plenum Publishers, New York, 2003

Table 1. Concentration of total protein in cell-free supernatant

Strain	Total protein in culture liquid under different growth temperatures, mg/ml	
	28 °C	37 °C
EV line NIIEG	0.013	0.011
EV11MpFSK3	0.461	0.001

Indirect (passive) hemagglutination test was performed with the broth cultures, supernatants, and bacterial pellets to determine comparative serologic activity (Table 2).

Table 2. F1 antigen serological activity

Strain	Indirect hemagglutination test (reciprocal titers)					
	28 °C			37 °C		
	Broth culture	Supernatant	Pellet	Broth culture	Supernatant	Pellet
EV line NIIEG	4	-	-	64	64	64
EV11MpFSK3	8192	8192	65536	32	64	512

Cell-free F1 antigen was precipitated directly from culture supernatants by adjusting the pH to its isoelectric point, 4.1. The insoluble pellets were suspended in 0.9% NaCl solution, pH 7.2, (one fiftieth of the initial supernatant volume). This F1 antigen preparation was evaluated in denaturing PAGE (data not shown) and an indirect hemagglutination test (Table 3), which confirmed the advantage of producing the antigen from the recombinant strain in comparison with wild type *Y. pestis*.

Table 3. Total yields of F1 from supernatants of broth cultures of *Y. pestis* strains

Source of F1	F1 yield, mg/liter		Relative value of serologic activity [a]	
	28 °C	37 °C	28 °C	37 °C
EV line NIIEG	13.4	11	9.3×10^{-3}	9.0×10^{-2}
EV11MpFSK3	230.6	1.4	9.0×10^{4}	2.8

[a] Expressed as F1 units per milligram of protein. One F1 unit is serological activity of 1 mg/ml solution of commercial preparation of F1 (Stavropol' Research Anti-Plague Institute, Russia) in indirect hemagglutination assay.

Strain EV11MpFSK3 was able to produce significant amounts of F1 antigen even in minimal nutritional media supplemented with glucose 0.2% (data not shown).

3. CONCLUSION

Growth of *Y. pestis* EV11MpFSK3 at 28°C yields high levels of highly serologically active F1 antigen, which is a promising component of subunit plague vaccines.

ACKNOWLEDGEMENTS

This work was performed within the framework of the International Science and Technology Center Partner Project #1197p, supported by the Cooperative Threat Reduction Program of the US Department of Defense.

REFERENCES

Anisimov, A. P., Nikiforov, A. K., Eremin, S. A., and Drozdov, I. G., 1995, [Design of a *Yersinia pestis* strain with increased protection]. *Biull. Eksp. Biol. Med.* **120**: 532-534.

Lowry, O. H., Rosenbrough, N. R., Farr, A. L., and Randall, R. J., 1951, Protein measurement with the folin phenol. *J. Biol. Chem.* **193**: 115-119.

Meyer, K. F., Hightower, J. A., and McCrumb F. R., 1974, Plague immunization. VI. Vaccination with the fraction I antigen of *Yersinia pestis*. *J. Infect. Dis.* **129**: S41-S45.

Simpson, W. J., Thomas, R. E., Schwan, T. G., 1990, Recombinant capsular antigen (fraction 1) from *Yersinia pestis* induces a protective antibody response in BALB/c mice. *Am. J. Trop. Med. Hyg.* **43**: 389-396.

Picture 28. Andrei Anisimov proudly posing beside his poster.

Chapter 84

Vaccination with Plasmid DNA Expressing the *Yersinia pestis* Capsular Protein F1 Protects Mice Against Plague

Haim GROSFELD, Tamar BINO, Yehuda FLASHNER, Raphael BER, Emanuelle MAMROUD, Shlomo LUSTIG, Baruch VELAN, Avigdor SHAFFERMAN and Sara COHEN
Department of Biochemistry and Molecular Genetics, Israel Institute for Biological Research, P.O.Box 19, Ness-Ziona, Israel 74100

1. INTRODUCTION

The fraction 1 capsular protein (F1) is considered an important but not essential virulence factor unique to *Y. pestis* (Welkos *et al.*, 1995). Immunization with the F1 protein has been shown to protect mice against subcutaneous challenge with wild type *Y. pestis* (Andrews *et al.*, 1996) and a combined formulation containing F1 and V antigen confers protection against airborne infection (Williamson *et al.*, 1997). The protein has been associated with eliciting protective immune response in humans as well.

The observation that genetic immunization is able to elicit a protective immunity has fostered a new generation in vaccine development. The *caf*1 gene, which codes for the F1 protein, was previously used as DNA vaccine. In this study inbred mice were found to be non responsive, and outbred mice responded by a weak anamnestic response (Brandler *et al.*, 1998). The advent in genetic vaccination and the accumulating information on factors modulating the extent of response to DNA vaccines led us to re-examine genetic vaccination based on F1 antigen.

Here we compare three F1 DNA derivatives carrying different signals for cellular localization and demonstrate that one such genetic derivative, which presumably targets expression to the cytosol induces an effective antibody response and confers protection against high doses of infective *Y. pestis.*

The Genus Yersinia, Edited by Skurnik *et al.*
uwer Academic/Plenum Publishers, New York, 2003

2. RESULTS AND CONCLUSIONS

Plasmid DNA expressing the *Y. pestis* capsular F1 antigen in conjunction with different signals for cellular location were constructed. These included plasmids harboring either the full length *caf1* gene, *caf1* gene fused downstream to the coding sequences of the Semliki Forest Virus E3 signal polypeptide, or the *caf1* gene devoid of its signal sequences. Intramuscular vaccination of mice with these F1 expressing vectors led to different levels of induction of humoral response in mice. The most effective construct in eliciting high levels of specific anti-F1 antibodies was the one, expressing the F1 protein devoid of its signal peptide (pCI-deF1). This response was not limited to one specific mouse strain, or to the mode of DNA administration. Nevertheless gene gun mediated vaccination of mice with pCI-deF1 was by far more potent than that attained by intramuscular DNA injection.

Three rounds of gene gun vaccination with pCI-deF1 elicits high titers of anti-F1 antibodies (up to 10^5 ELISA units) similar to those obtained by vaccination with F1 protein. Such pCI-deF1 DNA vaccination protects mice against challenge with virulent *Y. pestis* Kimberley53 strain (Ben-Gurion and Hertman, 1958) as high as 4000 LD_{50}. Thus, targeting F1 expression to the cytosol through truncation of the bacterial signal is required for generation of an effective F1 DNA vaccine.

REFERENCES

Andrews, G.P., Heath, D.G., Anderson, G.W. Jr., Welkos, S.L., and Friedlander, A.M., 1996, Fraction 1 capsular antigen purification from *Yersinia pestis* CO92 and from an *E. coli* recombinant strain and efficacy against lethal plague challenge. *Infect. Immun.* **64**: 2180-2187.

Ben-Gurion, R., and Hertman, I., 1958, Bacteriocin-like material produced by *Pasteurella pestis*. *J. Gen Microbiol.* **19**: 289-297.

Brandler, P., Sakh, K.U., Heath, D., Friedlander, A., and Urlich , R.G., 1998, Weak anamnestic responses of inbred mice to *Yersinia* genetic vaccine are overcome by boosting with F1 polypeptide while outbred mice remain nonresponsive. *J. Immunol.* **161**: 4195-4200.

Welkos, S.L., Davis, K.M., Pitt, L.M., Worsham, P.L., and Friedlander, A.M., 1995, Studies on the contribution of the F1-capsule-associated plasmid pFra to the virulence of *Yersinia pestis*. *Contrib. Microbiol. Immunol.* **13**: 299-305.

Williamson, E.D., Eley, S.M., Stagg, A.J., Green, M., Russell, P., and Titball, R.W., 1997, A sub-unit vaccine elicits IgG in serum, spleen cell culture and bronchial washing and protects immunized animals against pneumonic plague. *Vaccine* **15**: 1079-1084.

Chapter 85

Evaluation of Protective Immunity Induced by *Yersinia enterocolitica* Type-III Secretion System Mutants

Emanuelle MAMROUD, Yehuda FLASHNER, Avital TIDHAR, Raphael BER, David GUR, Moshe AFTALION, Shirley LAZAR, Baruch VELAN, Avigdor SHAFFERMAN and Sara COHEN
Department of Biochemistry and Molecular Genetics, Israel Institute for Biological Research, P.O.Box 19, Ness-Ziona, Israel 74100

1. INTRODUCTION

Yersinia enterocolitica is an enteric animal pathogen able to infect humans after being ingested in contaminated food or water. The infection results in a spectrum of diseases ranging from enterocolitis to fatal septicemia.

Two classes of attenuated *Y. enterocolitica* mutant strains have been evaluated as candidates for live vaccine in the orally infected mice model. The first includes mutants in which basic physiological functions were mutated such as *aroA, htrA* and *ompR* (Bowe *et al.*, 1989; Li *et al.*, 1996; Dorrell *et al.*, 1998). Mice immunized with a single immunization dose of the *aroA* mutant were not protected against a lethal wild-type infection whereas the *htrA* and *ompR* mutants conferred partial protection.

The second class of attenuated *Y. enterocolitica* strains evaluated as vaccine candidates includes mutants defective in specific virulence-associated genes. Three *Y. enterocolitica* attenuated strains mutated in *yadA sodA* and *irp1* were constructed (Roggenkamp *et al.*, 1995; Roggenkamp *et al.*, 1997; Pelludant *et al.*, 1998). All mutants were found to protect mice against a lethal wild-type challenge (Igwe *et al.*, 1999).

The type III secretion system, highly conserved among Gram-negative pathogens, is an essential virulence component of pathogenic *Yersinia*

he Genus Yersinia, Edited by Skurnik *et al.*
ıwer Academic/Plenum Publishers, New York, 2003

encoded on the common 70kb plasmid (Cornelis 2000). The system consists of structural components of the secretion apparatus (Ysc) and of a set of exported proteins (Yops). The YscL and YscU proteins were found to be essential for secretion and translocation of Yops virulence determinants into eukaryotic cells (Michiels *et al.,* 1991, Allaoui *et al.,* 1994). Mutant strains having transposon insertions in these genes were identified as attenuated for virulence by two *Yersinia* Signature-Tagged Mutagenesis screens (Darwin and Miller, 1999, Mecsas *et al.,* 2001). In the present study, two *Y. enterocolitica* strains mutated in the *yscL* and *yscU* genes were characterized for their level of attenuation and evaluated for their ability to induce protection.

2. RESULTS AND CONCLUSIONS

2.1 Characterization of *Y. enterocolitica* WA*yscL* and WA*yscU* mutants

The highly pathogenic *Y. enterocolitica* 0:8 WA strain, was used as a platform for generation of a limited library (according to Hensel *et al.,* 1995) consisting of several hundred clones of tagged mutants. Screening of this library by intraperitoneal (i.p) infection of mice led to the identification of attenuated *Y. enterocolitica* strains, among which were the WA*yscL*::km2 and WA*yscU*::km2 mutants. These mutant strains were submitted to further characterization as follows.

2.1.1 Yops secretion

WA*yscL* and WA*yscU* mutant strains were grown at 37°C in a Ca^{2+} deficient medium and proteins secreted into the culture medium were analyzed by SDS-polyacrylamide gel electrophoresis. As expected, Yops secretion was totally inhibited in both mutant strains as compared to wild-type strain (Figure 1A). To further characterize these cultures, supernatant samples and their corresponding cell pellets were subjected to immunoblot analyses with specific antisera against V antigen. As observed (Figure 1B), lower levels of the V antigen were produced in both WA*yscL* and WA*yscU* mutants as compared to the wild type strain, in agreement with previous observations that Yops expression is down regulated when secretion is blocked (Straley *et al.,* 1993). Although being produced, the V antigen protein is not secreted by both mutant strains, as evident by immunoblot analysis of the secreted proteins (Figure 1C).

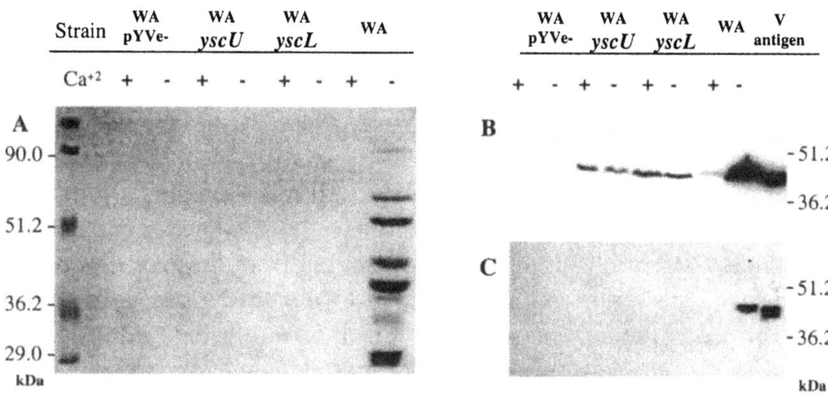

Figure 1. Functional analysis of the type III secretion system in the WA*yscL* and WA*yscU* mutant strains. **A:** SDS-PAGE (12.5% polyacrylamide gel) coomassie blue stain of ammonium-sulfate precipitated proteins from supernatants of *Y. enterocolitica* wild-type WA strain, WA*yscL* and WA*yscU* mutant strains and the WA(pYVe-) plasmidless strain (serving as a negative control). Bacteria were cultivated at 37°C in BHI broth supplemented with either 2.5 mM $CaCl_2$ (+ Ca^{2+}) or 20mM sodium oxalate and 20 mM $MgCl_2$ (-Ca^{2+}) according to Cornelis *et al.,* 1989. **B and C:** immunoblot analyses of V antigen from *Y. enterocolitica* strains, cultivated as described above. **B**- total cell extracts (equal number of bacteria were loaded in each lane) and **C**- secreted proteins prepared as described above. Purified V antigen (1 μg), was used as reference protein and as a source for production of the specific mouse polyclonal anti-V antigen antibodies used in the immunoblots.

2.1.2 Virulence in mice

The attenuation level of the *yscL* and *yscU* mutant strains was determined in BALB/c mice by two routes of infection. Using the i.p. route, these mutants were found to be severely attenuated with LD_{50}'s of more than 10^6 CFU while the LD_{50} of the wild type strain was 10^3 CFU. The observed attenuation was not due to a general growth defect since the mutant strains growth rates *in vitro* were similar to that of the wild type strain. Oral infection studies indicated that the mutant strains were non-lethal with an infection dose of more than 10^8 CFU, while the LD_{50} of the wild type strain was $5x10^6$-10^7 CFU.

2.1.3 Persistence of *Y. enterocolitica* WA*yscU* mutant strain in tissues

The course of i.p. infection with the wild type and WA*yscU* strains was analyzed by quantifying viable *Yersinia* cells from spleens, livers and lungs (recovered on BIN selective agar medium, Ber *et al.*, this Proceeding book) on days 1, 3, 7 and 10 post infection. Groups of mice (ten per group) were infected with 10^6 CFU of each strain, and two mice were sacrificed for analysis at each time point (experiments were carried out in duplicates).

On day 1 post infection the wild type and mutant strain were present in all the organs at about similar levels (Figure 2). However there were significant differences 3 days post infection. All mice infected with the wild type strain had high bacterial load in all the organs, whereas mice infected with the mutant strain began to clear the bacteria from these tissues and had 1000 fold lower bacterial counts compared to the wild type. By day 7 post infection, all mice inoculated with the wild type strain died while all mice infected with the mutant strain survived without any symptoms. Yet the mutant strain was still able to persist in the organs up to ten days post infection (limit of detection ≤5 CFU). Mice were monitored for 30 days and appeared healthy without any signs of disease. We note that the plasmidless WA(pYVe-) strain was rapidly cleared from internal organs (data not shown).

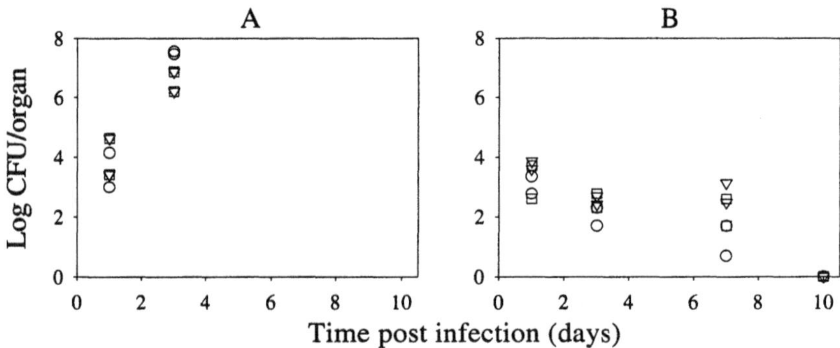

Figure 2. Colonization and persistence of *Y. enterocolitica* strains in mice organs. BALB/c mice were i.p. infected with 10^6 CFU of WA (A) or WAyscU (B) strains, respectively. Each point represents number of bacteria (CFU) present in spleen, liver and lungs (circles, triangles and squares, respectively) of an individual mouse.

2.2 Evaluation of *Y. enterocolitica* WA*yscL* and WA*yscU* mutant strains for inducing protective immunity

The WA*yscL* and WA*yscU* mutant strains ability to confer protection against a lethal WA strain challenge was determined using oral and i.p. immunization routes. Immunization via the oral route using $2x10^8$ CFU/mouse with each of the mutant strains failed to confer protection against an oral challenge of $20LD_{50}$ wild type strain. Intraperitoneal immunization with the attenuated strains was done at various doses (Table 1). Six weeks post immunization, mice were challenged i.p. with 300LD_{50} of wild type strain. High survival rates of 88% and 86% were observed in mice immunized with a single dose of either WA*yscL* or WA*yscU* strains respectively ($4x10^6$ CFU/mouse). Moreover, even at a ten-fold lower immunization dose, significant survival rates (60% and 40%) were observed. This protection was achieved although attenuated strains did not induce any detectable anti-V antigen antibodies (<20 ELISA units). We note that none of the mice immunized with the plasmidless strain ($4x10^6$ CFU/mouse) survived the challenge dose (Table 1). Thus, protection of mice immunized with the mutant strains might result from expression of plasmid-encoded regulators or antigens, other than Yops. In conclusion, the data suggests that a functional type III secretion apparatus is not important for protection via the i.p. route whereas in the oral route the system is essential.

Table 1. Protection conferred by *Y. enterocolitica* WA*yscU* and WA*yscL* strains

Immunization		Humoral response	Protection	
Strain	Dose[a] (CFU)	ELISA[b] GMT	% Survival[c] (Live/Total)	MTTD[d] (days)
WA*yscL*	$4x10^6$	1000	88 (7/8)	16
	$4x10^5$	500	60 (3/5)	15±0.5
WA*yscU*	$4x10^6$	800	86 (6/7)	16
	$4x10^5$	800	40 (2/5)	15±0.5
WA(pYVe-)	$4x10^6$	1000	0 (0/10)	7.8±0.5
Control	-	<20	0 (0/5)	5.5±0.5

a- Mice were immunized i.p, once. b- Whole cell *Y. enterocolitica* enzyme linked immunosorbent assay (ELISA) titer was determined five weeks post immunization. GMT-Geometric Mean Titer. c- Six weeks post immunization, mice were challenged i.p. with 300LD_{50} ($3x10^5$CFU/mouse) of *Y. enterocolitica* WA strain. Survival was followed for 4 weeks .d- MTTD - Mean Time To Death.

REFERENCES

Allaoui, A., Woestyn, S., Sluiters, C., and Cornelis, G. R., 1994, YscU, a *Yersinia enterocolitica* inner mambrane protein involved in Yop secretion. *J. Bacteriol* **176:** 4534-4541

Bowe, F., O'Gaora, P., Maskell, D., Cafferkey, M., and Dougan, G., 1989, Virulence, persistence, and immunogenicity of *Yersinia enterocolitica* 0:8 *aroA* mutants. *Infect. Immun.* **57:** 3234-3236

Cornelis, G. Sluiters, C., Lambert de Rouvroit, C., and michiels, T., 1989, Homology between VirF, the transcriptional activator of the *Yersinia* virulence regulon, and AraC, the *Escherichia coli* arabinose operon regulator. *J. Bacteriol.* **171:** 254-262

Cornelis, G. and Gijsegen, F. V., 2000, Assembly and function of type III secretory system. *Annu. Rev. Microbiol.* **54:**735-74.

Darwin, A.J. and Miller, V.L., 1999, Identification of *Yersinia enterocolitica* genes affecting survival in an animal host using signature-tagged transposon mutagenesis. *Mol. Microbiol.* **32:** 51-62

Dorrell, N., Li S., Everest, P.H., Dougan, G., and Wren, B.W., 1998, Construction and characterization of a *Yersinia enterocolitica* 0:8 *ompR* mutant. *FEMS Microbiol. Lett.* **165:** 145-151

Hensel, M., Shea, J. E., Gleeson, C., Jones, M. D., Dalton, E., and Holden, D.W., 1995, Simultaneous identification of bacterial virulence genes by negative selection. *Science* **269:** 400-403

Igwe, E.I., Russmann, H., Roggenkamp, A., Noll A., and Autenrieth, I.B., 1999, Rational live oral carrier vaccine design by mutating virulence-associated genes of *Yersinia enterocolitica. Infect. Immun.* **67:** 5500-5507

Li, S., Dorrell, N., Everest, P.H., Dougan, G., and Wren, B.W., 1996, Construction and characterization of a *Yersinia enterocolitica* 0:8 high-temperature requirement (*htrA*) isogenic mutant. *Infect. Immun.* **64:** 2088-2094

Mecsas, J., Bilis, I. and Falkow, S., 2001, Identification of attenuated *Yersinia pseudotuberculosis* strains and characterization of an orogastric infection in BALB/c mice on day 5 postinfection by signature-tagged mutagenesis. *Infect. Immun.* **67:** 2779-2787

Michiels, T., Vanooteghem, J.-C., Lambert de Rouvroit, C., China, B., Gustin, P., Boudry, P., and Cornelis, G.R., 1991, Analysis of *virC* , an operon involved in the secretion of Yop proteins by *Yersinia enterocolitica. J. Bacteriol.* **173:** 4994-5009

Pelludat, C., Rakin, A., Jacobi, C.A., Schubert, S., and Heesemann, J., 1998, The yersiniabactin biosynthetic gene cluster of *Yersinia enterocolitica*: organization and sidrophor-dependent regulation. *J. Bacteriol.* **180:** 538-546

Roggenkamp, A., Neuberger, H.R, Flugel, A., Schmoll, T., and Heesemann, J., 1995, Substitution of two histidine residues in YadA protein of *Yersinia enterocolitica* abrogates collagen binding, cell adherence and mouse virulence. *Mol. Microbiol.* **16:** 1207-1219

Roggenkamp, A., Bottner, T., Leitritz, L., Sing, A., and Heesemann, J., 1997, Contribution of the Mn-cofactored superoxid dismutase (SodA) to the virulence of *Y. enterocolitica* serotype 08. *Infect. Immun.* **65:** 4705-4710

Straley, S.C., Plano, G. V., Skrzypek, E., Haddix, P. L., Fields, K. A., 1993, Regulation by Ca^{2+} in the *Yersinia* low-Ca^{2+} response. *Mol. Microbiol.* **8:** 1005-1010

Chapter 86

Epidemiology and Diagnostics of *Yersinia*-Infections

Heinrich K.J. NEUBAUER[1] and Lisa D. SPRAGUE[2]
[1]Institute for Microbiology, German Federal Armed Forces, Neuherbergstr. 11, 80937 München, Germany; [2]Klinik und Poliklinik für Strahlentherapie und Radiologische Onkologie der TU-München, Klinikum rechts der Isar, Ismaningerstr. 22, 81675 München, Germany

1. INTRODUCTION

The genus *Yersinia* currently consists of eleven species of which three are pathogenic for humans and warm-blooded animals: *Y. enterocolitica* and *Y. pseudotuberculosis* causing Yersiniosis usually manifested as enteritis or enterocolitis and *Y. pestis*, the etiological agent of plague (Aleksic and Bockemühl, 1990; Perry and Featherstone, 1997). In the past two thousand years plague has had a disastrous effect on the development of human civilisation and is therefore still present in the consciousness of modern day people. This is probably the reason why *Yersinia* became such a well characterised model for the investigation of pathogenicity in bacteriological and immunological research (Cornelis *et al.*, 1998). Yet, the lack of knowledge on the epidemiology of Yersiniosis or infections with so called 'non-pathogenic' *yersiniae* e.g. *Y. kristensenii*, *Y. intermedia* or *Y. frederiksenii* is remarkable. To define a 'species' by the standards of diagnostic microbiology or scientific taxonomy the relatedness of isolates must be known.

2. EPIDEMIOLOGY OF YERSINIOSIS

The clinical picture of Yersiniosis in humans has been very well documented and described (Tauxe *et al.*, 1987; Hoogkamp-Korstanje and de Koning, 1990; Aleksic and Bockemühl, 1995): *Y. enterocolitica* causes acute

he Genus Yersinia, Edited by Skurnik *et al.*
ıwer Academic/Plenum Publishers, New York, 2003

but self-limiting enteritis or enterocolitis mostly in children. The clinical presentation may include diarrhoea with sometimes bloody stools, abdominal pain due to mesenteric lymphadenitis or terminal ileitis, 'pseudoappendicitis' in adults, fever, nausea with or without vomiting but also throat infections. Generally, adults overcome the disease in two weeks, children in four. Post infectious sequelae are reported to be reactive arthritis, erythema nodosum, uveitis and thyreoiditis.

Yersiniosis causes high economic losses due to hospitalisation, medication and time needed to care for paediatric patients (Tauxe *et al.*, 1987). In Europe *Y. enterocolitica* serovars O:3, O:5,27 and O:9 as well as *Y. pseudotuberculosis* serovars O:1, O:2 and O:3 are associated with human disease.

2.1 Human Yersiniosis: Example Germany

In contrast to the minuscule clinical data, actual and detailed epidemiological data on the occurrence of Yersiniosis in the EU are missing as illness or isolation of the agent is or was not mandatory in many countries. In Western Germany seroprevalence studies for anti *Yersinia* antibodies were carried out in healthy adult blood donors in 1988 and 1986 (Wenzel *et al.*, 1988; Mäki-Ikola *et al.*, 1997). These investigations showed that more than 40% of these volunteers had antibodies of the IgG, IgA and IgM subclasses indicating that many blood donors had had recent contact to pathogenic *Y. enterocolitica*. More then 10 years later Neubauer *et al.*, (2000a) demonstrated that the seroprevalence is still as high as 40% in healthy blood donors. The Western Blot technique based on native or recombinantly produced *Yersinia* outer proteins (YOP) was applied as a sensitive and specific tool for the identification of anti YOP antibodies (Wenzel *et al.*, 1988; Neubauer *et al.*, 2000a). By using this technique infections caused by all pathogenic *Y. enterocolitica* and *Y. pseudotuberculosis* isolates can be detected independently of their serovar thereby providing the exact prevalence. It is noteworthy that no data for children who show the highest rate of clinical manifestation are available. Consequently, the seroprevalence data available are not representative for the whole population of Germany and should be used and interpreted with caution. The high seroprevalence is to be contrasted with the small number of *Yersinia* isolates characterised by the national *Yersinia* reference laboratory every year (Aleksic *et al.*, 1995; Aleksic and Bockemühl, 1996). In the former German Democratic Republic where a mandatory notification system was introduced in 1978 approximately 30.000 isolations were registered in the following eleven years (Kiesewalter, 1992).

In a case control study in Belgium, Tauxe *et al.*, (1987) showed that the contamination of meat during the slaughtering of pigs and the consumption of raw pork or raw pork products can contribute to the incidence of human Yersiniosis. Therefore, it is believed that the reservoir for *Y. enterocolitica* is the slaughter pig, which carries the agent in its tonsils. However, this was never proved for other countries where slaughtering practise and consumption habits may vary markedly from those in Belgium.

The high prevalence of *Y. enterocolitica* in pigs is attributed to the development of fattening units where the agent can easily spread from one animal to the other after having been introduced by e.g. one piglet (Verhaegen *et al.,* 1998). Obviously, international trade of live stock, pork or pork products may also contribute to a global spread even to tropical or subtropical areas.

Fredriksson-Ahomaa *et al.,* (2001) demonstrated that contamination of pork with pathogenic *Y. enterocolitica* is currently at least a regional problem in southern Germany and pork may therefore be a source for human infection. Additionally, a high seroprevalence of anti YOP antibodies was found in meat juice samples of Bavarian slaughter pigs using the already described YOP Western blot technique and a commercially available kit using five recombinant proteins on a single blot strip (Nikolaou *et al.,* 2001). The investigators found anti *Yersinia* antibodies in 970 of 1014 samples. Only blots with three or more detectable signals were considered as being positive in order to exclude false positives. More than 82% of the pig farms were shown to have had a previous history of *Yersinia* infection.

However, these data only suggest that pork may be involved in human infection but the final proof that pork indeed is the 'missing link' is missing. It is also not clear which pathogenic *Y. enterocolitica* serovars are involved and whether *Y. pseudotuberculosis* may play the same important role in Yersiniosis as in other European countries such as Finland (Niskanen *et al.,* 2002). Some clues do lead to the assumption that many other factors may contribute to the epidemiology of Yersiniosis in Germany. In a recent survey, Nikolaou *et al.,* (2002) demonstrated that the seroprevalence of anti *Yersinia* antibodies is also very high in goats in the state of Schleswig-Holstein. 70 % of the animals investigated had antibodies. The prevalence of *Yersiniae* has also to be considered as high in the German cattle population. *Yersinia* infections became regularly detected (Hartung, 2000) during the course of *Brucella* surveillance when cross-reacting antibodies against the O:9 LPS antigen were detected (Garin-Bastuji *et al.,* 1999). Consequently, *Yersiniae* may be acquired through the consumption of meat, cheese and milk of ruminants, too. The contact with infected cows or their products has already been associated with human disease in France (Gourdon *et al.,* 1999). Surprisingly, many investigators failed to isolate pathogenic *Yersiniae* from food or even uncooked meat. In rural areas environmental risk factors have to be taken into consideration, e.g. contaminated drinking water (Kapperud *et al.,* 1995).

Considering all epidemiological data available to date and the many questions so far unanswered the dogma of '*Yersinia enterocolitica* infections and pork: the missing link′ (Tauxe *et al.,* 1987) has to be re-evaluated. The influences of modern animal and food production, legislation and changed consumer habits have to be included in new hypotheses on the epidemiology of Yersiniosis. It is more plausible that Yersiniosis is a multicausal illness with reservoirs in most animal populations used for the production of food and feed. High prevalence in humans and animals must be accepted as a consequence of industrialisation and globalisation. The dogma of Tauxe *et al.,* must also be questioned in view of countries such as the UK with high pork production and - consumption where infections in humans or isolation of pathogenic *Yersinia* strains are extremely seldom (Prentice *et al.,* 1991). A further possibility may be that clones have developed during the last decades with a certain species specificity. A human epidemic *Y. enterocolitica* clone may exist having its natural reservoir in chronic infected symptomatic or asymptomatic persons. Spread is facilitated by direct faecal-oral transmission. Or else bovine or caprine clones may exist which are accidentally introduced into human population. Current molecular epidemiological tools such as RFLP, RAPD etc. do not have the capacity to distinguish these clones, yet.

Only national and international surveillance and interdisciplinary research programmes are able to elucidate the 'truth about Yersiniosis' and are an urgent demand for the future.

2.2 Animal Yersiniosis

The occurrence of *Y. enterocolitica* has been documented for many animal species. Frequently isolation of the pathogenic agent was not linked with symptoms. Usually, clinical presentation is associated with systemic disease and / or enteritis in young animals. However, compared to well documented human Yersiniosis, animal Yersiniosis has to be considered as not investigated (Review: Neubauer *et al.,* 2001).

Pigs are believed to be the asymptomatic carriers of pathogenic *Y. enterocolitica* strains although experimentally infected piglets do develop enteritis or enterocolitis with anorexia and diarrhoea. Tonsillitis with the development of micro-abscesses seems to be a common finding in 'healthy' slaughter pigs. Nothing is known about the impact of these persistent infections on the increase of body mass or possible immunosuppressive effects.

The presence of *Yersiniae* in cattle seems to be a common finding worldwide. Clinical presentation in calves is reported to be diarrhoea and hyperplasia of the GALT (gut associated lymphoid tissue). In cows mastitis, mesenteric lymphadenitis, abortion and endocarditis are sometimes observed. This also seems to be the case for small ruminants. However, in sheep the 'apathogenic' serovar O:6,30 biovar 1A strains can provoke placentitis in pregnant ewes causing abortion. Hares, rabbits and chinchillas develop acute or granulomatous disease. Deaths are observed regularly.

3. LESSIONS TO BE LEARNED FROM THE 16S rRNA GENE SEQUENCE: TAXONOMY

The species *Y. enterocolitica* sensu stricto was established in 1980 by applying DNA relatedness studies and phenotypic characteristics. A species was defined as that its strains must have a DNA-DNA relatedness of more than 70%. This standard is still valid for the genus *Yersinia* today. In the past years it was noted that the species *Y. enterocolitica* consists of biochemically and serologically heterogeneous strains: so called 'European' and 'American' bioserotypes named after the continent of their first isolation. By sequencing the 16S rRNA gene of these isolates it became obvious that members of both bioserotypes also had type specific 16S rRNA gene regions. This lead to the postulation that both 'types' might be species of their own. The correlation of DNA relatedness and 16S rRNA gene sequence similarity has been clarified. A 97 % sequence identity corresponds to a 70 % DNA - DNA relatedness. No evidence for the existence of two distinct species based on 16S rRNA gene sequences could be found by sequencing various strains of both bioserotypes. Isolates of both 16S rRNA gene types had sequence identities of more than 97 %. However, we could demonstrate the presence of three DNA - DNA relatedness groups within the species *Y. enterocolitica* represented by the 'American' bioserotypes, by the enteropathogenic 'European' strains and by the non-enteropathogenic 'European' strains.

The lowest official taxonomic rank is the subspecies. It is characterised by 'minor but consistent phenotypic variations within the species or on genetically determined clusters of strains within the species'. Considering the presence of three relatedness clusters and the 'minor but consistent phenotypic variation' i.e. the highly conserved 16S rRNA gene sequence of European and American isolates, the division of the species *Y. enterocolitica* into two subspecies was justified. We proposed the names *Yersinia enterocolitica* subsp. *enterocolitica* for strains belonging to the 16S rRNA gene type represented by strain ATCC 9610 of American origin and *Yersinia enterocolitica* subsp. *palearctica* for strains belonging to the 16S rRNA gene type of the strain DSMZ 13030 of European origin (Neubauer *et al.*, 1999; Neubauer *et al.*, 2000b).

The technique of sequencing the 16S rRNA gene to search for new species and subspecies in a current phenotypic species has been applied to a total of 269 isolates belonging to ten *Yersinia* species (except *Y. pestis).* Twenty different 16S rRNA gene types could be identified within the genus. Described sequence clusters of the species *Y. frederiksenii* were confirmed. New 16S rRNA gene sequence clusters were detected for the species *Y. frederiksenii, Y. intermedia, Y. mollaretii, Y. aldovae, Y. kristensenii*, and *Y. rohdei* (Neubauer *et al.*, 2000c). For the species *Y. kristensenii* and *Y. pseudotuberculosis* the presence of DNA-DNA clusters below 70% relatedness representing new species were confirmed (unpublished data).

4. DIAGNOSTICS OF *YERSINIA*

The gold standard for the final biochemical identification of a *Yersinia* isolate is still classical tube testing (Aleksic and Bockemühl, 1990). Commercially available identification systems such as the API 20E, API rapid 32 IDE, the GNI card of the VITEK system (all bioMerieux) or the Micronaute E (Merlin Diagnostika) lack sensitivity and specificity at the species level (Neubauer *et al.*, 1998; Linde *et al.*, 1999). Optimising the Micronaute E system layout by including the biochemical key reactions needed for the differentiation of species and *Y. enterocolitica* biovars resulted in an increase of the overall sensitivity up to 98 % (Neubauer *et al.*, 2000d). The practicability of this miniaturised semi automated system was also tested for field use. The definite identification of *Y. enterocolitica* isolates can also be achieved by sequencing the 16S rRNA gene (Neubauer *et al.*, 2000b). A commercially available PCR kit and a dip stick probe assay for the identification of *Y. enterocolitica* isolates based on 16S rRNA gene sequences proved to be specific either for 'European' or 'American' isolates (Neubauer *et al.*, 1999). These tests cannot be recommended for routine diagnostic use.

For the assessment of the presumptive pathogenicity of an *Y. enterocolitica* isolate a variety of empiric virulence markers are used (Aleksic and Bockemühl, 1990). These assays can easily be replaced by PCR assays targeting plasmid genes (Neubauer *et al.*, 2000e). The evaluation of molecular systems in terms of routine diagnostic procedures, however, is still missing.

Typically, a combination of various PCR assays targeting the *Y. pestis* genes i.e. Fraction 1 capsular antigen gene, plasminogen activator / coagulase gene, the gene of the V-antigen, and the 16S rRNA gene are used for identification and characterization of *Y. pestis* isolates. These assays are not affected by the presence of DNA of *Rattus norvegicus* or from fleas (*Xenopsylla cheopis* and *Ctenocephalides* spp.) in samples (Neubauer *et al.*,

2000f). Real time PCR such as the Light Cycler ™ technique will be used in future to fasten diagnosis.

ACKNOWLEDGEMENTS

I would like to thank all the colleagues and friends who have contributed to my work on *Yersinia* by supplying materials, time and knowledge. The superb technical assistance of C. Lodri, R. Schneider and S. Scholz of my lab group is appreciated.

REFERENCES

Aleksic, S., and Bockemühl, J., 1990, Mikrobiologie und Epidemiologie der Yersiniosen. *Immun. Infekt.* 18: 178-185.

Aleksic, S., Bockemühl, J., and Wuthe, H.H., 1995, Epidemiology of *Y. pseudotuberculosis* in Germany, 1983-1993. *Contrib. Microbiol. Immunol.*, 13: 55-58.

Cornelis, G.R., Boland, A., Boyd, A.P., Geuijen, C., Iriarte, M., Neyt, C., Sory, M.-P., and Stainier, I., 1998, The virulence plasmid of *Yersinia*, an anti host genome. *Microbiol. Mol. Biol. Rev.* 62: 1315-1352.

Fredriksson-Ahomaa, M., Bucher, M., Hank, C., Stolle, A., Korkeala, H., 2001, High prevalence of *Yersinia enterocolitica* 4:O3 on pig offal in southern Germany: a slaughtering technique problem. *Syst Appl Microbiol.* 24: 457-63.

Garin-Bastuji, B., Hummel, N., Gerbier, G., Cau, C., Pouillot, R., Da Costa, M., Fontaine, J.-J., 1999, Non specific serological reactions in the diagnosis of bovine brucellosis: experimental oral infection of cattle with repeated doses of *Yersinia enterocolitica* O:9. *Vet. Microbiol.* 66: 223-233.

Gourdon, F., Beytout, J., Reynaud, A., Romaszko, J.-P., Perre, D., Theodore, P., Soubelet, H., and Sirot, J., 1999, Human and animal epidemic of *Yersinia enterocolitica* O:9, 1989-1997, Auvergne, France. *Emerg. Infect. Dis.* 5: 719-721.

Hartung, M., 2000, Mitteilungen der Länder über *Y. enterocolitica*-Nachweise in Deutschland. In: bgvv- Hefte 08/2000 (ed.: bgvv, Berlin), 133-136.

Hoogkamp-Korstanje, J.A.A., and de Koning, J., 1990, Enteropathogene Yersinien: Pathogenitätsfaktoren und neue diagnostische Methoden. *Infekt. Immun.* 18: 186-191.

Kiesewalter, J., 1992, Klinische und epidemiologische Bedeutung von *Yersinia enterocolitica* für Mensch und Tier. *Bundesgesundhbl.* pp. 495-500.

Kapperud, G., Ostroff, S.M., Nesbakken, T., Hutwanger, L.C., Bean, N.H., Lassen, J., and Tauxe, R.V., 1995, Risk factors for sporadic *Yersinia enterocolitica* infections in Norway: A case control study. *Contr. Microbiol. Immunol.* 13: 25-28.

Linde, H-J., Neubauer, H., Meyer, H., Aleksic, S., and Lehn, N., 1999, Identification of *Yersinia* species by the Vitek GNI card. *J. Clin. Microbiol.* 37, 211-214.

Mäki-Ikola, O., Heesemann, J., Toivanen, I., and Granfors, K., 1997, High frequency of *Yersinia* antibodies in healthy populations in Finland and Germany. *Rheumatol. Int.*, 16: 227-229.

Neubauer, H., Sprague, L.D., Scholz, H., and Hensel, A., 2001, *Yersinia enterocolitica*-Infektionen: 1. Bedeutung bei Tieren. *Berl. Münch. Tierärztl. Wochenschrift*, 114: 8-12.

Neubauer, H., Rasolomaharo, M., Brooks, T., Chanteau, S., and Splettstößer, W.D., 2000a, IgG / IgM ELISA and immunoblot for the serodiagnosis of human plague. *Epidemiology and Infection,* 125: 593-597.

Neubauer, H., Hensel, A., Aleksic, S., Finke, E.-J., and Meyer H., 2000b, *Yersinia enterocolitica* 16S rRNA gene types belong to the same genospecies but form three different homology clusters. *Int. J. Med. Microbiol.*, 290: 61-64.

Neubauer, H., Hensel, A., Aleksic, S., Meyer, H., 2000c, Identification of *Yersinia enterocolitica* within the genus *Yersinia. System. Appl. Microbiol.,* 23: 58-62.

Neubauer, H., Molitor, M., Rahalison, L., Aleksic, S., Bakes, H., Chanteau, S., Meyer, H., 2000d, A semiautomated system for identification of *Yersinia* species within the genus *Yersinia., Clin. Lab.*, 46: 561-567.

Neubauer, H., Hensel, A., Aleksic, S., and Meyer, H., 2000e, Specific detection of plasmid bearing *Yersinia* isolates by PCR. *Clin. Lab.*, 46: 583-587.

Neubauer, H., Splettstösser, W., Prior, J., Aleksic, S., and Meyer, H. , 2000f, A comparison of different PCR assays for the rapid and presumptative diagnosis of *Yersinia pestis*. *J. Vet. Med., Series B*, 47: 573-580.

Neubauer, H., Reischl, U., Köstler, J., Aleksic, S., Finke, E.-J., and Meyer, H., 1999, Variations in the 16S rRNA gene sequence of *Yersinia enterocolitica* isolates influence the specificity of molecular identification systems. *Zentralbl. Bakt.*, 289: 329-337.

Neubauer, H., Sauer, T., Becker, H., Aleksic, S., and Meyer, H., 1998, Comparison of systems for the identification and differentiation of the species within the genus *Yersinia. J. Clin. Microbiol.* 36: 3366-3368.

Nikolaou, K., Hensel, A., Meyer, H., Czerny, C.P., and Neubauer, H., Seroprevalence of anti-*Yersinia* antibodies in meat juice samples of pigs. In: Proceedings of the 4th International Symposium on the epidemiology and control of *Salmonella* and other food borne pathogens in pork, Leipzig 2001, 159-161.

Nikolaou, K., Hensel, A., Rösler, U., Ganter, M., Arnold, T., and Neubauer, H., Seroprävalenz von anti-*Yersinia* Antikörpern in deutschen Ziegenbeständen. In: DVG Tagung Fachgruppe Bakteriologie und Mykologie, Hannover 2002.

Niskanen, T., Fredriksson-Ahomaa, M., Korkeala, H., 2002, *Yersinia pseudotuberculosis* with limited genetic diversity is a common finding in tonsils of fattening pigs. *J. Food. Prot.*, 65: 540-545.

Perry, R.D., and Fetherston, J.D., 1997, *Yersinia pestis*: the etiologic agent of plague. *Clinical Microbiology Reviews*, 10: 35-66.

Prentice, M., Cope, D., and Swann, R.A., 1991, The epidemiology of *Yersinia enterocolitica* infection in the British Islands 1983-1988. *Contr. Microbiol. Immunol.* 12: 17-25.

Tauxe, R.V., Vandepitte, J., Wauters, G., Martin, S.M., Gossens, V., De Mol, P., Van Noyen, R., and Thiers, G., 1987, *Yersinia enterocolitica* infections and pork: the missing link. *Lancet* 1: 1129-1132.

Verhaegen J., Chalier J., Lemmens P., Delmee M., Van Noyen R., Verbist L., Wauters G. (1998): Surveillance of human *Yersinia enterocolitica* infections in Belgium: 1967-1998 CID **27**, 59-64.

Wenzel, B.E., Heesemann, J., Wenzel, K.W., and Scriba, P.C., 1988, Antibodies to plasmid-encoded proteins of enteropathogenic *Yersinia* in patients with autoimmune thyroid disease. *Lancet* 2: 56.

Chapter 87

Susceptibility to Plague of the Rodents in Antananarivo, Madagascar

Lila RAHALISON[1], Michel RANJALAHY[1], Jean-Marc DUPLANTIER[2], Jean-Bernard DUCHEMIN[1], Jocelyn RAVELOSAONA[3], Lala RATSIFASOAMANANA[4] and Suzanne CHANTEAU[1]
[1]Institut Pasteur de Madagascar Centre Collaborateur OMS, BP 1274, Antananarivo, BP 1274 - Antananarivo, Madagascar; [2] Programme RAMSE, IRD Madagascar ; present address : IRD, BP 1386, Dakar, Sénégal; [3]Bureau Municipal d'Hygiène, Antananarivo, Madagascar; [4]Programme National de Lutte contre la Peste, Ministère de la Santé, Antananarivo, Madagascar; [5]Present address : CERMES, BP 10887, YN 034, Niamey, Niger

1. INTRODUCTION

Plague reached Antananarivo city, the capital of Madagascar in 1921 and caused severe outbreaks during fifteen years. After annual campaigns of mass vaccination, treatment of patients with sulfamides and streptomycin, flea control using insecticides, no human case was detected from 1949 until 1978, date of the re-emergence of plague in the capital. This re-emergence coincided with a dramatic reduction of the socioeconomic conditions of the population and therefore low hygiene and salubrity conditions. Since the 90's, 100 to 150 sporadic human cases are yearly notified, among them 20 to 30 patients are confirmed using bacteriology or F1 ELISA (Chanteau *et al.*,2000). No epidemic was observed despite an intense transmission of *Yersinia pestis* among the rodent populations. Indeed, in 1995 and 1996, 90% of the rats were seropositive to F1 antibody and 10% were *Y. pestis* carriers in a whole market of the city. At present, the incidence of human plague has remained stable.

When plague was introduced in the capital, *Rattus rattus* was the main rodent. *Rattus norvegicus* appeared later, probably during the 50's (Rakotondravony, 1983). Its settlement and expansion has been favoured by

The Genus Yersinia, Edited by Skurnik *et al.*
·ıwer Academic/Plenum Publishers, New York, 2003

urbanization and at present this species has almost completely replaced *R. rattus* making it the main reservoir of *Y. pestis.* The flea *Xenopsylla cheopis* is the unique vector involved in the epidemiological cycle of the disease in Antananarivo (1.5 million inhabitants).

The aim of this study was to follow up the transmission of plague in the natural rodent populations in the city and to test their resistance to *Y. pestis* in laboratory conditions.

2. METHODOLOGY

From 1998 to 2001, a monthly surveillance of the rats was undertaken in 9 quarters of the city. The indicators used were the *X. cheopis* flea index, the isolation of *Y. pestis* from spleen homogenates and pool of fleas (fleas collected from one animal represented a pool). The seroprevalence for anti-F1 IgG was determined by indirect ELISA.

The LD_{50} (lethal dose killing 50% of the animals) was determined on wild anti-F1 seronegative *R. rattus* and *R. norvegicus* from Antananarivo city and the F1 generation. As controls, *R. rattus* and *R. norvegicus* from non endemic plague area in Madagascar, and laboratory Wistar white rats were used. A virulent reference strain of *Y. pestis* (LD_{50} < 10 cfu on mice) was subcutaneously injected at increasing gradual doses on 6 batches of 6 rats. The mortality of the animals was followed during 7 weeks. Anti-F1 serology and isolation of *Y. pestis* from spleens were systematically performed for all the animals.

3. RESULTS

3.1 Plague in rodent population in Antananarivo city

Table 1. Annual indicators of rodent plague surveillance in Antananarivo from 1998 to 2001

Year	No. trapped rats	% R. norvegicus	Global Sp[1] [min-max]	Cheopis index [min-max]	% *Y. pestis*+ fleas pool
1998	1288	94.4	25.5 [7.4-87.7]	4 [0.9-9]	4.2 (19+/453)
1999	1514	96.2	23 [2-91]	2.6 [0.4-6.5]	6.4 (28+/437)
2000	1595	99.6	25.8 [1.5-62]	3.5 [0.13-6.8]	7 (42+/598)
2001	888	100	9.8 [1.8-17]	1.6 [0.35-3]	1.2 (3+/245)

[1]Sp, seroprevalence

A total of 5285 rats were trapped from 1998 to 2001 (Table 1). The annual global seroprevalence was respectively 25.5 [7.4-7.7], 23 [2-91], 25.8 [1.5-62], 9.8 [1-17] in 1998, 1999, 2000 and 2001. It reflects a high circulation of *Y. pestis* among the rodent populations. According to the quarter of the city, the seroprevalence ranged from 1% to 91%. The Cheopis Index (CI) was from 1.2 to 7 which was higher than the theorical threshold of risk for an outbreak (CI=1). A significant proportion of pools of fleas was found infected with *Y. pestis*. All these indicators are in agreement with a high transmission of *Y. pestis* among the rat populations.

3.2 Resistance of *R. norvegicus* and *R. rattus* in experimental conditions

The LD_{50} found was respectively 10^3 cfu for *R. norvegicus* and 10^5 cfu for *R rattus* (wild and F1 animals), while the LD_{50} of the rats from non endemic areas in Madagascar was < 100 cfu for both species. In comparison, the LD_{50} of the white Laboratory Wistar rats was < 10 cfu. Eighty (4/5) to 100% (12/12) of the rats that survived from the infection developed anti-F1 antibody but no *Y. pestis* strain was isolated from their spleen.

4. CONCLUSIONS

R. novegicus has gradually replaced *R. rattus* in Antananarivo city after its introduction in the 50's. In 2001, all the rats trapped were *R. norvegicus* suggesting that this species is nowadays the main reservoir of plague in Antananarivo city. Experimentally, we have demonstrated the natural resistance to *Y. pestis* of the two populations of rats in Antananarivo (*R. norvegicus* and *R. rattus*), as compared to the white laboratory rats and to the rats from non endemic areas. The longitudinal surveillance of the rodents during 4 years, has showed their high seroprevalence for the anti-F1 IgG antibody and a high flea index. The survey has also evidenced *Y. pestis* in pools of fleas confirming the transmission of the plague bacillus among these rodents. It is likely that many of the rodents get infected but survive and further develop anti-F1 antibodies or remain carriers of the plague bacillus. Indeed rat mortality phenomenon is not observed in the capital, conversely to what is seen in the rural villages or in the harbor of Mahajanga. This natural resistance may explain the maintenance of plague in this city. The absence of epizootic and subsequently the lack of free fleas probably explain the sporadic transmission of plague to human.

REFERENCES

Chanteau S., Ratsitorahina M., Rahalison L., Rasoamanana B., Chan F., Boisier P., Rabeson D. and Roux J. 2000. Current epidemiology of human plague in Madagascar . *Microbes and Infection*, 1: 1-7.

Rakotondravony A.D.S. 1983. Etude comparée de trois rongeurs des milieux malgaches: *R. norvegicus*, *R. rattus* et *Eliurus sp*. Biologie et dynamique des populations. *Thèse Doctorat de 3ème cycle*. Université d'Antananarivo.

Chapter 88

Food-PCR

Validation and Standardization of Diagnostic PCR for Detection of Yersinia enterocolitica *and Other Foodborne Pathogens*

Mathilde H. JOSEFSEN[1], Susanne THISTED LAMBERTZ[2], Stefan JENSEN[1] and Jeffrey HOORFAR[1]
[1]*Danish Veterinary Institute, Copenhagen, Denmark;* [2]*National Food Agency, Uppsala, Sweden*

1. INTRODUCTION

Molecular methods, using nucleic acid diagnostics, are receiving increasing attention for testing the microbiological safety of food. Nucleic acid molecules can, via sequence combinations, precisely prescribe the phenotypic characteristics of a microorganism, and detection of specific sequences unique to a particular species can obviate any requirement for confirmatory tests. Furthermore, detection of nucleic acids is very rapid compared to conventional culture-based analysis. The most widely used nucleic acid diagnostic test is based on the powerful PCR technology (Mullis *et al.*, 1986).

The enthusiasms of scientists, and the enormous bulk of publications presenting convincing data, have encouraged many diagnostic laboratories to implement PCR-based methods for pathogen detection (Scheu *et al.*, 1998). However, the results of tests developed or published by one laboratory can sometimes be difficult to reproduce by other laboratories. Although this relates to most laboratory techniques, lack of reproducibility is more pronounced in molecular techniques due to sensitive reagents, complex equipment and the need for personnel with specific skills. Proper validation based on consensus criteria is an absolute prerequisite for successful adoption of PCR-based diagnostic methodology.

The Genus Yersinia, Edited by Skurnik *et al.*
·wer Academic/Plenum Publishers, New York, 2003

2. FOOD-PCR: A EUROPEAN EFFORT

Recognizing this, in 1999 the European Commission approved the research project, FOOD-PCR (www.PCR.dk), with the aim of validating and standardizing the use of diagnostic PCR for detection of bacterial pathogens in foods.

An intention of FOOD-PCR was to devise standardized PCR-based detection methods for *Yersinia enterocolitica, Salmonella enterica*, thermophilic *Campylobacter* spp., enterohemorrahgic *Escherichia coli* (EHEC) and *Listeria monocytogenes*. The methods would focus on three sample types from primary food production: poultry-carcass rinse, pig-carcass swab, and milk.

The early tasks of the 3-year project included: Production of certified DNA material by researchers working in expert laboratories, preparation of a standard for thermal cycler performance criteria (Anon, 2002(a)), selection of promising candidate PCR methods, and optimization and testing of these for accuracy against comprehensive collections of reference strains. The final selected PCR for detection of *Y. enterocolitica* was based on the *ail* gene. The accuracy and robustness of the PCR's were evaluated through multi center inter-laboratory trials.

The project also dealt with sample pretreatment. Here, methods were developed based on current ISO pre-enrichment procedures (Anon., 1999(a); Anon., 2000), adapting then-existing procedures where necessary to allow PCR to replace conventional post-enrichment and/or detection. Another important area has been automated detection, including real-time PCR.

Finally the complete procedure comprising sample pretreatment and PCR will be subjected to inter-laboratory trials, to provide validated PCR-based pathogen detection protocols.

Amongst the outcomes of the project is the production of a guideline and a biochemical method for validation of different types and brands of cyclers. In addition four PCR standardization documents are drafted in collaboration with the European Committee on Standardization (CEN). One general standard deals with the definition of terms and test controls (Anon, 2002(b)). A separate standard is devoted to sample preparation (Anon, 2002(c)), and another to amplification/detection (Anon, 2002(d)). The reference DNA material, including *Y. enterocolitica*, has been produced by the Joint Research Center for wider use. Different methods of DNA purification have been evaluated, and a simple and suitable method selected.

3. THE PERSPECTIVE

The experience gained through the FOOD-PCR project and activities of various international working-groups suggests that a very basic aspect of PCR standardization is elaboration of a vision. The vision adopted by the PCR working group of CEN is that diagnostic PCR will have the same status as conventional bacteriological culture techniques by the year 2010 (Figure 1).

Vision:
Diagnostic PCR is implemented as alternative to culture methods by year 2010

↓

Strategy:
Rapid publication of general guidelines based on validated approaches with specific instructions as appendix

↓

Principles:
Non-commercial effort
European standardization
Scientific documentation
Cross-country validation

Figure 1. Proposed vision, strategy and principles for standardization of diagnostic PCR for detection of foodborne pathogens

The strategy, or long-term plan to achieve that vision, is *rapid* publication of a few basic guidelines, including protocols for specific pathogens as appendices. The principles of the work will be based on non-commercial and international effort, including extensive multi-center ring-trials, with the aim of providing end-users with non-exclusive protocols. The emphasis in the strategy is the speed and simplicity of the publications, avoiding protocols outdated by the rapid pace of method development.

4. INTERPRETATION OF RESULTS

Due to the risk of false positive PCR cases, a European working group is currently considering inclusion of a solution hybridization step for confirmation of amplicons (Anon., 2001). However, in our experience, if primer sequences that amplify genetic regions unique to a species have been carefully chosen, tested and validated, post-PCR confirmation is unnecessary and will negate the advantage of rapidity, although it may look convincing on a paper for standard protocol. In addition, it could be too complicated for many small laboratories, and exclude commercial kits based solely on gel electrophoresis.

Possible false-positive results can be revealed through further culturing of the enriched microorganisms from PCR-positive cases, although one should have in mind the sub-optimal diagnostic sensitivity of most culture methods. PCR can be in particular useful for detection of *Y. enterocolitica*, which has a lengthy culture protocol (Anon., 1996), and can result in isolation of the so-called environmental strains, such as *Y. frederiksenii*.

5. TEST CONTROLS

In PCR DNA is amplified, while culture methods isolate live bacteria, in some cases leaving injured target bacteria behind. Many workers have addressed this issue by spike-in experiments that demonstrate a detection limit of one target bacterium in a 25-g sample. However, the applicability to real-life situations of spiked studies using fresh cultures of "healthy" inoculates can be questioned.

Although the application of PCR appears to be quite straightforward, the importance of including proper test controls may be easily overseen.

Thus, a PCR cannot be given diagnostic status before it includes, as a minimum, an internal positive control.

6. VALIDATION STRATEGY

The occurrence of an incorrect sequence of events is seen with some of the traditional culture techniques. When many workers begin to face the challenge of quality control requirements it is realized that it is actually necessary to go several steps back and begin with proper validations. The writing of PCR standards is thus based on validation studies and experience gained through ring trials of FOOD-PCR and other validation trials.

The choice and testing of primers was the first phase in the production of standards. The expert laboratory, SLV, Uppsala, Sweden, assessed the selectivity of the *ail*-based PCR on 175 strains of *Y. enterocolitica* and closely related bacteria (Lambertz *et al.*, submitted). DNA was extracted from each strain, and analyzed in a series of PCR's containing the primer sets to be evaluated. Each PCR contained reagents, which were identical as far as possible (supplier, batch etc.), and the thermal cyclers used were routinely and checked. The initial evaluation was performed through a limited inter-laboratory trial including at least three expert laboratories as partners. The criteria for successful evaluation were strong and specific amplification of correct target sequences and no others. One primer set was chosen to take forward into the next rounds of the validation process. The PCR was then optimized. Hereafter the detection limit of the PCR in terms of the number of cells it can detect with 99 % probability (Knutsson *et al.*, 2002) was established.

The second phase took the form of a large-scale inter-laboratory trial to confirm the specificity of the PCR's. This trial involved 12 partners (Anon., 1999), in addition to the organizing laboratory. Each participating laboratory received a Standard Operating Procedure (SOP), samples of DNA from the strain list established previously, and sufficient reagents (once again as identical as possible e.g. supplier and batch) to perform PCR's in duplicate upon the DNA. The DNA samples were blind, i.e. their identity known only to the organizing laboratory, and coded. The participants performed the PCR's and reported the results to the organizing laboratory. The percentage ratio of true positive results to false positive, and true negative results to false negatives was recorded. Substantial data analysis was performed according to Langton *et al.*, (2002).

The Phase 2 ring trial participants also performed the PCR's using *non*-identical reagents, hereby the ultimate robustness of the methods were evaluated (Malorny *et al.*, 2002).

The final PCR-based detection protocol will contain a simple and universal sample treatment of the matrix in combination with a PCR-compatible (pre) enrichment broth prior to amplification. This will give a degree of familiarity to the method, which should encourage end users to replace conventional methodology with it.

7. FINAL PHASE

The next and final phase therefore involves validation of the complete PCR-based method, as a comparison with the equivalent conventional method. Again, this is done as an inter-laboratory trial involving 10 to 12 partners. Samples of matrix spiked at various target cell densities are sent to each participant by the organizing laboratory. In the NORDVAL guideline (Anon., 2001(a)), these levels are set to be zero, 1-10 cells, and 10 – 100 cells per 25 g sample. The participants should then incubate (pre-enrich) the samples and apply the PCR according to the SOP, and record the results. Concurrently, they are to perform a conventional detection procedure (plating etc) after the pre-enrichment, and record the results. The organizing laboratory should compare the results, or responses, of each method. Results from at least 8 laboratories with valid results must be available, if the comparison is to be thorough (Anon., 1999). The diagnostic specificity (relating to the number of false positives), diagnostic sensitivity (relating to the number of true positives), and the overall accuracy of each method should be determined and compared. The findings from this final phase will be made available at www.pcr.dk.

A robust PCR-based method should be at least as accurate as the conventional method.

REFERENCES

Anon. (1996) *Yersinia enterocolitica*. Detection in food. Nordic Committee on Food Analysis, Method no. 117, 3rd Edition. Oslo, Norway.

Anon. (1999(a)) Microbiology of food and animal feeding stuff – Carcass sampling for microbiological analysis. ISO/CD 17604. International Organization for Standardization, Geneva, Switzerland.

Anon. (1999) Microbiology of food and feeding stuffs – Protocol for the validation of alternative methods. Document established by the Joint Group of MicroVal (WG 6) and CEN (WG 6/TAG 2). ISO/DIS 16140, Geneva, Switzerland.

Anon. (2000) Microbiology of food and animal feeding stuff – Preparation of test samples, initial suspension and decimal dilution for microbiological examination. Doc CEN/TC 275/ WG 6 N 114. Published by AFNORE, Paris, France.

Anon. (2001(a)) Protocol for the validation of alternative microbiological methods. NV-DOC.D-2001-04-25. Published by NordVal. Copenhagen, Denmark.

Anon. (2002(a)) Microbiology of food and animal feeding stuffs – Polymerase chain reaction (PCR) for the detection of foodborne pathogens – Performance criteria for thermal cyclers. European Committee for standardization (CEN), CEN/TC 275/WG 6/TAG 3 N 46. Published by DIN, Berlin, Germany.

Anon. (2002(b)) Microbiology of food and animal feeding stuffs – Polymerase chain reaction (PCR) for the detection of foodborne pathogens – General method specific requirements. International Standard Organization, Geneva, Switzerland.

Anon. (2002(c)) Microbiology of food and animal feeding stuffs – Polymerase chain reaction (PCR) for the detection of foodborne pathogens – Requirements for sample preparation for qualitative detection. European Committee for standardization (CEN), CEN/TC 275/WG 6/TAG 3 N 45. Published by DIN, Berlin, Germany.

Anon. (2002(d)) Microbiology of food and animal feeding stuffs – Polymerase chain reaction (PCR) for the detection of foodborne pathogens – Requirements for amplification and detection for qualitative methods. European Committee for standardization (CEN), CEN/TC 275/WG 6/TAG 3 N 47. Published by DIN, Berlin, Germany.

Knutsson, R., Blixt, Y., Grage, H., Borch, E., and Rådström, P. (2002) Evaluation of selective enrichment PCR procedures for *Yersinia enterocolitica*. *Int. J. Food Microbiol.* **73**, 35-46.

Lambertz, S. Thisted., Knutsson, R., Hoorfar, J., Cook, N., Rådström, P. Multi centre validation of a PCR assay for detection of pathogenic *Yersinia enterocolitica*: towards an international standard. *Submitted for Appl. Environ. Microbiol.*

Langton, S.D., Chevennement, R., Nagelkerke, N., Lombard, B. (2002). Analyzing collaborative trials for qualitative microbiological methods: accordance and concordance. *Int. J. Food Microbiol.* 79:175-181.

Malorny, B., Tassios, P., Rådström, P., Cook, N., Wagner, M., Hoorfar, J. (2002). Standardization of diagnostic PCR for detection of foodborne pathogens. *Int. J. Food Microbiol.* (in press).

Mullis, K., Faloona, F., Scharf, S., Siaki, R., Horn, G., Erlich, H. (1986) Specific enzymatic amplification of DNA in vitro: The polymerase chain reaction. *Cold Spring Harbor Symposia on Quantitative Biology* **51(1),** 263-273.

Scheu, P.M., Berghof K., and Stahl U. (1998) Detection of pathogenic and spoilage micro-organisms in food with the polymerase chain reaction. *Food Microbiol.* **15**, 13-31.

Chapter 89

A Multiplex PCR-Detection Assay for *Yersinia enterocolitica* Serotype O:9 and *Brucella* spp. Based on the Perosamine Synthetase Gene

Application to Brucella *Diagnostics*

Peter S. LÜBECK[1], Mikael SKURNIK[2], Peter AHRENS[1] and Jeffrey HOORFAR[1]

[1]*Department of Bacteriology, Danish Veterinary Institute, Copenhagen, Denmark;* [2]*Department of Bacteriology and Immunology, Haartman Institute, University of Helsinki and Department of Medical Biochemistry and Molecular Biology, University of Turku, Finland.*

1. INTRODUCTION

The gradual increase during 1990s in herds infected with *Yersinia enterocolitica* serotype O:9 (YeO9) in Official Brucellosis Free countries has created an international problem in laboratory diagnosis of brucellosis (Fenwick *et al.,* 1996, Cheasty *et al.,* 1998). Although several recent enzyme immunoassays and Western blot techniques have increased the diagnostic specificity of serology, the problem of cross-reactions remains unsolved (Kittelberger, 1998).

The cross-reactivity is caused by the identical structures of the O-side chains (O-antigen) of the smooth lipopolysaccharide (LPS) of *Brucella* and YeO9. LPS is the major component of the outer membrane of Gram-negative bacteria. It is the bacterial outermost structure, which makes it the immuno-dominant epitope, e.g., in the course of *Brucella* infections (Nielsen *et al.,* 1989).

The conventional culture protocols can take up to 2-4 weeks to complete and are not YeO9 specific. Since polymerase chain reaction (PCR) is a rapid and sensitive detection method widely in use in microbial diagnostics we are developing a differentiative multiplex

The Genus Yersinia, Edited by Skurnik *et al.*
ıwer Academic/Plenum Publishers, New York, 2003

PCR-detection assay based on the amplification of the perosamine synthetase (*per*) gene specific DNA sequences of *Brucella* and YeO9.

2. METHODS AND RESULTS

The *per* gene of YeO9 was cloned and sequenced. This sequence together with the *per* genes of *Vibrio cholerae, V. anguillarum, E. coli* O157:H7 and *Brucella melitensis* formed the basis for development of the PCR assay. The sequences (50-60 % identical) were used to design primer pairs to specifically amplify *per* gene fragments of YeO9 or *Brucella* (Table 1).

Table 1. YeO9 (per-8 - per-15) and *Brucella* (bruc-1 - bruc-5) *per*-gene specific primers.

Name	Sequence
per-8	5'-CCAATATATCAGCCTAGCCTAGG
per-9	5'-TCTTTCTGCGAACATCCTGTAT
per-10	5'-AAGAATCCTTCTCCAAATATATAGG
per-11	5'-AACTATTATTGTCGCAAAGAGATTT
per-12	5'-GCAGACGGGGGCAAAAGTA
per-13	5'-TCACGGGAGGCCGATACA
per-14	5'-CACCTTGATTTTTTAAATGGCAT
per-15	5'-CATGATGTATGGAGAGAGGGAGAT
bruc-1	5'-CGGTTTATGTGGACTCTCTCG
bruc-2	5'-CTTGAGGATTGCGCGCTAG
bruc-3	5'-CGAAAAAATCGCTACTGCTCG
bruc-4	5'-TAACGTCTCCCACAGGGGTG
bruc-5	5-CAGTATTCTCGTGTAGGCGAAGTA

Fifty *Y. enterocolitica* strains of different serotypes (O:3; O:4; O:4,33; O:5,27; O:8; O:9; O:11; O:12; O:12,25; O:14; O:15; O:16; O:20; O:28; O:35; O:57), four other *Yersinia* spp., and 18 *Brucella* spp strains (Table 2) were tested with different combinations of the primers. Among the studied primer pairs, one primer set (bruc-1 + bruc-5, 322 bp fragment) was found to be specific for *Brucella* spp. and another set (per-12 + per-14, 312 bp fragment), for YeO9 strains. The primer pairs are specific for *Brucella* and YeO9 respectively, but because of the similar size the fragments are difficult to separate in gel electrophoresis. Therefore, a further development for Real-Time PCR with TaqMan probes is currently in progress to allow correct identification between *Brucella* and YeO9. The primer pairs will be further evaluated against several *Brucella* species and compared with an international referred *Brucella* specific PCR method to validate the diagnostic accuracy.

Table 2. Specificity of the YeO9 and *Brucella* specific primer pairs. *Different serotypes, see text.

Species/type	Number of strains	**O:9 primers** per12 + per14	***Brucella* primers** bruc1 + bruc5
Y. enterocolitica spp*.	25	neg	neg
Y.e. O:3	9	neg	neg
Y.e. O:8	7	neg	neg
Y.e. O:9	**9**	**pos**	**neg**
Y. frederiksenii	1	neg	neg
Y. kristensenii	1	neg	neg
Y. ruckeri	1	neg	neg
Y. pseudotuberculosis	1	neg	neg
Brucella suis 2	7	neg	pos
B. suis 1	1	neg	pos
B. abortus (1+3+4+6)	4	neg	pos
B. ovis 1	1	neg	pos
B. melitensis (1+2+3)	3	neg	pos
B. canis 1	1	neg	pos
B. neotomae	1	neg	pos

ACKNOWLEDGMENTS

This work was supported by the Danish Council of Science (SJVF) and the Academy of Finland.

REFERENCES

Cheasty T., Said B., Jiggle B., Threlfall J. 1998. *Yersinia* species associated with diarrhoeal disease in the United Kingdom during the period 1990-1996. Nederlands Tijdschrift voor Medische Microbiologie 6, Suppl. II, S11.

Fenwick S.G. 1996. *Yersinia enterocolitica* strains recovered from domestic animals and people in New Zealand. FEMS Immunol. Med. Microbiol. 16:241-5.

Kittelberger R, Bundesen PG, Cloeckaert A, Greiser-Wilke I, Letesson JJ. 1998. Serological cross-reactivity between *Brucella abortus* and *yersinia enterocolitica* 0:9: IV. Evaluation of the M- and C-epitope antibody response for the specific detection of *B. abortus* infections. Vet Microbiol 60:45-57.

Nielsen K., Cherwonogrodzky J.W., Duncan J.R., Bundle D.R. 1989. Enzyme-linked immunosorbent assay for identification of the antibody response of cattle naturally infected with *Brucella abortus* or vaccinated with strain 19. Am. J. Vet. Res. 50:5-9.

Picture 29. Arve Saebo and Laura Franzin

Chapter 90

Vascular Endothelial Growth Factor in Yersiniosis

A study on 157 military recruits

Arve SAEBO[1] and Jørgen LASSEN[2]

[1]*Department of Surgery, Molde hospital, Molde; and the* [2]*National Institute of Public Health, Oslo, Norway*

1. INTRODUCTION

Yersinia enterocolitica causes a great variety of clinical syndromes such as intestinal inflammation, rheumatic disease, hepatitis, nephritis, thyreoiditis etc. Vascular endothelial growth factor (VEGF) is implicated in the pathological angiogenesis of such non-neoplastic disorders (Dvorak *et al.*, 1995).

2. AIM OF THE STUDY

The aims of the study were:

1. to elucidate whether chronic manifestations of *Y. enterocolitica* infection might be associated with elevated VEGF levels; and
2. whether VEGF level might be related to antibody titers against *Y. enterocolitica.*

3. MATERIALS AND METHODS

In 1987, serum samples from 755 male military recruits (mean age 20 years) were examined for: IgG, IgM and IgA antibody response to *Y. enterocolitica* O:3 using an ELISA. The recruits answered a questionnaire

The Genus Yersinia, Edited by Skurnik *et al.*
·ıwer Academic/Plenum Publishers, New York, 2003

regarding clinical complaints as recurrent diarrhoea, steatorrhea, or arthralgy (Saebo *et al.*, 1994).

In 2000, the frozen serum samples from 157 recruits were examined for VEGF levels, using the Quantikine immunoassay. Among them, 24/66 with antibody response, and 37/91 without antibody response had reported clinical complaints.

4. RESULTS

VEGF levels: The highest mean VEGF level was recorded among seropositive recruits with clinical complaints; the lowest among seropositive recruits without complaints (Table 1). Seronegative recruits presented with intermediate levels.

VEGF related to immunoglobulin levels: Seropositive recruits with clinical complaints presented with significantly lower IgA response than did seropositive recruits without clinical complaints; i.e. an inverse relationship as compared with the observed VEGF levels (Table 1).

Table 1. VEGF serum levels and antibody response (ELISA) against *Y. enterocolitica* among seropositive recruits, with or without clinical complaints.

Complaints	Antibody response by ELISA						VEGF (pg/ml)	
	IgG Mean	SD	IgM Mean	SD	IgA Mean	SD	Mean	SD
Yes (N= 24)	0.215	0.18	0.055	0.08	0.069*	0.10	344.3¤	183
No (N= 42)	0.278	0.27	0.080	0.13	0.156*	0.22	218.9¤	158

* $p = 0.03$, ¤ $p < 0.01$ (two-sample, two-sided student's t-test)

5. CONCLUSION

The observed positive correlation between high VEGF levels and clinical complaints may indicate that VEGF may be involved in development of chronic conditions related to *Y. enterocolitica* infection. Further studies are required to elucidate the possible connection between VEGF, IgA, and chronic sequels of infectious diseases.

REFERENCES

Dvorak, H..F, Detmar, M., Claffey, K.P., Nagy, J.A., van de Water, L., and Senger, D.R., 1995, Vascular permeability factor / vascular endothelialgrowth factor: an important mediator of angiogenesis in malignancy and inflammation. *Int Arch Allergy Immunol* 107: 233-5.

Saebo, A., Kapperud, G., Lassen, J., and Waage, J., 1994, Prevalence of antibodies to *Yersinia enterocolitica* O:3 among Norwegian military recruits: association with risk factors and clinical manifestations. *Eur J Epidemiol* 10: 749-55.

Chapter 91

Pathogenic Role of a Superantigen in *Yersinia pseudotuberculosis* Infection

Jun ABE[1], Hirotsugu KANO[1], Hiroko NOGAMI[1], Shinichi MATSUMOTO[2], Kiyosi BABA[3], Hirohisa SAITO[1] and Takao KOHSAKA[4]
[1]Dep. of Allergy and Immunology, National Research Institute for Child Health and Development, Tokyo, [2]Yamaga Municipal Hospital, Kumamoto, [3]Division of Pediatrics, Heart Institute, Kurashiki Central Hospital, Okayama, [4]Dep. of Gstroenterology, National Center for Child Health and Development, Tokyo, Japan

1. INTRODUCTION

Yersinia pseudotuberculosis infection is accompanied by multiple systemic symptoms. In the acute phase, not only gastrointestinal symptoms such as abdominal pain, diarrhea, and vomiting, but also fever, scarlatiniform skin rash, conjunctivitis, and lymphadenopathy are the main symptoms of the disease. Besides, in some patients, extra-gastrointestinal complication such as coronary aneurysms, reactive arthritis, or acute renal failure develops following the acute illnesses.

Y. pseudotuberculosis produces a superantigen, YPM, which selectively stimulates T cells bearing Vβ3, 9, 13.1, and 13.2 gene segments. It was reported that the *ypm* genes are not present in all strains and that the distribution of the *ypmA* gene positive strains is much more frequent in Japan than in Europe, which correlated to the differences in the frequency and severity of the systemic manifestations of the disease. In order to study a pathological role of YPM in this infection, we analyzed the proinflammatory cytokines and anti-YPM antibody levels in patients. We also showed that YPM stimulates human PBMC to secrete IL-12 and to express skin-homing receptor, cutaneous lymphocyte-associated antigen (CLA).

The Genus Yersinia, Edited by Skurnik *et al.*
'ıwer Academic/Plenum Publishers, New York, 2003

2. MATERIALS AND METHODS

Thirty-six patients (22 males and 14 females, ages ranged from 1 year 10 months to 14 years, median 5 years 2 months) with *Y. pseudotuberculosis* infection were studied. Eight patients were complicated with acute renal failure and seven patients had aneurysms of the coronary arteries or pericardial effusion during the acute phase of the illness. Sera were obtained 3-10 days and 7-33 days after the onset of the disease for analysis of cytokines and anti-YPM antibodies, respectively. Quantitation of IL-6, IL-12p40, and IFN-γ in sera was performed by sandwich ELISA kit (BD Biosciences) and anti-YPM antibody was measured by ELISA using a purified rYPM-coated plates and peroxidase-conjugated $F(ab')_2$ fraction of rabbit anti-human IgG antibodies.

For in vitro stimulation of human PBMC, cells ($1x10^6$/ml) were cultured with either PHA (1 μg/ml), SEB (100 ng/ml), rYPM (1 μg/ml), or medium alone for 48 hr and the culture supernatant was measured for IL-12p40 by ELISA. PBMC were also cultured with PHA or rYPM for 3 days, stained with PerCP-conjugated anti-human CD3, FITC-conjugated anti-CLA, and PE-conjugated anti-TCR-Vβ3 antibodies and analyzed by a flow cytometer.

3. RESULTS

Among the patients who developed a renal complication later in the sub-acute phase, serum levels of IL-12p40 and IFN-γ were significantly elevated before the renal failure took place. Their anti-YPM IgG titers were also higher than in the patients without renal complications during the convalescent phase of the disease. On the other hand, among the patients with a heart complication, serum level of IL-6 was elevated but the levels of IL-12p40, IFN-γ and anti-YPM antibody was not significantly elevated.

Table 1. Cytokine levels in sera from patients with heart or kidney complications

Complications	IL-6 (pg/ml)	IL-12 (pg/ml)	IFN-γ (pg/ml)	anti-YPM IgG (O.D. units)
Heart	119.3 ± 51.5[a]	122.9 ± 52.6	33.4 ± 12.4	0.35 ± 0.06
Kidney	79.8 ± 41.2	225.7 ± 80.0[b]	50.2 ± 23.8[c]	0.75 ± 0.19[d]
None	33.4 ± 6.6	97.2 ± 16.4	11.0 ± 2.6	0.32 ± 0.04

a; p=0.03, b; p=0.03, c; p=0.01, d; p=0.002 compared to "None" patients

Of 36 patients, 24 patients had erythematous skin rash during the acute phase of illnesses. When the cytokine levels and anti-YPM titer in sera were compared between the two groups of patients, IL-12p40 and anti-YPM antibodies were significantly elevated in patients with skin rash.

Table 2. Cytokine levels in sera from patients with or without skin rash

Skin Rash	IL-6 (pg/ml)	IL-12 (pg/ml)	IFN-γ (pg/ml)	anti-YPM IgG (O.D. units)
+	59.9 ± 17.4	168.9± 33.3[a]	30.2 ± 10.0	0.52 ± 0.08[b]
-	54.4 ± 23.0	62.1 ± 16.8	13.0 ± 3.9	0.25 ± 0.05

a; p=0.005, b; p=0.004 compared to "-" patients

It was reported that staphylococcal and streptococcal pyrogenic superantigens induce the expansion of skin-homing CLA+ T cells in an IL-12-dependent manner (Leung, *et al.,* 1995). In patients with atopic dermatitis, the superantigen reactive T cells were more enriched in the CLA+ subsets in PBMC (Strickland, *et al.,* 1998). Therefore, we tried to determine whether YPM is able to induce PBMC to produce IL-12 and to expand CLA+ T cells. In *in vitro* culture of human PBMC with PHA, SEB, rYPM, or medium alone, IL12p40 concentrations in the supernatants were 95.8 ± 30.6 pg/ml, 102.2 ± 16.1 pg/ml, 237.7 ± 70.0 pg/ml, and 13.8 ± 8.5 pg/ml, respectively. After stimulation of human PBMC with PHA or rYPM for 3 d, the percentages of CLA+/Vβ3+ cells among T cells was 1.3 ± 0.8% and 10.1 ± 3.8%, respectively. Thus, YPM stimulated the IL-12 production and the expansion of CLA+ T cells.

4. CONCLUSION

Among the various systemic manifestations in *Y. pseudotuberculosis* infection, renal complication such as acute renal failure, and an erythematous skin rash were more strongly related to the elevation of anti-YPM titer. Acute renal failure might be caused by the T cells that was stimulated and matured under the Th1 type inflammatory cytokine milieu produced by YPM during the acute phase of the disease. Patients with erythematous skin rash had an elevated level of serum IL-12p40 during the acute phase. The *in vitro* experiments indicated that YPM could contribute to the pathogenesis of skin rash through the enhanced production of IL-12 and the expansion of the skin-homing receptor CLA+ T cells in patients.

ACKNOWLEDGEMENTS

This work was supported in part by a grant from the Ministry of Health, Labour, and Welfare.

Chapter 92

Chronic Infection with *Yersinia enterocolitica* in Patients with Clinical or Latent Hyperthyroidism

Björn E. WENZEL[1], Thea M. STRIEDER[2], Erzsébet GÁSPÁR[1] and Wilma M. WIERSINGA[2]
[1]*Cell & Immunobiological Laboratory, Department of Medicine I, Medizinische Universität Lübeck, Germany;* [2]*Department of Endocrinology and Metabolism, Academic Medical Centre, University of Amsterdam, The Netherlands*

1. INTRODUCTION

A role for *Y. enterocolitica* in the pathogenesis of various thyroid diseases has been proposed on the basis of the presence of cellular (Bech *et al.*, 1978) and humoral immunity (Lidman *et al.*, 1976) in patients with Graves' disease (GD), Hashimoto's thyroiditis (HT), non-toxic goiter, and other thyroid disorders. Although the epidemiological distribution of YE serotypes is geographically different, the use of antibodies against the plasmid encoded YOP (YOP-Ab) is a very specific and reliable marker for chronic sub clinical YE-infections independently from the YE-serotype distribution (Heeseman *et al.*, 1987). The best-studied YE-infection coinciding with thyroid disorders is that in GD.

Wenzel *et al.*, 1988 found a higher prevalence of YOP antibodies in GD and HT than in healthy blood donors and patients with non-toxic goiter using Western blots for the detection of YOP-Ab.

The present study was undertaken, in order to re-assess the prevalence of YOP-Ab in patients with autoimmune and <u>non</u>-autoimmune thyroid disorders, with reference to the thyroid hormone status. A standardized and quantifiable Western blot method was used and well-defined sex and age matched control groups were investigated.

The Genus Yersinia, Edited by Skurnik *et al.*
Kluwer Academic/Plenum Publishers, New York, 2003

2. SUBJECTS AND METHODS

Control groups: i) Healthy blood donors (from Lübeck, Hamburg, Berlin and Amsterdam), male and female, age 18 -65 years, n=452; ii) 100 females, age 20-69, selected for reference values of endocrine function tests. They had no history of thyroid diseases and were self-proclaimed in good health. All sera were collected during the same time period.

Patients with thyroid disorders: Sera of sex and age matched patients were collected over a time period of 1 year according to their abnormal thyroid function tests; namely thyrotropin (TSH), free Thyroxin (fT4), TSH-Receptor Antibodies (TSH-R) and antibodies to the thyroid peroxidase (TPO-Ab). Reference values: fT4 = 9.3-20.1 pmol/l Delfia, Turku, Finland, TSH = 0.03 - 4.0 mU/l (Delfia), TSH-R = 1.0 – 4.1 IU/l (CT RRA, B.R.A.H.M.S. Berlin, Germany), (Zophel *et al.,* 2000); TPO = >60 kU/l (LUMITEST, B.R.A.H.M.S., Berlin). Non-autoimmune hyperthyroidism (Thyroid Autonomy), n=41 was defined as: i) TSH-R<0.3 IU/l, ii) TPO-Ab<60 kU/l, iii) TSH<0.005 mU/l. Autoimmune hyperthyroidism (GD), n=58 was defined as: TSH-R>1.0 IU/l. There is a gray zone of TSH-R >0.3<1.0 IU/L and TPO-Ab<100 kU/l, where a diagnosis is only relying on clinical assessment.

Specific IgG and IgA antibodies against purified plasmid-encoded virulence associated YOPs of YE serotype O9 (LCR) were detected in sera by immunoblotting and related to the intensity of bands (AID, Straßberg, Germany) (Cremer *et al.,* 1993). In short, YOPs (25,34,36,37,39,40,46,48 kDa) are blotted onto nitrocellulose. Sera are diluted 1:51 and incubated with the antigen-coated nitrocellulose strips overnight at 22°C. The IgG /IgA antibody-antigen complexes formed are quantified with the AID-Scan-System. References are included in each assay run, using human acute sera (culture-positive YE infection) containing antibodies to the YOPs. Test sera are judged positive, if at least three bands (IgG) or two band (IgA) are seen in immunoblotting at a level >10% (IgG) or >5% (IgA) of reference standards.

The YOP-Ab assay was performed without prior knowledge of thyroid function tests or the presence of TPO-Ab in the serum samples. The inter assay variance was <3%. The significance of differences between groups was analyzed with the Fisher exact test; p-values are two tailed.

3. RESULTS

There is an unexpected difference of YOP-Ab prevalence between a random control cohort (blood donors) and the "thyroid normal standard" (sex/age matched females of a normal laboratory standard cohort for thyroid function tests); YOP-IgG 36% versus 24%, (p<0.05), respectively.

Patients with an overt or latent autoimmune hyperthyroidism (TSH<0.005 IU/l) have a significant increased prevalence of YOP-Ab (IgG=76%/ IgA=48%; p<0.0001) and with TSH-R antibodies (IgG=68%/ IgA=38%; p<0.001).

Patients with "autonomous" hyperthyroidism (low TSH<0.3 IU/l) but clearly no signs for autoimmune processes (no thyroid antibodies) also show a high prevalence of YOP-Ab (IgG=66%/IgA=22%, p<0.01).

Table 1. Prevalence of YOP antibodies in patients with hyperthyroidism and in healthy controls

Cohort	Gender	Age /m	Range	Thyroid/Status	YOP-IgG (%)	YOP-IgA (%)
Blood Donors	♀/♂	42	18-57	Normal/no Ab	116/336 =36	40/336 =12
Controls/ matched	♀	45	20-65	Normal/clin.+ lab. tests	24/100 =24	13/100 =13
Hyper/ matched	♀	45	22-65	****TSH<0.005	44/58 =76	28/58 =48
Hyper/ matched	♀			***TSH<0.005 + TSH-R	26/38 =68	14/37 =38
Hyper/ matched	♀			**TSH<0.005 -TSH-R	27/41 =66	9/41 =22

**p<0.01, ***p<0.001, ****p<0.0001

4. CONCLUSIONS

In hyperthyroidism a high prevalence of YOP-Ab, of IgG and IgA class was detected. There are always a minimum number (>20%) of patients without any YOP-Ab. This may be due to inconsistencies in the diagnosis, in the definition of cohorts or to other factors/agents involved; i.e.: hormone status, committing immune factors or other infectious agents.

YOP-Ab appear solely dependent on low TSH (hyperthyroidism); But not on the autoimmune status (TSH-R or TPO-Ab) of patients, which, however, untreated is accompanied by hyperthyroidism. This suggests that *Yersinia* virulence properties do not directly induce autoimmunity. In contrast, thyroid hormone status or unknown thyroid factors may trigger *Yersinia* virulence, which through it's cross reactivity with thyroid antigen, may contribute to the chronic thyroid hyperthyroidism (GD).

In light of this report, of direct and indirect evidence that *Y. enterocolitica* as well as *Y. pseudotuberculosis* share antigen homologies with the thyroid epithelial cell (Tomer and Davies, 1993; Weetman and McGregor, 1994), of the reported coincidence of the onset of GD with the expression of YOP-Ab (Wenzel *et al.*, 1990), and of the prevalence of YOP-Ab in female first degree relatives of patients with autoimmune thyroid disorders (Strieder *et al.*, 2002) further investigation will have to focus on: i) the genetic predisposition to acquire chronic infections with *Y. enterocolitica* of families with a history of thyroid disorders ii) direct interactions of thyroid hormone with the enterobacteria.

Finally, the difference between the two normal control groups, namely blood donors and matched female controls, may reflect a gray zone of individuals with sub clinical thyroid disorders excluded in our sex/age matched thyroid control group.

ACKNOWLEDGMENTS

„Margarete Markus Charity Foundation", MMMG2000, supported the study. We thank Ms. A. Oldörp and Mr. F. Holst for their excellent technical assistance.

REFERENCES

Cremer J, Putzker M, Faulde M, Zöller L; 1993. Immunoblotting of *Yersinia* plasmid-encoded released proteins: A tool for serodiagnosis., *Electrophoresis*, 14: 952-959.

Heesemann J, Eggers C, Schröder J, 1987. Serological diagnosis of Yersiniosis by immunoblot technique – using virulence associated antigens of enteropathogenic *Yersinae. Contr Microbiol Immunol,*): 285.

Strieder TGA, Wenzel BE, Prümmel MF, Tijssen JGP, Wiersinga WM, 2002. Increased prevalence of antibodies to enteropathogenic *Yersinia enterocolitica* virulence proteins in relatives of patients with autoimmune thyroid disease. *Clin Exp Immunol,* in press.

Tomer Y and Davies T; Infection, Thyroid Disease, and Autoimmunity, 1993. *Endocrine Reviews,*; 14/1: 107-120.

Weetman and McGregor, 1994. Autoimmune Thyroid disease: Further development in our understanding. *Endocrine Reviews,*15/6:788-830.

Wenzel BE, Heesemann J, Wenzel KW, Scriba PC, 1988. Antibodies to plasmid-encoded proteins of enteropathogenic *Yersinia* in patients with thyroid disease. *Lancet,* 1:56.

Wenzel BE, Franke TF, Heufelder AE, Heesemann J, 1990. Autoimmune thyroid disease and enteropathogenic *Yersinia enterocolitica. Autoimmunity,* 7/4:295-304.

Zophel K, Wunderlich G, Koch R, Franke WG, 2000. Measurement of thyrotropin receptorantibodies (TRAK) with a second generation assay in patients with Graves' disease. *Nuklearmedizin,* 39/4:113-20.

Chapter 93

A New Selective Medium Provides Improved Growth and Recoverability of *Yersinia pestis*

Raphael BER, Emanuelle MAMROUD, Moshe AFTALION, David GUR, Avital TIDHAR, Yehuda FLASHNER, and Sara COHEN.
Department of Biochemistry and Molecular Genetics, Israel Institute for Biological Research, P.O.Box 19, Ness-Ziona, Israel 74100

1. INTRODUCTION

Current WHO regulations for isolation of *Y. pestis* recommend the use of brain-heart infusion agar (BHIA), sheep blood agar and MacConkey agar. These recommendations are best suited for isolation from clinical samples that are otherwise sterile (such as blood, lymph node, and cerebrospinal fluïd), where pure culture is expected to be obtained. However, when the clinical disease form requires the use of samples such as sputum, respiratory tract swabs or washings, skin swabs or skin scrapping, the isolation of *Y. pestis* may become complicated due to presence of competing background flora. This is even more pronounced when environmental samples are tested for the presence of *Y. pestis*, or in tissues of decaying carcasses originating from animals suspected to have died from plague. In such cases, efficient isolation of *Y. pestis* can be achieved using selective solid media. Among the above media, only MacConkey possesses some selectivity, mainly due to presence of crystal violet (inhibits Gram-positive organisms) and bile salts (inhibits non-enteric bacteria). However, this medium is widely used for isolation of many enteric bacteria and other Gram-negative organisms in general. Moreover, the slow growth rate exhibited by *Y. pestis* on MacConkey restricts its suitability as selective medium for mixed cultures. The use of *Yersinia*-selective agar (CIN) agar has been proposed as an alternative for isolation of *Y. pestis* (Rasoamanana *et al.*, 1996). Though *Y. pestis* can generally tolerate the levels of selective substances used in CIN,

'he Genus Yersinia, Edited by Skurnik *et al.*
'wer Academic/Plenum Publishers, New York, 2003

colonies on this medium vary in size, and most important, only a small portion of the plated bacteria grows to form colonies (Russel *et al.,* 1997). It thus appears that the level of selective agents in both MacConkey and CIN agar causes a decrease in growth rate and result in low recovery of *Y. pestis*. The aim of this study was the formulation of a medium that would provide both high selectivity for isolation of *Y. pestis* as well as better recovery compared to the currently available selective media.

2. RESULTS AND CONCLUSIONS

We developed a simple new selective solid medium (BIN) that is based on introduction of irgasan, cholate salts, crystal violet and nystatin into BHIA. Components concentration was optimised for growth performance of *Y. pestis* using the plating-efficiency technique. The formulated BIN was found to provide better growth rate and higher recovery levels compared to the conventional MacConkey and CIN, using virulent (Kimberley53) and attenuated (EV76 and A1122) strains as shown in Table 1.

Table 1. Relative recovery on BIN, CIN and MacConkey compared to BHIA after 48 hours

Y. pestis Strain	BIN (% ±SD)	CIN (% ±SD)	MacConkey (% ±SD)
Kimberley53	82 ± 11.5	21 ± 1.5	19 ± 14.1
EV76	95 ± 4.6	2.4 ± 1.6	38 ± 15.3
A1122	92 ± 12.4	18 ± 12.4	<0.1

The selectivity of BIN to Gram-negative strains appeared to be similar to that of CIN, and none of the tested Gram-positive bacteria showed any growth on this medium. The advantage of BIN has been also demonstrated in decaying carcasses of infected mice, where it was found to be superior to CIN, MacConkey, BHIA or Blood agar for recovering *Y. pestis* from the mixed bacterial populations found in the spleens. It thus seems that for isolation of *Y. pestis*, the rich BIN is advantageous to other selective media, especially for samples expected to be problematic, such as complex clinical cases or environmental samples.

REFERENCES

Rasoamanana, B., Rahalison, L., Raharimanana, C. and Chanteau, S., 1996, Comparison of *Yersinia* CIN agar and mouse inoculation assay for the diagnosis of plague. *Trans. Roy. Soc. Trop. Med. Hyg.* **90:** 651.

Russel, P., Nelson, M., Whittington, D., Green, M., Eley, S.M. and Titball, R.W., 1997, Laboratory diagnosis of plague. *British J. Biomed. Sci.* **54:** 231-236.

Chapter 94

Evaluation of a One-Step Biochemical Screening Test to Determine Pathogenic Strains of *Yersinia enterocolitica*

Bruce CIEBIN, Jeremy WAN and Frances JAMIESON
Ministry of Health and Long-Term Care, Laboratories Branch, National Centre for Yersinia, Etobicoke, Ontario, Canada

1. BACKGROUND

Yersinia enterocolitica is associated with a variety of clinical syndromes in humans including enterocolitis, pseudoappendicitis, septicemia and reactive arthritis. Laboratory protocols for the isolation and identification of this organism are well established; however, further tests required to determine the potential pathogenicity of an isolate are not always convenient for a diagnostic laboratory to perform. In this study, the inability of pathogenic *Y. enterocolitica* strains to utilize D-arabitol was evaluated as an alternative to other more complex, expensive and time consuming tests for determining pathogenicity.

2. METHODS

A total of 2127 *Y. enterocolitica* isolates received by the National Centre for Yersinia between January 1998 to April 2002 were used in the evaluation and included submissions from across Canada, U.S.A. and U.K. Each isolate was subcultured to Blood Agar and MacConkey Agar and examined for purity of growth after 18–24 hours incubation at 28 °C. The identity of each isolate was confirmed as *Y. enterocolitica* using conventional biochemical testing (Gray, 1995). Biogrouping and serotyping of each *Y. enterocolitica*

The Genus Yersinia, Edited by Skurnik *et al.*
'uwer Academic/Plenum Publishers, New York, 2003

isolate was performed according to the scheme of Wauters *et al.*, (Wauters, 1987). A small loopful of *Y. enterocolitica* growth from the Blood Agar purity plate was used to inoculate a tube containing 5 ml of a 1% D (+) arabitol peptone water broth (Sigma Chemical Co. St. Louis, MO) and incubated at 28°C up to 48 hours. No evidence of acidity in the arabitol broth culture (as observed by lack of development of a yellow colour) indicates that D-arabitol was not utilized, and therefore, the isolate of *Y. enterocolitica* was considered potentially pathogenic.

3. RESULTS AND DISCUSSION

There was a 100% correlation between an arabitol negative reaction and 1692 isolates of *Y. enterocolitica* belonging to 12 different recognized pathogenic biogroup/serotypes (Table 1).

Table 1. Pathogenic strains of *Y. enterocolitica* tested

Biogroup	Serotype	Number of Isolates tested
1B	O:21	1
1B	O:4	2
1B	O:8	22
1B	O:9	13
2	O:5,27	26
2	O:9	4
2	non-typeable	1
3	O:1, 2, 3	11
3	O:3	4
3	O:5, 27	11
3	non-typeable	4
4	O:3	1593
Total		1692[1]

[1]100% of isolates were arabitol negative

Only 14 (3.2%) of 435 isolates of 16 different non-pathogenic biogroups/serotypes of *Y. enterocolitica* failed to utilize arabitol (Table 2). The 100% sensitivity and 96.8% specificity rates of the arabitol test to recognize pathogenic isolates of *Y. enterocolitica* in this study are an improvement over commonly used screening tests such as the salicin fermentation esculin hydrolysis and pyrazinamidase (Table 3).

Table 2. Non-pathogenic Strains of *Y. enterocolitica* tested

Biogroup	Serotype	Number of isolates tested
1A	O:12,25	1
1A	O:14	4
1A	O:16	1
1A	O:31	1
1A	O:34	14
1A	O:36	9
1A	O:40	1
1A	O:41,42	27
1A	O:41,43	51
1A	O:5	76
1A	O:5,27	3
1A	O:6,30	42
1A	O:6,31	26
1A	O:7,8	60
1A	O:7,13	26
1A	non-typeable	93
Total		435[1]

[1]14 (3.2%) of isolates were arabitol negative

Table 3. Comparison of pathogenicity screening tests

Test	Sensitivity	Specificity
Arabitol utilization[1]	100.00%	96.8%
Salicin fermentation, esculin hydrolysis[2]	100.00%	92.0%
Pyrazinamidase[2]	95.0%	92.0%

[1]results from current study
[2]Farmer *et al.,* 1992

4. CONCLUSIONS

The use of D-arabitol can be considered a reliable, inexpensive, one-step procedure for screening pathogenicity of *Y. enterocolitica* isolates. We believe that diagnostic laboratories will find this test to be more convenient and easier to interpret than the existing pathogenicity screening methods currently available. All suspect *Y. enterocolitica* isolates, whether determined to be potentially pathogenic or non-pathogenic by screening methods, should be referred to a reference laboratory for further testing. Biochemical confirmation, biotyping, serotyping, pathogenicity testing and molecular characterization performed by reference laboratories can provide useful epidemiological information and allow for early recognition of outbreaks.

ACKNOWLEDGEMENTS

We thank the National Microbiology Laboratory, Health Canada, for their support and the Instructional Media Centre, Ministry of Health and Long-Term Care for their assistance.

REFERENCES

Gray, L. 1995. *Escherichia, Salmonella, Shigella* and *Yersinia*. p454–456. In Murray P. et al (ed), Manual of Clinical Microbiology, 6th ed. ASM, Washington, D.C.

Farmer, J., G. Carter, V. Miller, S. Falkow and I. Wachsmuth, 1992. Pyrazinamidase, CR-MOX agar, salicin fermentation esculin hydrolysis, and D-xylose fermentation for identifying pathogenic serotypes of *Yersinia enterocolitica*. J. Clin. Microbiol. 30: 2589–2594.

Wauters, G., K. Kandolo and M. Janssens 1987. Revised biogrouping scheme of *Yersinia enterocolitica*. Contrib. Microbiol. Immunol. 9: 14–21.

Chapter 95

Production of Polyvalent *Yersinia enterocolitica* Bacteriophage Preparation for Medical Prophylactic Use

Marina DARSAVELIDZE, Zhana KAPANADZE, Temo CHANISHVILI, Tamuna SULADZE, Maia ELIZBARASHVIL and Megi DVALISHVILI
G. Eliava Institute of Bacteriophage, Microbiology and Virology, Georgian Academy of Sciences, Tbilisi Georgia.

1. INTRODUCTION

Since the late 60's increased numbers of human infections caused by *Yersinia enterocolitica* have been recorded. Between 0.4 and 21% of the intestinal infections are caused by *Y. enterocolitica.*This gives high priority to investigation of *Y. enterocolitica* infections, especially since disease incidence is increasing in the whole world.

Y. enterocolitica strains of serovars 03, 05, 06, 08 and 09 cause most human infections. Domestic animals and vegetable products are the main sources of human infections. The major transmission route is food, although in the last years transmission via direct contact, water or within hospitals have been reported. Therefore, probability to encounter *Y. enterocolitica* has increased in all countries of the world. One reason for this the widely developed industrial food production, cold storage of food raw materials and public food enterprises.

The treatment of yersiniosis is restricted to wide spectrum antibiotics, the resistance towards which is increasing. Therefore there is an urgent need to produce of antibacterial preparation with wide spectrum of lytic activity that would be applicable for sanitation, prevention and treatment of the infection.

The aim of our study was to isolate and investigate the main biological properties of virulent *Y. enterocolitica* - phages against the most frequent

The Genus Yersinia, Edited by Skurnik *et al.*
..wer Academic/Plenum Publishers, New York, 2003

homologous strains of enteric *yersiniae*, and to develop technology for the production of polyvalent bacteriophages for medicinal-prophylactic use.

2. BIOLOGICAL PROPERTIES OF Y. ENTEROCOLITICA BACTERIOPHAGES

We have isolated 16 bacteriophage clones active against *Y. enterocolitica* serovars O3, O5,25, O6,30, O8 and O9. Of these, 13 *Yersinia*-phage clones were isolated from the sewage waters; one clone from *Y. enterocolitica* strain #164-568-77, and two clones from the polyvalent *S. flexneri* and *E. coli* bacteriophage preparation.

The plaque morphology of the 16 bacteriophages on the corresponding standard strains was variable, however, the bacteriophages were divided into four groups according to this property. The phage neutralization test using homologous and heterologous anti-phage antisera allowed division of the *Yersinia* phages into 8 groups between which no cross-reactions took place.

Electrom microscopic analysis revealed that all the phages were tailed and could be classified into *Myoviridae* or *Podoviridae*. Many of the *Podoviridae* phages were divided into serological groups 2,3,4 and 5. The phages belonging to serological groups 1, 6 and 7 could be attributed to the *Myoviridae* group (Figure 1).

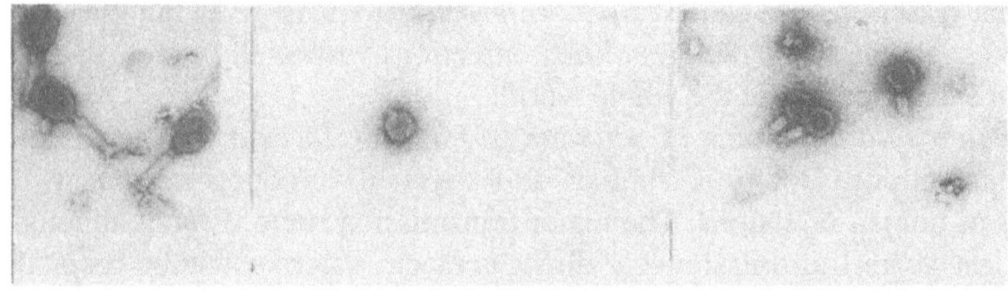

Figure 1. Electron micrographs of *Y. enterocolitica* -specific bacteriophages. M10 X 200 000

The host range of the bacteriophages is usually limited within the respective group of microorganisms; rarely one encounters phages with specificity restricted to a single strains of a homologous species. The host range spectrum of the isolated *Yersinia* -phages was tested on homologous and heterologous strains both at 24°C and 37°C. No difference were found in the lytic activity of the phages between these two temperatures.

The results obtained clearly demonstrated that the serogroup 1, 2 and 5 phages are specific only infecting *Y. enterocolitica* while the serogroup 3, 4 and 6 phages are polyvalent, and some strains of *Shigella, Salmonella* and *E. coli* were lysed as well. The phages with a wide host range were used as construction material for production of the polyvalent *Y.enterocolitica* phage preparation.

3. PRODUCTION OF POLYVALENT PHAGE

We established the conditions for the production of the specific polyvalent bacteriophage preparation intended for preventive use in medicine. The optimal plurality of infection ensuring maximal yield of phages from the infected cells was 1:8 to 1:10, the optimal incubation temperature, 20±2°C, and the the bacterial lysis took place after 4±1 hours. After production of lysates of individual phages (each specific against different serovars of *Y. enterocolitica*), the lysates were mixed and sterilized by filtration.

The resulting polyvalent phage preparation was tested against a large number of bacterial strains (Figure 2). It was found out that *Y. enterocolitica* polyvalent phage preparation was active against a wide range of homologic and heterologic strains. The preparation was stably lytic and had the highest lytic activity (with titers of 10^{-5} - 10^{-6}) on *Y. enterocolitica* serovar 03 and 09 strains, and a little lower (10^{-3} - 10^{-4}), on *Y. enterocolitica* serovar 05,27; 06,30 and 08 strains as determined by the method of Appelman.

4. CONCLUSIONS

A polyvalent *Y. enterocolitica* bacteriophage preparation has been produced which has several potential applications:
- Treatment and prophylaxis of yersiniosis patients and sanitation.
- Treatment and prophylaxis of farm animals.
- Treatment of hospital quarters, places of public catering, farms, vegetable storehouses

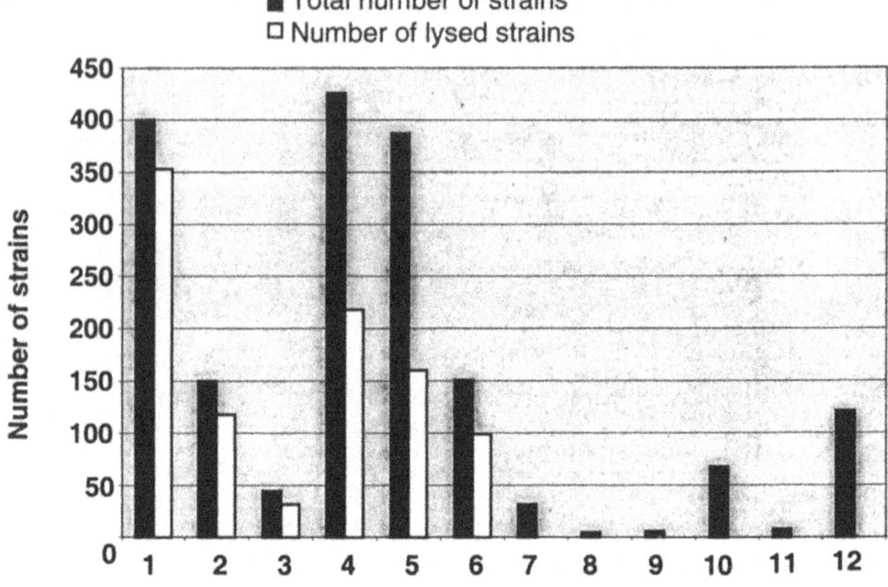

Figure 2. The host specificity of the polyvalent *Y. enterocolitica* phage preparation. Columns: 1, *Y. enterocolitica;* 2, *Y. pseudotuberculosis;* 3, *Y. pestis;* 4, *Shigella;* 5, *Salmonella;* 6, *E. coli;* 7, *Klebsiella;* 8, *Hafnia;* 9, *Citrobacter;* 10, *Proteus;* 11, *Enterobacter;* 12, *Pseudomonas aeruginosa.*

REFERENCES

Aglamova V.T. The role of the *Yersinias* in the human infection pathology. Tashkent, 1980

Kawaoka J. Characteristics of the *Y. enterocolitica* bacteriophages.Abt. la. 253, 1982

Khairullina, R.G., Cherepanova, G.V. and Nateev A.A. Isolation of *Yersinia enterocolitica* from objects of the environment in 1994-1999. Ulyanovsk, 2002.

Yutsenko G.V. The Intestinal Yersinioses. Moscow, 1977

Chapter 96

New Approaches to Detect and Assess the Pathogenicity of Clinical Strains of *Yersinia* spp. Based on Molecular Biology Techniques

Galina Y. TSENEVA[1], Airat R. MAVZUTOV[2], Alena V. SVARVAL[1], Andrey H. BAYMIEV[2], Ekaterina A. VOSKRESSENSKAYA[1] and Igor V. SMIRNOV[1]
[1]*St.Petersburg Pasteur Institute;* [2] *State Medical University of Bashkortostan*

1. INTRODUCTION

Infections caused by members of the genus *Yersinia, Y. pseudotuberculosis* and *Y. enterocolitica*, are a real concern both in Russia and in many other areas of the world. This situation results from their widespread prevalence and broad polymorphism of clinical manifestations that varies from the generalized process involving different systems and organs due to *Y. pseudotuberculosis* to more frequent disorders of alimentary canal due to *Y. enterocolitica.* However, laboratory diagnostics of these infections, especially those by *Y. enterocolitica* presents a serious problem.

2. METHODS AND RESULTS

The aim of the present study was to develop new molecular biology methods for the detection and assessment of etiological significance of *Y. pseudotuberculosis* and *Y. enterocolitica.*

The following tasks have been achieved:

The Genus Yersinia, Edited by Skurnik *et al.*
'uwer Academic/Plenum Publishers, New York, 2003

2.1 Solid phase invasin specific EIA

The 103 kD invasin protein was synthesized and used to prepare a hyper-immune serum that served to obtain a conjugate for solid-phase immunoenzyme assay (IEA for the detection of *Y. pseudotuberculosis* O:1).

The developed IEA method was shown to have a sensitivity of $5x10^5$-10^6 cell/ml as determined by testing with standard dilutions (McFarland) from 10^9 to 10^5. The system has highly diagnostic efficacy and allows diagnosis within 1-5 days of disease in at least 70 % of patients in comparison with the bacteriological method (the growth at the differential diagnostic plating medium) that allows diagnosis within 10-14 days (Figure 1).

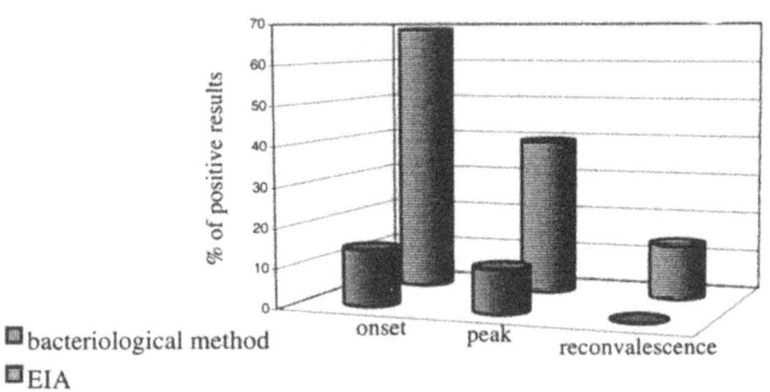

Figure 1. Efficacy of detection of *Y. pseudotuberculosis* in stools of patients during the different periods of pseudotuberculosis infection by EIA and bacteriological method.

2.2 Slide agglutination with hyper-immune sera

Hyper-immune sera specific for virulent *Yersinia* were obtained against the *Y. enterocolitica* virulence plasmid (pYV) encoded outer membrane proteins (SVY-1) and against the outer membrane proteins encoded by both the plasmid and chromosomal genes (SVY-2). To eliminate cross-reactions between SVY-1 and SVY-2, heterologous antibodies were adsorbed from the sera by plasmid-free and mutant *Y. enterocolitica* strains.

SVY-1 was used to analyse by slide agglutination reaction human and rodent isolates of *Y. enterocolitica* to assess the etiological significance of these *Y. enterocolitica* isolates. The results showed that one third of the studied isolates gave positive reaction and were considered as pathogenic

(Table 1). Similar analysis with SVY-2 allowed further differentiation and classified 6,7 % of strains as virulent.

Table 1. Identification of virulent *Yersinia* isolates using SVY slide agglutination.

Species		*n*	Serovars						
			03	04,33	05,27	06,30	07,8	08	09
Y. enterocolitica		214 (human)	43	12	113	12	18	0	0
(n = 328)		114 (rodents)	23	4	19	55	9	3	1
	pYV^+	92 (28 %)	47	3	25	14	3	0	0
Y. pseudotuberc.		33	I	II	III	IV	V		
	pYV^+	33 (100 %)	22	5	3	4	5		

2.3 Enterotoxin PCR

All available sequences of the genes that control enterotoxin production in *Enterobacteriaceae*, including those of *Yersinia* were analyzed. A conserved 30 bp region has been found at the 3'-end in ST-genes. It served to develop a specific DNA probe and PCR primers (Table 2).

Table 2. Primer pairs used in PCR analysis

No	Gene / group of genes	Nucleotide sequences in primers	Expected size of product
1	Sta	5'-acaggcaggattacaaca-3' 5'-ccgtgaaacaacatgacg-3'	240
2	Stb	5'-ttcacctttccctcaggatg-3' 5'-gcacccggtacaagcaggatt-3'	169

PCR of the conserved region of the thermostable enterotoxin gene of *Enterobacteriaceae* additionally revealed 20 % of etiologically significant *Y. enterocolitica* isolates that were found to be negative by other pathogenicity tests (SVY-1, SVY-2, pyrazinamidase test, pYV detection).

3. CONCLUSIONS

The molecular biological preparations and assays that we have developed have opened a new perspective of more precise and early diagnostics of *Yersinia* spp. infections.

For identification of *Y. pseudotuberculosis* O:1 strains from the outbreaks and from sporadic cases of human, rodent, environmental and food origin the IEA method on the basis of the invasin protein gave better results and allowed diagnosis approximately 10 days earlier in comparison the bacteriological method.

The pathogenicity of *Y. enterocolitica* (invasion, reproduction in the cells, enterotoxigenicity) varies considerably not only within the species but within a serotype. Also it is known that *Y. enterocolitica* cause a large part of diarrhoea syndrome. Gastritis, enteritis and colitis of *Yersinia* etiology make no less than 2/3 of cases. Thus detection of the conserved region of the thermostable enterotoxin gene of *Enterobacteriaceae* by PCR in the virulent strains could be an interesting tool.

Index

GPSR Compliance
The European Union's (EU) General Product Safety Regulation (GPSR) is a set of rules that requires consumer products to be safe and our obligations to ensure this.

If you have any concerns about our products, you can contact us on

ProductSafety@springernature.com

In case Publisher is established outside the EU, the EU authorized representative is:

Springer Nature Customer Service Center GmbH
Europaplatz 3
69115 Heidelberg, Germany

www.ingramcontent.com/pod-product-compliance
Ingram Content Group UK Ltd.
Pitfield, Milton Keynes, MK11 3LW, UK
UKHW022321190726
13856UKWH00001B/130